SYSTEM THEORY

A Hilbert Space Approach

This is a volume in
PURE AND APPLIED MATHEMATICS
A Series of Monographs and Textbooks
Editors: SAMUEL EILENBERG AND HYMAN BASS

A list of recent titles in this series appears at the end of this volume.

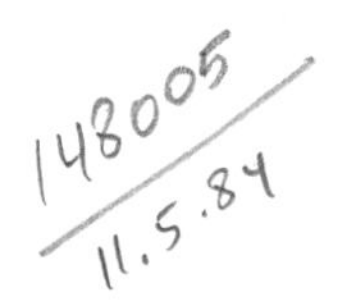

SYSTEM THEORY
A Hilbert Space Approach

Avraham Feintuch
Department of Mathematics
Ben Gurion University of the Negev
Beer Sheva, Israel

Richard Saeks
Department of Electrical Engineering
Texas Tech University
Lubbock, Texas

1982

ACADEMIC PRESS
A Subsidiary of Harcourt Brace Jovanovich, Publishers

New York London
Paris San Diego San Francisco São Paulo Sydney Tokyo Toronto

ACADEMIC PRESS, INC.
111 Fifth Avenue, New York, New York 10003

United Kingdom Edition published by
ACADEMIC PRESS, INC. (LONDON) LTD.
24/28 Oval Road, London NW1 7DX

Library of Congress Cataloging in Publication Data

Feintuch, Avraham.
System theory.

(Pure and applied mathematics ;)
Bibliography: p.
Includes index.
1. System analysis. 2. Hilbert space. I. Saeks, R.
II. Title. III. Series: Pure and applied mathematics (Academic Press) ;
QA3.P8 [QA402] 510s [003] 82-1816
ISBN 0-12-251750-4 AACR2

PRINTED IN THE UNITED STATES OF AMERICA

82 83 84 85 9 8 7 6 5 4 3 2 1

To our parents

Contents

Part IV STOCHASTIC SYSTEMS

Preface

Although one can trace the heritage of system theory back through the centuries, the field did not truly come into its own until the post–World War II era. By the mid-1960s, however, the field had expanded to the point where communication between its various branches had begun to break down, with some practitioners applying the new state space concepts while others employed the traditional frequency domain techniques. Moreover, the systems community was beginning to encounter new classes of distributed, time-varying, multivariate, and discrete time systems, and a multiplicity of new concepts associated therewith. As such, with the goal of bringing some order into this mounting chaos, a number of researchers began to search for a unified approach to linear system theory. Although many approaches were tried with varying degrees of success, the present work represents the culmination of one such search, in which Hilbert space techniques are used to formulate a unified theory of linear systems.

Unlike the classical applications to mathematical physics, however, the formulation of a viable theory of linear systems required the development of a modified Hilbert space theory in which a "time structure" is adjoined to the classical Hilbert space axioms. The present book thus represents an exposition of the resultant "theory of operators defined on a Hilbert resolution space" together with the formulation of a unified theory of linear systems based thereon.

Interestingly, however, essentially the same theory evolved independently in the pure mathematics community. Indeed, the theory was developed simultaneously by two different research groups working independently and motivated by different problems. One such group developed the theory of nest algebras in the context of a study of non–self-adjoint operator algebras, while the second developed the theory of triangular operators models in the context of an effort to extend the Jordan canonical form to an infinite-dimensional setting. Although different terminology and notation were employed, the three theories proved to be essentially identical; as such, concepts developed in each of the three theories are consolidated in the present work within the Hilbert resolution space setting.

In order to make it accessible to the system theory community at large, only a single course in Hilbert space techniques is assumed; the text is otherwise self-contained. In particular, the book can be employed in a second-year graduate course for students interested in either operator theory or system theory. Indeed, the authors have taught such courses to both mathematics and electrical engineering students.

The text is divided into four parts dealing with

I. operator theory in Hilbert resolution space,
II. state space theory.
III. feedback systems, and
IV. stochastic systems.

As such, it can be used for a second course in operator theory in which Part I is covered in detail together with a sampling of topics from Parts II–IV. Alternatively, one can gloss over Part I with the emphasis on the latter parts for a course on linear systems.

Although it is impossible to acknowledge everyone who has contributed to this book in one way or another, the authors would like to express their sincerest thanks to the students who served as guinea pigs in our classes, to our colleagues who have served as sounding boards for our ideas, and to the numerous individuals who have read the several drafts of the manuscript and commented thereon. To list but a few, we thank Gary Ashton, Roman DeSantis, John Erdos, Maria Fuente, Israel Gohberg, Dave Larson, Phil Olivier, Lon Porter, Al Schumitzky, Leonard Tung, and George Zames. Finally, we would like to express our sincerest thanks to Mrs. Pansy Burtis, who painstakingly typed the several drafts of the manuscript and too many revision thereof to count.

Introduction

Intuitively, a system is a black box whose inputs and outputs are functions of time (or vectors of such functions). As such, a natural model for a system is an operator defined on a function space. This observation and its corollary to the effect that system theory is a subset of operator theory, unfortunately, proved to be the downfall of early researchers in the field. The projection theorem was used to construct optimal controllers that proved to be unrealizable, operator factorizations were used to construct filters that were not causal, and operator invertibility criteria were used to construct feedback systems that were unstable.

The difficulty lies in the fact that the operators encountered in system theory are defined on spaces of time functions and, as such, must satisfy a *physical realizability* (or *causality*) condition to the effect that the operator cannot predict the future. Although this realizability condition usually takes care of itself in the analysis problems of classical applied mathematics, it must be externally imposed on the synthesis problems that are central to system theory.

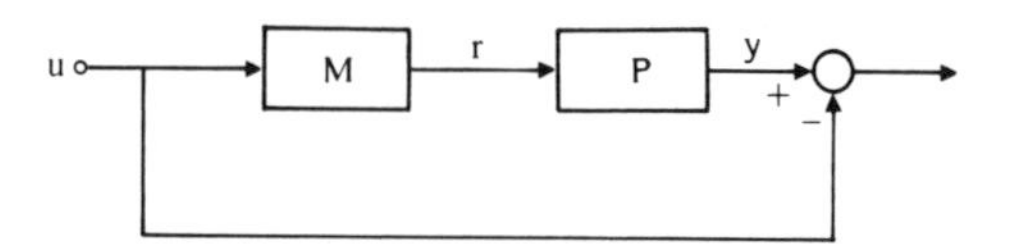

Fig. 1. Block diagram for the optimal control problem.

The problem is readily illustrated by the *optimal control problem*, illustrated in Fig. 1. Here, the black box represented by the operator P is termed the *plant*; it could be an aircraft, an electric generator, or a chemical process, for example. To control the plant we desire to build a compensator, represented by the operator M, which generates a plant input r designed to cause the resultant plant output y to track a reference input u. Since large plant inputs are not acceptable, M is typically chosen to minimize the *performance measure*

$$J(M) = \|y - u\|^2 + \|r\|^2$$

which measures both the magnitude of the plant input and the deviation between plant output and reference input.

From the block diagram of Fig. 1

$$r = Mu \qquad \text{and} \qquad y - u = (PM - 1)u$$

Hence for any given M

$$\begin{aligned} J(M) &= \|y - u\|^2 + \|r\|^2 = \|(PM - 1)u\|^2 + \|Mu\|^2 \\ &= ((PM - 1)u, (PM - 1)) + (Mu, Mu) \\ &= ([(PM - 1)^*(PM - 1) + M^*M]u, u) \\ &= ([M^*(1 + P^*P)M - M^*P^* - PM + 1]u, u) \\ &= ([(1 + P^*P)^{1/2}M - (1 + P^*P)^{-1/2}P^*]^* \\ &\quad \times [(i + P^*P^{1/2}M - (1 + P^*P)^{-1/2}P^*]u, u) \\ &\quad + ([P^*(1 + PP^*)^{-1}P]u, u) \end{aligned}$$

where we have used the fact that the positive definite hermitian operator $1 + P^*P$ admits a positive definite hermitian square root. Now, the term $([P^*(1 + PP^*)P]u, u)$ is independent of M, while the term

$$([(1 + P^*P)^{1/2}M - (1 + P^*P)^{-1/2}P^*]^* \times [(1 + P^*P)^{1/2}M - (1 + P^*P)^{-1/2}]u, u)$$

is nonnegative. Hence the performance measure will be minimized by choosing an M that makes this latter term zero. That is,

$$M_0 = (1 + P^*P)^{-1}P^*$$

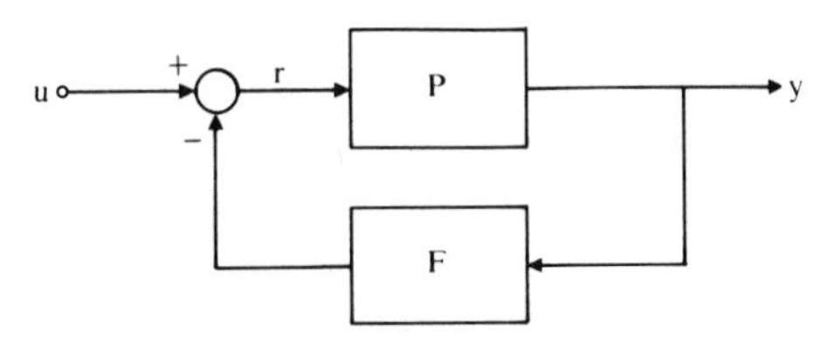

Fig. 2. Feedback system.

Although this may at first seem to be a complete solution to our optimal control problem, a closer investigation will reveal that M_0 may fail to be physically realizable. Indeed, consider the case where P is the *ideal delay* $\tilde{D}$ defined on $L_2(-\infty, \infty)$ by

$$[\tilde{D}f](\dagger) = f(\dagger - 1)$$

Since $\tilde{D}$ is unitary, its adjoint is its inverse, the *ideal predictor* $\tilde{P}$, defined on $L_2(-\infty, \infty)$ by

$$[\tilde{P}f](\dagger) = f(\dagger + 1)$$

Thus

$$M_0 = (1 + P^*P)^{-1}P^* = \tfrac{1}{2}\tilde{P}$$

which cannot be implemented.

The difficulty lies not with our mathematics but with the formulation of the problem, which should have included some type of physically realizability constraint. Although such a constraint can be readily formulated in $L_2(-\infty, \infty)$, causality is not well defined in an abstract operator theoretic setting, so the methods of classical operator theory are not immediately applicable, as one might have expected, to the optimal control problem.

As a second example consider the *feedback system*, illustrated in Fig. 2. Here

$$r = u - Fy \qquad \text{and} \qquad y = Pr$$

Hence when the appropriate inverse exists the operator mapping u to y defined by the feedback system takes the form $y = Hu$, where

$$H = (I + PF)^{-1}P$$

Now consider the case where $P = I$ is the identity on $L_2(-\infty, \infty)$ and $F = \tilde{D} - I$, where $\tilde{D}$ is the ideal delay. Then

$$H = [I + I(\tilde{D} - I)]^{-1}I = [\tilde{D}]^{-1} = \tilde{P}$$

is once again the ideal predictor. It would seem that we have constructed a device for predicting the future using physically realizable components. In fact, the feedback system is *unstable* and cannot be implemented. The precise formulation of the concept of stability and its relationship to causality, however, once again requires additional structure not available in a classical operator theoretic setting.

In an effort to alleviate these and similar problems encountered in the design of regulators, passive filters, and stochastic systems, the theory of operators defined on a Hilbert resolution space was developed in the mid-1960s. In essence, a *Hilbert resolution space* is simply a Hilbert space to which a time structure has been axiomatically adjoined, thereby allowing one to define such concepts as causality, stability, memory, and passivity in an operator theoretic setting. The present text is therefore devoted to an exposition of the theory of operators defined on a Hilbert resolution space and the formulation of a theory of system based thereon.

Although the development of the theory of operators defined on a Hilbert resolution space was motivated by its potential applications to system theory, it is interesting to observe that the resultant theory is closely allied with two parallel developments in abstract operator theory, also dating to the mid-1960s: *nest algebras* and *triangular operator models.* Indeed, the algebras of causal operators we construct are in one-to-one correspondence with the nest algebras, and both theories define a class of triangular operator models. Accordingly, Part I, summarizing the theory of operators defined on a Hilbert resolution space, simultaneously serves as an exposition of these theories.

The remaining three parts of the text are devoted to the formulation of a theory of systems defined in a Hilbert resolution space setting. In Part II the concept of memory is formalized by the development of a state space theory for an operator defined on a Hilbert resolution space. The design of feedback systems is the primary topic of Part III, where stability is defined in a resolution space setting and algorithms for the optimal and/or asymptotic design of stable feedback systems are developed. Finally, stochastic systems are investigated in Part IV. A Hilbert resolution space–valued random variable is adopted as a model for a stochastic process and used to derive a theory for stochastic estimation and control.

PART I

Operator Theory in Hilbert Resolution Space

The purpose of Part I is to survey the theory of operators defined on a Hilbert resolution space, on which the system theory developed in the succeeding parts is based. We assume that the reader is familiar with the elementary theory of operators defined on a Hilbert space and concentrate on those aspects of the theory that are unique to the resolution space setting. In Chapters 1 and 2 we define the serveral classes of causal and hypercausal operators, which will be used throughout the text, and develop their fundamental properties. With this foundation Chapters 3–5 are then devoted to the study of the three fundamental problems in the theory of operators defined on a Hilbert resolution space: operator decomposition, operator factorization, and operator invertibility.

1

Causality

A. HILBERT RESOLUTION SPACE

1. Remark: In the sequel H will denote a complex Hilbert space, which we shall assume, for the sake of convenience, to be separable. While this assumption is not necessary for most of the results, all of the examples considered are separable, and there thus seems to be no real advantage to proving results in their ultimate generality.

2. Definition: Suppose that H is a (complex separable) Hilbert space and $\$$ is a linearly ordered time set with minimum and maximum elements $\dagger_0$ and $\dagger_\infty$, respectively. A family $\mathscr{E} = \{E^\dagger : \dagger \in \$\}$ of orthogonal projection on H is called a *resolution of the identity* if

(i) $E^\ddagger(H) \supset E^\dagger(H)$ whenever $\ddagger \geq \dagger$, $\ddagger, \dagger \in \$$;
(ii) $E^{\dagger 0} = 0$, $E^{\dagger\infty} = I$;
(iii) E is closed under strong limits; i.e., if $\{E^\dagger n\} \in \mathscr{E}$ such that $E^{\dagger n} \rightarrow E$ strongly, then $E \in \mathscr{E}$.

3. Definition: If H is a Hilbert space and $\mathscr{E}$ is a resolution of the identity on H then the pair $(H, \mathscr{E})$ is called a *Hilbert resolution space.*

4. Remark: The motivation for defining a Hilbert resolution space is to equip an abstract Hilbert space with a time structure. To this end we interpret the subspace $H^\dagger = E^\dagger(H)$, $\dagger \in \$$, as the part of H that comes before time $\dagger$, and we interpret the subspace $H_\dagger = E_\dagger(H)$ as the part of H that comes after time $\dagger$. Here $E_\dagger = I - E^\dagger$ and thus $H_\dagger$ is the orthogonal complement of $H^\dagger$. These interpretations will become clear with the following classical examples.

5. Example: Let $H = L_2(-\infty, \infty)$, for each $f \in L_2(-\infty, \infty)$ and real $\dagger$ define

$$[E^\dagger f](s) = \begin{cases} f(s), & s \leq \dagger \\ 0, & s > \dagger \end{cases}$$

and let $E^{-\infty} = 0$ and $E^{\infty} = I$. It is a simple exercise to check that $\mathscr{E} = \{E^\dagger\}$ is a resolution of the identity and thus that $(H, \mathscr{E})$ is a resolution space. Indeed, with this resolution structure $E^\dagger f$ represents the part of f that comes before $\dagger$ in a natural manner.

6. Remark: In the preceding example we could just as well have formulated an L_2 space over an arbitrary, linearly ordered, locally compact set $\$$ relative to a σ-finite Borel measure on $\$$ with values taken in a Hilbert space. As before, the resolution of the identity given in Example 5 defines a natural time structure in the L_2 space, which we term its *usual time structure.*

7. Example: A special case of this L_2 space are the *sequence spaces* $l_2(-\infty, \infty)$ and $l_2[0, \infty)$ composed of square summable sequences $\{a_k\}_{-\infty}^{\infty}$ and $\{a_k\}_0^{\infty}$, respectively, with the usual time structure defined by

$$E^n\{a_k\}_{-\infty}^{\infty} = \{a_k\}_{-\infty}^{n}$$

with $E^{-\infty} = 0$ and $E^{\infty} = I$ and similarly for $l_2[0, \infty)$.

8. Example: Although the usual resolution structure for a function space corresponds to our most natural physical interpretation of time, alternative structures can be formulated and, indeed, may be appropriate for certain applications.[1] For instance, consider the *reproducing kernel Hilbert space* associated with the *Wiener process.* Here H is composed of absolutely continuous functions defined on $[0, \infty)$ such that

(i) $f(0) = 0$,
(ii) $\int_0^\infty \|\dot{f}\|^2 \, dt < \infty$,

together with the norm

$$\|f\|^2 = \int_0^\infty \|\dot{f}\|^2 \, dt$$

Now, in this space the usual resolution of the identity is undefined since the truncate of an absolutely continuous function may not be absolutely continuous. On the other hand, we may define a new resolution structure by

$$[E^\dagger f](s) = \begin{cases} f(s), & s \leq \dagger \\ f(\dagger), & s > \dagger \end{cases}$$

for any $\dagger \geq 0$ with $E^\infty = I$. Relative to the topology of our reproducing kernel Hilbert space this is a perfectly well-defined resolution of the identity, though this would not be the case in an L_2 space. In fact, this is a rather natural resolution structure for this Hilbert space,[1] for if f is interpreted as a sample function generated by the Wiener process, then $E^\dagger f$ is our "best estimate" of f based on observations of the sample function taken before time $\dagger$.

9. Remark: The above resolution of the identity can be generalized to an arbitrary reproducing kernel Hilbert space[1] by defining $E^\dagger f$ to be that element of the space minimizing $\|E^\dagger f\|$ under the constraint that

$$[E^\dagger f](s) = f(s), \qquad s \leq \dagger$$

Such an element exists and defines a resolution of the identity by virtue of the projection theorem.

10. Example: An alternative method for associating a resolution structure with the Wiener process is as follows. Let $\{y_\dagger : \dagger \in [0, \infty)\}$ be a Wiener process for each $\dagger$ and let $H_y^\dagger = V\{y_s : 0 \leq s \leq \dagger\}$ and $H_y = V\{y_s : 0 \leq s < \infty\}$. We may now define a resolution of the identity on H_y by letting $E_y^\dagger$ be the orthogonal projection of H_y onto $H_y^\dagger$, $0 \leq \dagger < \infty$. Finally, we may define a unitary transformation of H_y onto $L_2[0, \infty)$ that maps the projections $E_y^\dagger$ into a correpsonding resolution structure on $L_2[0, \infty)$:

$$\tilde{E}_y^\dagger f = \int_0^\dagger \frac{d}{ds}(f, y_s)_{H_y} \, dy_s$$

where the left-hand side of the equation is a *Wiener integral*. The Wiener process thus defines three natural resolution spaces; one in a reproducing kernel Hilbert space, one in H_y, and one in $L_2[0, \infty)$.

11. Remark: Although we are interested in Hilbert space in the present context, occasionally we find it necessary to work with an algebraic resolution space in which a time structure is introduced into a vector space without the aid of an inner product.

12. Definition Suppose that V is a vector space and $\$$ is a linearly ordered time set with minimum and maximum elements $\dagger_0$ and $\dagger_\infty$, respectively. A family, $\mathcal{F} = \{F^\dagger : \dagger \in \$\}$ of projections on V is called an *algebraic resolution of the identity* if

(i) $F^\ddagger(V) \supset F^\dagger(V)$ whenever $\ddagger > \dagger$, $\ddagger, \dagger \in \$$; and
(ii) $F^{\dagger 0} = 0$, $F^{\dagger\infty} = I$.

13. Definition: If V is a vector space and $\mathcal{F}$ is an algebraic resolution of the identity on V, then the pair $(V, \mathcal{F})$ is termed an *algebraic resolution space.*

14. Example: Let $V_\dagger$, $\dagger \in \$$, be a family of vector spaces parametrized by a time set $\$$ and let

$$V = \prod_\dagger [V_\dagger]$$

Now, if f is a section of V we may define $F^\dagger f$ by

$$[F^\dagger f](s) = \begin{cases} f(s), & s \leq \dagger \\ 0, & s > \dagger \end{cases}$$

which together with $F\dagger_0 = 0$ and $F^{\dagger\infty} = I$ defines an algebraic resolution of the identify $\mathcal{F}$ and an algebraic resolution space $(V, \mathcal{F})$.

15. Remark: Of course, every Hilbert resolution space is also an algebraic resolution space.

B. CAUSAL OPERATORS

1. Remark: We are now able to define the notion of a causal operator on a Hilbert resolution space $(H, \mathcal{E})$. To motivate our definition let us look at an example where the time structure is obvious, $(L_2(-\infty, \infty), \mathcal{E})$, where $\mathcal{E}$ is the usual resolution structure for a function space defined in Example A.5. Let A be an operator on this space and let f and g be L_2 functions such that $f = Ag$, i.e., $[Ag](\dagger) = f(\dagger)$ for almost all real $\dagger$. In classical system theory[3] such an operator is said to be causal if the value of f at time $\dagger$ does not depend on the values of f after time $\dagger$ but only on the values of f at and/or before time $\dagger$; i.e., the present output is dependent only on the present and past inputs, and hence the operator does not predict the future. This can be stated mathematically by requiring that

$$[Af_1](s) = [Af_2](s), \qquad s \leq \dagger$$

whenever

$$f_1(s) = f_2(s), \qquad s \leq \dagger$$

Recalling the definition for the usual resolution structure for $L_2(-\infty, \infty)$ where $E^\dagger$ "kills" that part of f that comes after $\dagger$, the above implication can be reformulated in our resolution space notation by requiring that

$$E^\dagger A f_1 = E^\dagger A f_2$$

whenever

$$E^\dagger f_1 = E^\dagger f_2$$

Extrapolating this equality to the general resolution space setting we have

2. Definition: An operator A on $(H, \mathscr{E})$ is *causal* if for all $E^\dagger \in \mathscr{E}$ and any $x, y \in H$ the equality $E^\dagger x = E^\dagger y$ implies that $E^\dagger A x = E^\dagger A y$.

3. Remark: In Definition 2 no mention was made of the linearity of A. In fact, this definition will be used in the notes at the end of Part I for nonlinear as well as linear operators. Within the main text, however, we shall restrict ourselves to the linear theory, wherein causality has a natural geometric interpretation.

4. Property: Let A be a linear operator on $(H, \mathscr{E})$. Then the following are equivalent:

(i) A is causal,
(ii) $E^\dagger A = E^\dagger A E^\dagger$ for all $\dagger \in \$$,
(iii) $A E_\dagger = E_\dagger A E_\dagger$ for all $\dagger \in \$$, and
(iv) the subspaces $H_\dagger = E_\dagger(H) = [I - E^\dagger](H)$ are invariant under A.

Proof: Suppose that A is causal. Since $E^\dagger$ is an orthogonal projection,

$$E^\dagger([I - E^\dagger]x) = E^\dagger x - E^\dagger x = 0 = E^\dagger 0$$

for all $x \in H$. Thus since A is causal

$$E^\dagger A[(I - E^\dagger]x) = E^\dagger A 0 = 0, \qquad \dagger \in \$$$

or equivalently,

$$E^\dagger A x = E^\dagger A E^\dagger x, \qquad \dagger \in \$, \quad x \in H$$

verifying (ii). Now, substituting $E^\dagger = I - E_\dagger$ into the last equality and invoking the linearity of A yields $AE_\dagger = E_\dagger AE_\dagger$, verifying (iii). This is, however, just the classical criterion for $H_\dagger = E_\dagger(H)$ to be an invariant subspace of A, and therefore (iv) is verified.

Finally, suppose that $E^\dagger x = E^\dagger y$ for some $x, y \in H$ and $\dagger \in \$$. Then $E^\dagger x - E^\dagger y = E^\dagger(x - y) = 0$, which implies that $E_\dagger(x - y) = [I - E^\dagger](x - y) = x - y$ or, equivalently, that $x - y \in H_\dagger$. Hence (iv) implies that $A(x - y) \in H_\dagger$, and $E^\dagger A(x - y) = 0$. Equivalently,

$$E^\dagger Ax = E^\dagger Ay$$

implying that A is causal and completing the proof. //

5. Remark: Of the various characterizations of the linear causal operators given by Property 4, (iv) is by far the most important. Indeed, the invariance of the causal operators on the *nest of subspaces* $\{H_\dagger : \dagger \in \$\}$ implies that these operators form a *nest algebra* and are thus equipped with the powerful geometric structure associated with such algebras. Some of the more elementary geometric properties associated with such a nest algebra will be derived in the remainder of this section, and a summary of the deeper results on nest algebras will be formulated in the remainder of Part I. We begin by verifying that the causal operators in fact form an algebra.

6. Property: The set of causal bounded linear operators on a resolution space $(H, \mathscr{E})$ is a weakly closed Banach algebra with identity.

Proof: Since the nest of subspaces $\{H_\dagger : \dagger \in \$\}$ is invariant for any two causal operators A and B, it is also invariant for AB and $A + B$ while the identity is clearly invariant on such a nest. Now assume that $\{A_\pi\}$ is a weakly convergent net of causal operators with limit A. Let $x \in H_\dagger$ and $y \in H^\dagger$ for any $\dagger \in \$$, in which case $A_\pi x \in H_\dagger$, implying that $(A_\pi x, y) = 0$. Since A is the weak limit of A_π, this implies that $(Ax, y) = 0$, which in turn implies that $Ax \in H_\dagger$ since y is an arbitrary element from $H^\dagger$. Since x is an arbitrary element from $H_\dagger$, this implies that $H_\dagger$ is invariant for A, $\dagger \in \$$, and hence that A is causal. Finally, since the algebra of causal operators is weakly closed it is also uniformly closed and hence is a Banach algebra (since it is a uniformly closed subalgebra of the Banach algebra $\mathscr{B}$ made up of all bounded linear operators on H). //

7. Remarks: In the sequel the causal operators on a Hilbert resolution space $(H, \mathscr{E})$ will be denoted $\mathscr{C}$ with the particular resolution space understood. Of course, $\mathscr{C} \subset \mathscr{B}$.

8. Example: Let H be of finite dimension n and let $\mathscr{E}$ be composed of n distinct projections:

$$0 < E^1 < E^2 < \cdots < E^n = I$$

An operator A is causal on $(H, \mathscr{E})$ if and only if it leaves the subspaces H_i invariant. Therefore, if we construct an orthonormal basis $\{e_1, e_2, \ldots, e_n\}$ for H such that $E^i(H)$ is the span of $\{e_1, e_2, \ldots, e_i\}$, then A is causal if and only if its matrix representation with respect to this basis is lower triangular. More generally if $\mathscr{E}$ contains less than n projectors, A will be causal if and only if it is block lower triangular relative to an appropriate basis.

9. Example: Let $H = l_2(-\infty, \infty)$ with the usual resolution of the identity, in which case a linear operator A is causal if and only if it leaves invariant the subspaces $l_2(n, \infty)$ for all $n \in Z$. Thus if $\{e_i\}_{-\infty}^{\infty}$ is the standard basis for l_2, A is causal if and only if its matrix representation with respect to this basis is lower triangular.

10. Example: We now consider the continuous case. At this point we have no analog to a matrix representation for an operator without first choosing an orthonormal basis. Unfortunately, if $H = L_2(-\infty, \infty)$, there is no natural orthonormal basis consistent with the usual time structure. We therefore restrict ourselves to a class of operators that do generalize the matrix idea, the *integral operators*. Consider the operator defined by

$$[Af](\dagger) = \int_{-\infty}^{\infty} G(\dagger, \ddagger) f(\ddagger)\, d\ddagger$$

where $G(\dagger, \ddagger)$ is a (possibly distributional) kernel that defines a bounded linear operator on H. For this operator the triangular characterization of causality seen in the previous examples manifests itself by the requirement that $G(\dagger, \ddagger)$ be supported on the triangular half-space $\{(\dagger, \ddagger) \in R^2 : \dagger \geq \ddagger\}$. Equivalently,

$$G(\dagger, \ddagger) = 0, \qquad \dagger < \ddagger$$

To verify this we must show that A leaves the subspaces $L_2(s, \infty)$ invariant if and only if the equality holds. If $f \in L_2(s, \infty)$, then $f(\ddagger) = 0$ for $\ddagger \leq s$. Thus if $\dagger \leq s$,

$$\begin{aligned}
[Af](\dagger) &= \int_{-\infty}^{\infty} G(\dagger, \ddagger) f(\ddagger)\, d\ddagger \\
&= \int_{-\infty}^{\dagger} G(\dagger, \ddagger) f(\ddagger)\, d\ddagger + \int_{\dagger}^{\infty} G(\dagger, \ddagger) f(\ddagger)\, d\ddagger \\
&= \int_{\dagger}^{\infty} G(\dagger, \ddagger) f(\ddagger)\, d\ddagger
\end{aligned}$$

since $f(\ddagger) = 0$ in the interval $-\infty < \ddagger < \dagger$. Thus $[Af](\dagger) = 0$ if and only if $\int_{\dagger}^{\infty} G(\dagger, \ddagger) f(\ddagger)\, d\ddagger = 0$. Since f is arbitrary over the interval of integration, this will be the case if and only if $G(\dagger, \ddagger) = 0$, $\dagger < \ddagger$, as was to be shown.

11. Remark: If in the above example $G(\dagger, \ddagger)$ is of the form $G(\dagger - \ddagger)$, the integral operator becomes a *convolution operator* while our causality criterion reduces to the requirement that $G(\dagger) = 0$ for $\dagger < 0$, i.e., the convolutional kernel has support on the half-line.

12. Example: If the above-described convolution operator is formulated in $l_2(-\infty, \infty)$ rather than $L_2(-\infty, \infty)$, we obtain

$$a_i = [Ab]_i = \sum_{j=-\infty}^{\infty} G_{i-j} b_j$$

which, as before, is causal relative to the usual time structure on $l_2(-\infty, \infty)$ if and only if $G_k = 0$ for $k < 0$. On taking Fourier transforms our discrete convolution on $l_2(-\infty, \infty)$ is transformed into a multiplication operator on $L_2(T)$ in the form

$$\hat{a} = \hat{G}\hat{b}$$

Here the operator is bounded if and only if $\hat{G} \in L_\infty(T)$, and the causality condition implies that the operator will be causal if and only if $\hat{G} \in H_\infty(T)$. Hence multiplication by an $H_\infty(T)$ function may be viewed as a causal operator within the class of bounded operators defined by multiplication with an $L_\infty(T)$ function.

13. Remark: From the point of view of applying these results, $H_\infty(T)$ functions interpreted as causal operators are most naturally viewed as a representation of the discrete convolution operators that are causal relative to the usual resolution structure on $l_2(-\infty, \infty)$. We could, however, define a family of orthogonal projections on $L_2(T)$ by FE^nF^{-1} where $F: l_2(-\infty, \infty) \to L_2(T)$ is the Fourier transform operator, in which case multiplication by an $L_\infty(T)$ function would be causal relative to $(L_2(T), F\mathscr{E}F^{-1})$ if and only if the function lies in $H_\infty(T)$.

Although most of our theory is applicable to any Hilbert resolution space, it is occasionally convenient to work with a restricted class of resolution spaces. These restrictions are most readily formulated in terms of the nest of subspaces $\{H_\dagger : \dagger \in \$\}$. For this purpose we define

$$H_{\dagger -} = V\{H_s : s < \dagger\}$$

and let $E_{\dagger -}$ be the orthogonal projection onto $H_{\dagger -}$. Using this notation we then make the following definitions.

14. Definition: A Hilbert resolution space is said to be *continuous* if $H_{\dagger-} = H_\dagger$ for all $\dagger \in \$$. It is said to be *maximal* if $H_{\dagger-} \backslash H_\dagger$ has dimension 0 or 1 for all $\dagger \in \$$. Finally, we say that the resolution space is *discrete* if each $\dagger \in \$$ admits an immediate predecessor $\ddagger$ and $H_\ddagger \backslash H_\dagger$ is finite dimensional.

15. Remark: In the examples of this section $L_2(-\infty, \infty)$ is continuous while $l_2(-\infty, \infty)$ is discrete and both are maximal.

C. ANTICAUSAL AND MEMORYLESS OPERATORS

1. Remark: In a resolution space we deal with two natural families of projections, the resolution of the identity $\{E^\dagger : \dagger \in \$\}$ and the projections $\{E_\dagger = I - E^\dagger : \dagger \in \$\}$. The latter family of projections, however, fails to be a resolution of the identity since the projections decrease rather than increase with $\dagger$. We may, however, define a *dual resolution of the identity* by reversing the linear ordering on $\$$. That is, we let $\$^*$ be the set $\$$ linearly ordered by $<^*$ where $\dagger <^* \ddagger$ if and only if $\ddagger < \dagger$, while $\dagger_0^* = \dagger_\infty$ and $\dagger_\infty^* = \dagger_0$:

$$\mathscr{E}^* = \{E_\dagger : \dagger \in \$^*\}$$

is thus a resolution of the identity relative to the time set $\$^*$ and $(H, \mathscr{E}^*)$ is a resolution space.

2. Definition: An operator A on $(H, \mathscr{E})$ is *anticausal* if it is causal on $(H, \mathscr{E}^*)$.

Of course, all of the properties developed for causal operators in the preceding section hold for anticausal operators and are summarized as follows.

3. Property: Let A be a linear operator on $(H, \mathscr{E})$. Then the following are equivalent:

(i) A is anticausal,
(ii) $E_\dagger A = E_\dagger A E_\dagger$,
(iii) $AE^\dagger = E^\dagger A E^\dagger$, and
(iv) the subspaces $H^\dagger = E^\dagger(H)$ are invariant under A.

4. Property: The set of anticausal bounded linear operators on a resolution space $(H, \mathscr{E})$ is a weakly closed Banach algebra with identity.

5. Remark: Of course, all of the examples given in Section B characterizing the causal operators of various types can be replicated for anticausal operators on replacing the lower triangular representations by upper triangular representations.

Consistent with the dual nature of the causal and anticausal operators we denote the anticausal operators by $\mathscr{C}^*$. Indeed, this notation is further justified by the following property.

6. Property: A linear operator A on $(H, \mathscr{E})$ is anticausal if and only if A^* is causal.

Proof: Since M is an invariant subspace of A if and only if its orthogonal compliment $M^{\perp}$ is invariant for A^*, the nest of subspaces $\{H_{\dagger} : \dagger \in \$\}$ is invariant for A if and only if the nest of subspaces $\{H^{\dagger} : \dagger \in \$^*\}$ is invariant for A^*. Thus A^* is anticausal if and only if A is causal. //

Since a causal operator cannot "remember" the future and an anticausal operator cannot remember the past, it is natural to refer to an operator that remembers neither the past or the future as memoryless.

7. Definition: An operator A on $(H, \mathscr{E})$ is *memoryless* if and only if it is both causal and anticausal.

8. Example: Suppose that A is the integral operator considered in Example B.10. Then A is causal if and only if $G(\dagger, \ddagger) = 0$ for $\dagger < \ddagger$ and it is anticausal if and only if $G(\dagger, \ddagger) = 0$ for $\dagger > \ddagger$. Thus a memoryless operator $G(\dagger, \ddagger) = 0$ for $\dagger \neq \ddagger$. In particular, for it to define a nonzero operator $G(\dagger, \ddagger)$ must be distributional and so take the form

$$G(\dagger, \ddagger) \in g(\dagger)\delta(\dagger - \ddagger)$$

Therefore

$$[Af](\dagger) = \int_{-\infty}^{\infty} g(\dagger)\delta(\dagger - \ddagger) f(\ddagger)\, d\ddagger = g(\dagger) f(\dagger)$$

is a multiplication operator.

9. Remark: Since the causal operators typically admit a lower triangular representation and the anticausal operators typically admit an upper triangular representation, it is not surprising that the memoryless operators are typically diagonal, as indicated by the above example. Moreover, the memoryless operators on $l_2(-\infty, \infty)$ with its usual time structure are represented by diagonal matrices, etc.

We denote memoryless operators on $(H, \mathscr{E})$ by $\mathscr{M}$. They are characterized by combining the properties already derived for the causal and anticausal operators as follows.

10. Property: Let A be a linear operator on $(H, \mathscr{E})$. Then the following are equivalent:

(i) A is memoryless,
(ii) the nest of subspaces $\{H_\dagger : \dagger \in \$\}$ are reducing subspaces for A,
(iii) the nest of subspaces $\{H^\dagger : \dagger \in \$^*\}$ are reducing subspaces for A,
(iv) $AE^\dagger = E^\dagger A$ for all $\dagger \in \$$, and
(v) $AE_\dagger = E_\dagger A$ for all $\dagger \in \$$.

11. Remark: M is a *reducing subspace* for A if $A(M) \subset M$ and $A(M^\perp) \subset M^\perp$. Unlike the causal and anticausal operators, the memoryless operators are closed under adjoints; i.e., if A is causal and anticausal, then A^* is anticausal since it is the adjoint of a causal operator and causal since it is the adjoint of an anticausal operator, and hence it is also memoryless. Accordingly, rather than simply obtaining a Banach algebra of memoryless operators as the intersection of two Banach algebras, we actually obtain a *von Newmann algebra* (a weakly closed Banach algebra of operators that is closed under adjoints).

12. Property: The set of memoryless bounded linear operators on a resolution space $(H, \mathscr{E})$ is a von Neumann algebra.

Rather than always working with the memoryless operators on a resolution space, we occasionally find one of its subalgebras to be more useful.

13. Definition: The *core* of a resolution space $(H, \mathscr{E})$ is the von Neumann algebra generated by the resolution of the identity $\mathscr{E} = \{E^\dagger : \dagger \in \$\}$.

14. Remark: The core of $(H, \mathscr{E})$ is denoted by $\mathscr{K}_\mathscr{E}$. Unlike $\mathscr{M}$ it is an abelian (commutative) algebra satisfying

$$\mathscr{K}_\varepsilon \subset \mathscr{M}$$

Moreover,

$$\mathscr{M} = \mathscr{K}'_\mathscr{E}$$

the *commutant* of $\mathscr{K}_\mathscr{E}$. As such $\mathscr{M}$ is abelian if and only if $\mathscr{K}_\mathscr{E}$ is maximal abelian, in which case $\mathscr{M} = \mathscr{K}_\mathscr{E}$. In general, $\mathscr{K}_\mathscr{E}$ is the *center* of $\mathscr{M}$, the set of operators in $\mathscr{M}$ that commute with every operator in $\mathscr{M}$.

D. THE INTEGRALS OF TRIANGULAR TRUNCATION

1. Remark: In the previous sections descriptions of the causal and anticausal operators were given in algebraic terms. At this point we formulate a family of operator-valued integrals that yield an analytic characterization of the causal operators by appropriate abstraction the "triangular nature" of the causal operators noted in the examples of Section B.

2. Definition: Let $\mathscr{E}$ be a resolution of the identity on H. *A partition* of $\mathscr{E}$ is a finite subset of $\mathscr{E}$ containing zero and the identity.

3. Remark: We denote a partition $\mathscr{P} = \{E^{\dagger i} : i = 0, 1, \ldots, n\}$, where the indices are chosen such that

$$0 = E^{\dagger 0} < E^{\dagger 1} < E^{\dagger 2} < \cdots < E^{\dagger n} = I$$

For the sake of notational brevity we let $E^i = E^{\dagger i}$, $i = 0, 1, \ldots, n$, and we let $\Delta^i = [E^i - E^{i-1}]$. Finally, the set of all partitions of $\mathscr{E}$ may be ordered by refinement and forms a *directed set* under this ordering.

4. Definition: Let $(H, \mathscr{E})$ be a Hilbert resolution space and let $F: \$ \to \mathscr{B}$ be a bounded operator valued function of $\$$. Then

$$(M) \int F(\dagger)\, dE(\dagger) = \lim_{\mathscr{P}} \left[\sum_{i=1}^{n} F(i) \Delta^i \right]$$

where the limit is taken over the directed set of partitions in the uniform operator topology and $F(i) = F(\dagger_i)$.

5. Remark: Note that $(M) \int F(\dagger)\, dE(\dagger)$ is only defined for those functions for which the limit exists. It is, however, a linear transformation from that set of functions into $\mathscr{B}$, though it is not bounded or continuous in any viable topology.[2] Because of the idempotent nature of the projection-valued measure defined by $\mathscr{E}$, double integrals are characterized by a powerful theory, which is the key to the tractability of $(M) \int F(\dagger)\, dE(\dagger)$.

6. Property: Let $F: \$ \to \mathscr{B}$ and $G: \$ \to \mathscr{B}$. Then

$$(M) \int G(\ddagger) \left[(M) \int F(\dagger)\, dE(\dagger) \right] dE(\ddagger) = (M) \int G(\dagger) F(\dagger)\, dE(\dagger)$$

whenever the integrals exist.

Proof: Let $\mathscr{P}'$ and $\mathscr{P}''$ be partitions for the two integrals on the left side of the equality and let $\mathscr{P} = \{E^i : i = 0, 1, \ldots, n\}$ be a partition that refines both $\mathscr{P}'$ and $\mathscr{P}''$. Using the partition $\mathscr{P}$ we then have

$$\sum_{j=1}^{n} G(j)\left[\sum_{i=1}^{n} F(i)\Delta^i\right]\Delta^j = \sum_{i=1}^{n} G(i)F(i)\Delta^i$$

since $\Delta^i\Delta^j = 0$, $i \neq j$, and $\Delta^i\Delta^i = \Delta^i$. As such, modulo a refinement that is valid since we are taking limits, the partial sums on both sides of the equality are equal, and hence so are their limits when these exist. //

7. Example: Let $A \in \mathscr{B}$ and interpret A as a constant function defined on $. Then

$$(M)\int A\, dE(\dagger) = \lim_{\mathscr{P}}\left[\sum_{i=1}^{n} A\Delta^i\right] = \lim_{\mathscr{P}}\left[A\sum_{i=1}^{n}\Delta^i\right]$$
$$= \lim_{\mathscr{P}}\,[A] = A$$

where we have used the fact that $[\sum_{i=1}^{n} \Delta^i] = I$.

8. Remark: Although their properties are similar, numerous variations can be defined on the integral of Definition 4. First, since $\mathscr{B}$ is noncommutative we may interchange the function and measure, obtaining

$$(M)\int dE(\dagger)\, F(\dagger) = \lim_{\mathscr{P}}\left[\sum_{i=1}^{n} \Delta^i F(i)\right]$$

when it exists. Alternatively, we may use a measure on both sides of the function to obtain

$$(M)\int dE(\dagger)\, F(\dagger)\, dE(\dagger) = \lim_{\mathscr{P}}\left[\sum_{i=1}^{n} \Delta^i F(i)\Delta^i\right]$$

if it exists.

Additionally, rather than evaluating $F(\dagger)$ at $\dagger_i$, we may choose to evaluate it at $\dagger_{i-1}$. While such a modification would not effect a scalar-valued integral,[2] it yields a distinct set of integrals in our operator-valued setting that prove to be fundamental to our theory. We may thus define

$$(m)\int F(\dagger)\, dE(\dagger) = \lim_{\mathscr{P}}\left[\sum_{i=1}^{n} F(i-1)\Delta^i\right]$$

$$(m)\int dE(\dagger)\, F(\dagger) = \lim_{\mathscr{P}}\left[\sum_{i=1}^{n} \Delta^i F(i-1)\right]$$

and

$$(m) \int dE(\dagger)\, F(\dagger)\, dE(\dagger) = \lim_{\mathscr{P}} \left[\sum_{i=1}^{n} \Delta^i F(i-1) \Delta^i \right]$$

when they exist.

Finally, we may take limits in the strong operator topology rather than in the uniform operator topology, obtaining six additional integrals, which we denote s $(M) \int F(\dagger)\, dE(\dagger)$ and similarly for the other integrals.

In all we have defined 12 distinct integrals, though they all have similar properties. They are defined *only* when the appropriate limits exist, and they map their domain linearly into $\mathscr{B}$ but are neither bounded nor continuous. They do, however, possess the double integral property whenever the projection valued measures associated with two integrals are "aligned"; for example,

$$\left[(M) \int F(\dagger)\, dE(\dagger)\right] \left[(M) \int dE(\ddagger)\, G(\ddagger)\right] = (M) \int F(\dagger)\, dE(\dagger)\, G(\dagger)$$

or

$$\text{s}\ (m) \int dE(\dagger) \left[\text{s}\ (m) \int F(\ddagger)\, dE(\ddagger) \right] dE(\ddagger) = \text{s}\ (m) \int dE(\dagger)\, F(\dagger)\, dE(\dagger)$$

9. Example: Let $F(\dagger) = E^\dagger$; in that case

$$\sum_{i=1}^{n} E^i \Delta^i = \sum_{i=1}^{n} E^i [E^i - E^{i-1}] = \sum_{i=1}^{n} [E^i - E^{i-1}] = I$$

Hence

$$(M) \int E^\dagger\, dE(\dagger) = \lim_{\mathscr{P}}\, [I] = I$$

On the other hand,

$$\sum_{i=1}^{n} E^{i-1} \Delta^i = \sum_{i=1}^{n} E^{i-1} [E^i - E^{i-1}] = \sum_{i=1}^{n} E^{i-1} - E^{i-1} = 0$$

Hence

$$(m) \int E^\dagger\, dE(\dagger) = \lim_{\mathscr{P}}\, [0] = 0$$

10. Example: Consider the resolution space $l_2^+ = l_2[0, \infty)$ with its usual resolution structure and let V be the *unilateral shift* $[Va]_i = a_{i-1}$, $i = 1, 2, \ldots,$ and $[Va]_0 = 0$. Because of the countable nature of the resolution of the identity, we may without loss of generality assume that our partitions take the form

$$\mathscr{P}_n = \{E^i : i = 0, 1, \ldots, n\}$$

where E^i is the projection onto $V\{e_0, e_1, \ldots, e_{i-1}\}$, $0 < i < n$, with $\{e_i\}_0^\infty$ the standard basis for l_2^+. Of course, $E^0 = 0$ and $E^n = I$, as required of any partition. Therefore E_i is the projection onto $V\{e_i, e_{i+1}, \ldots\}$ with $E_0 = I$ and $E_n = 0$, while Δ^i is the projection onto $V\{e_{i-1}\}$, $i = 1, 2, \ldots, n-1$, with Δ^n the projection onto $V\{e_n, e_{n+1}, \ldots\}$. To compute $\mathrm{s}(M) \int E_\dagger V \, dE(\dagger)$ we must work with the partial sums

$$\sum_{i=1}^{n} E_i V \Delta^i = \sum_{i=1}^{n-1} E_i V \Delta^i$$

where the last term in the sum has been dropped since $E_n = 0$. Now V maps e_{i-1} to e_i, and hence $E_i V \Delta^i = V\Delta^i$, $i = 1, 2, \ldots, n-1$, and

$$\sum_{i-1}^{n} E_i V \Delta^i = \sum_{i=1}^{n-1} E_i V \Delta^i = \sum_{i=1}^{n-1} V \Delta^i = V E^{n-1}$$

Finally, E^{n-1} converges strongly to the identity as n goes to ∞. Hence

$$\mathrm{s}\,(M) \int E_\dagger V dE(\dagger) = V$$

On the other hand, the sequence of operators E^{n-1} is not uniformly convergent, and since the unilateral shift is isometric, neither is the sequence VE^{n-1}. Therefore we conclude that $(M) \int E_\dagger V \, dE(\dagger)$ does not exist.

11. Remark: Although as defined the integrals map a space of $\mathscr{B}$-valued functions into $\mathscr{B}$, we would like to use them to define a mapping from $\mathscr{B}$ to $\mathscr{B}$ by identifying an operator $A \in \mathscr{B}$ with a function $F: \$ \rightarrow \mathscr{B}$ and then integrating the resultant function to obtain a new operator in $\mathscr{B}$. Since the integral of a constant function is trivial, we identify A with the functions $E^\dagger A$, $E_\dagger A$, $AE^\dagger$, and $AE_\dagger$, obtaining the following *integrals of triangular truncation*.

12. Property: For any bounded linear operator A on $(H, \mathscr{E})$ the following pairs of integrals coincide; i.e., one exists if and only if the other exists and when the exist they coincide:

(i)
$$(m)\int E_{\dagger} A\, dE(\dagger) = (M)\int dE(\dagger)\, AE^{\dagger}$$

(ii)
$$(M)\int E_{\dagger} A\, dE(\dagger) = (m)\int dE(\dagger)\, AE^{\dagger}$$

(iii)
$$(m)\int dE(\dagger)\, A\, dE(\dagger) = (M)\int dE(\dagger)\, A\, dE(\dagger)$$

(iv)
$$(M)\int dE(\dagger)\, A\, E_{\dagger} = (m)\int E^{\dagger} A\, dE(\dagger)$$

(v)
$$(m)\int dE(\dagger)\, AE_{\dagger} = (M)\int E^{\dagger} A\, dE(\dagger)$$

and similarly for the corresponding strongly convergent integrals.

Proof: (iii) is trivial since the integrand is constant. In each of the other four cases one can verify equality by showing that the partial sums for any given partition for the two integrals coincide. We illustrate the equivalence of the partial sums for (i). In particular, the partial sums for $(m)\int E_{\dagger} A\, dE(\dagger)$ take the form

$$\sum_{i-1}^{n} E_{i-1} A\Delta^{i} = \sum_{i=1}^{n} \sum_{j=1}^{n} \Delta^{j} A\Delta^{i} = \sum_{j\geq i} \Delta^{j} A\Delta^{i} = \sum_{j=1}^{n} \sum_{i-1}^{j} \Delta^{j} A\Delta^{i} = \sum_{j-1}^{n} \Delta^{j} AE^{j}$$

where the latter summation is just the typical partial sum for

$$(M)\int dE(\dagger)\, AE^{\dagger} \quad //$$

13. Remarks: As with the general integrals, each of the above integrals of triangular truncation is defined only for those A for which the appropriate limits exist. The resultant mapping from (a domain in) $\mathscr{B}$ to $\mathscr{B}$ is, however, linear but, as with the general integrals, unbounded. Indeed, it can be shown that the range space for each of the five integrals of triangular truncation is closed,[2] which together with the fact that they are unbounded and the open mapping theorem implies that they are not even closed operators when viewed as mappings from $\mathscr{B}$ to $\mathscr{B}$. One therefore cannot formulate a straightforward theory around these integrals. They are, however, fundamental to our theory, and the remainder of the present chapter and significant portions of Chapters 2–4 are devoted to characterizing their range spaces and domains. As in the general care the double integral property is the key to the tractability of the integrals of triangular truncation, manifesting itself as

14. Property: The integrals of triangular truncation are projections.

Proof: This follows immediately from the double integral condition of Property 6, and we shall therefore simply sketch the proof for one of the integrals of triangular truncation. Indeed,

$$(m)\int E^{\dagger}\left[(m)\int E^{\ddagger}A\,dE(\ddagger)\right]dE(\dagger) = (m)\int E^{\dagger}E^{\dagger}A\,dE(\dagger)$$

$$= (m)\int E^{\dagger}A\,dE(\dagger)$$

since $E^{\dagger}$ is a projection. //

We complete the section with a characterization of the relationship between the various integrals of triangular truncation, the proof of this relationship follows by direct comparison of the appropriate partial sums coupled with the equalities $A = (M)\int A\,dE(\dagger) = (m)\int A\,dE(\dagger)$.

15. Property: Let A be a bounded linear operator on $(H, \mathscr{E})$. Then

(i) $(m)\int E_{\dagger}A\,dE(\dagger) = (m)\int dE(\dagger)\,A\,dE(\dagger) + (M)\int E_{\dagger}A\,dE(\dagger)$

(ii) $(M)\int E^{\dagger}A\,dE(\dagger) = (M)\int dE(\dagger)\,A\,dE(\dagger) + (m)\int E_{\dagger}A\,dE(\dagger)$

(iii) $A = (M)\int E_{\dagger}A\,dE(\dagger) + (M)\int E^{\dagger}A\,dE(\dagger)$

(iv) $A = (m)\int E^{\dagger}A\,dE(\dagger) + (m)\int E_{\dagger}A\,dE(\dagger)$

(v) $A = (M)\int E_{\dagger}A\,dE(\dagger) + (m)\int dE(\dagger)\,A\,dE(\dagger) + (m)\int E^{\dagger}A\,dE(\dagger)$

wherever the appropriate integrals of triangular truncation exist, with similar results for the corresponding strongly convergent integrals. Moreover, if all but one of the integrals of triangular truncation in any of the above expressions exist, then they all exist.

E. INTEGRAL REPRESENTATION

1. Remark: The range spaces for the integrals of triangular truncation (i), (iii), and (iv) are just the causal, memoryless, and anticausal operators, respectively, as follows.

2. Property: The integrals of triangular truncation

$$(m)\int E_{\dagger} A\, dE(\dagger) = (M)\int dE(\dagger)\, AE^{\dagger}$$

and

$$\mathrm{s}\,(m)\int E_{\dagger} A\, dE(\dagger) = \mathrm{s}\,(M)\int dE(\dagger)\, AE^{\dagger}$$

are projections onto $\mathscr{C}$.

Proof: If A is causal, the partial sums for $(M)\int dE(\dagger)\, AE^{\dagger}$ are given by

$$\sum_{i=1}^{n} \Delta^i A E^i = \sum_{i=1}^{n} \Delta^i E^i A E^i = \sum_{i=1}^{n} \Delta^i E^i A = \sum_{i=1}^{n} \Delta^i A = \left[\sum_{i=1}^{n} \Delta^i\right] A = A$$

where we have used the properties $\Delta^i = \Delta^i E^i$ and $[\sum_{i=1}^{n} \Delta^i] = I$ in addition to the causality of A. Since each partial sum is equal to A, so is the limit independently of topology. Hence

$$(m)\int E_{\dagger} A\, dE(\dagger) = (M)\int dE(\dagger)\, AE^{\dagger} = A = \mathrm{s}\,(M)\int dE(\dagger)\, AE^{\dagger}$$

$$= \mathrm{s}\,(m)\int E_{\dagger} A\, dE(\dagger)$$

verifying that $\mathscr{C}$ is contained in the range of these integrals. To verify that $\mathscr{C}$ is the entire range of the integral operator we let $B = (M)\int dE(\dagger)\, AE^{\dagger}$ where $A \in \mathscr{B}$. Since this integral converges over the net of all partitions it also converges over the subnet defined by partitions $\mathscr{P}^{\dagger}$, which include $E^{\dagger}$, say $E^{\dagger} = E^k$, where $\dagger$ is a fixed element of $. Then

$$E^{\dagger}B = E^{\dagger}\left[\lim_{\mathscr{P}^{\dagger}}\left[\sum_{i=1}^{n} \Delta^i A E^i\right]\right] = \lim_{\mathscr{P}^{\dagger}}\left[\sum_{i=1}^{n} E^{\dagger}\Delta^i A E^i\right]$$

$$= \lim_{\mathscr{P}^{\dagger}}\left[\sum_{i=1}^{k} E^k\Delta^i A E^i\right] = \lim_{\mathscr{P}^{\dagger}}\left[\sum_{i=1}^{k} E^k\Delta^i A E^i E^k\right]$$

$$= \lim_{\mathscr{P}^{\dagger}}\left[\sum_{i=1}^{n} E^{\dagger}\Delta^i A E^i E^{\dagger}\right] = E^{\dagger}\left[\lim_{\mathscr{P}^{\dagger}}\left[\sum_{i=1}^{n} \Delta^i A E^i\right]\right]E^{\dagger}$$

$$= E^{\dagger}BE^{\dagger}$$

showing by Remark B.3 that B is causal. A parallel argument may be invoked for the strongly convergent integrals. //

3. Remark: The fact that both the uniformly convergent and strongly convergent integrals of triangular truncation define projections onto the causals should not be surprising since the causals were defined algebraically and are independent of topology. It should, however, be noted that the domains of these projections are dependent on the topology employed, with the domain of the strongly convergent integrals strictly larger than the domain of the uniformly convergent integrals since the strong topology is coarser than the uniform topology.

By combining the equalities,

$$(m) \int E_{\dagger} A \, dE(\dagger) = (M) \int dE(\dagger) \, AE^{\dagger} = A$$

which hold if and only if A is causal, with the equalities of Property D.15, a number of alternative characterizations for the causal operator are obtained:

4. Corollary: The following are equivalent:

(i) A is causal,

(ii) $$(m) \int E_{\dagger} A \, dE(\dagger) = (M) \int dE(\dagger) \, AE^{\dagger} = A$$

(iii) $$(M) \int dE(\dagger) \, AE_{\dagger} = (m) \int E^{\dagger} A \, dE(\dagger) = 0$$

(iv) $$\mathrm{s}\,(m) \int E_{\dagger} A \, dE(\dagger) = \mathrm{s}\,(M) \int dE(\dagger) \, AE^{\dagger} = A$$

(v) $$\mathrm{s}\,(M) \int dE(\dagger) \, AE_{\dagger} = \mathrm{s}\,(m) \int E^{\dagger} A \, dE(\dagger) = 0$$

The proof follows immediately from Properties 1 and D.15 and is left to the reader (Exercise 19).

By similar arguments one can formulate parallel characterizations for the anticausal and memoryless operators:

5. Property: The integrals of triangular truncation

$$(m) \int dE(\dagger) \, AE_{\dagger} = (M) \int E^{\dagger} A \, dE(\dagger)$$

and

$$\mathrm{s}\,(m)\int dE(\dagger)\,AE_{\dagger} = \mathrm{s}\,(M)\int E^{\dagger}A\,dE(\dagger)$$

are projections onto $\mathscr{C}^*$.

6. Corollary: The following are equivalent:

(i) A is anticausal,

(ii) $$(m)\int dE(\dagger)\,AE_{\dagger} = (M)\int E^{\dagger}A\,dE(\dagger) = A$$

(iii) $$(M)\int E_{\dagger}A\,dE(\dagger) = (m)\int dE(\dagger)\,AE^{\dagger} = 0$$

(iv) $$\mathrm{s}\,(m)\int dE(\dagger)\,AE_{\dagger} = \mathrm{s}\,(M)\int E^{\dagger}A\,dE(\dagger) = A$$

(v) $$\mathrm{s}\,(M)\int E_{\dagger}A\,dE(\dagger) = \mathrm{s}\,(m)\int dE(\dagger)\,AE^{\dagger} = 0$$

7. Property: The integrals of triangular truncation

$$(m)\int dE(\dagger)A\,dE(\dagger) = (M)\int dE(\dagger)\,dE(\dagger)$$

and

$$\mathrm{s}\,(m)\int dE(\dagger)A\,dE(\dagger) = \mathrm{s}\,(M)\int dE(\dagger)\,A\,dE(\dagger)$$

are projections onto $\mathscr{M}$.

8. Corollary: The following are equivalent:

(i) A is memoryless,

(ii) $$(m)\int dE(\dagger)\,A\,dE(\dagger) = (M)\int dE(\dagger)\,A\,dE(\dagger) = A$$

(iii) $$\mathrm{s}\,(m)\int dE(\dagger)\,A\,dE(\dagger) = \mathrm{s}\,(M)\int dE(\dagger)\,A\,dE(\dagger) = A$$

9. Example: Consider the unilateral shift V on $l_2[0, \infty)$ with its usual resolution structure and let $\mathscr{P}_n = \{E^i : i = 0, 1, \ldots, n\}$ be the partition defined in Example D.10. Then using the notation of that example and recalling that V maps e_{i-1} to e_i, which is in the range of E_{i-1}, we have

$$E_{i-1} V \Delta^i = V \Delta^i, \qquad i = 1, 2, \ldots, n$$

Hence

$$(m) \int E_\dagger V \, dE(\dagger) = \lim_{\mathscr{P}_n} \left[\sum_{i=1}^{n} E_{i-1} V \Delta^i \right] = \lim_{\mathscr{P}_n} \left[\sum_{i=1}^{n} V \Delta^i \right]$$

showing that V is causal. Of course, since the uniform limit converges, so does the strong limit, implying that

$$\text{s}\,(m) \int E_\dagger V \, dE(\dagger) = V$$

10. Remark: A comparison of Example 9 with Example D.10 will reveal that while

$$(m) \int E_\dagger V \, dE(\dagger) = V = \text{s}\,(m) \int E_\dagger V \, dE(\dagger)$$

independently of topology,

$$\text{s}\,(M) \int E_\dagger V \, dE(\dagger) = V$$

while $(M) \int E_\dagger V \, dE(\dagger)$ fails to exist. Thus, unlike the integrals of triangular truncation discussed in the present section,

$$(M) \int E_\dagger A \, dE(\dagger) = (m) \int dE(\dagger) \, A E^\dagger$$

and

$$\text{s}\,(M) \int E_\dagger A \, dE(\dagger) = \text{s}\,(m) \int dE(\dagger) \, A E^\dagger$$

do not project onto the same space of operators. The characterization of these range spaces and the subtle distinction between them is the primary topic of Chapter 2.

PROBLEMS

1. Show that a sequence of orthogonal projections is uniformly convergent if and only if it is eventually constant.

2. Let E be a projection on a Hilbert space with range $E(H)$. Show that $I - E$ is also a projection and that its range is the orthogonal complement of $E(H)$.

3. Show that the class of operators $E^{\dagger}$ defined in Example A.8 are, in fact, orthogonal projections relative to the reproducing kernel Hilbert space topology.

4. Repeat Problem 3 for the class of operators defined in Remark A.9.

5. For a nonlinear operator A on $(H, \mathscr{E})$ show that items (i) and (ii) of Property B.4 are equivalent and that items (iii) and (iv) of Property B.4 are equivalent. Furthermore, show that items (i) and (ii) imply items (iii) and (iv) but not conversely.

6. Show that a linear operator on $(H, \mathscr{E})$ is causal if and only if $\|E^{\dagger}Ax\| \leq \|AE^{\dagger}x\|$ for all $x \in H$ and $\dagger \in \$$.

7. Give an explicit representation for the resolution of the identity made up from the projections $FE^{n}F^{-1}$, $n \in Z$, defined on $L_2(T)$ in Remark B.13.

8. Show that the resolution spaces $L_2(-\infty, \infty)$ and $l_2(-\infty, \infty)$ are both maximal.

9. Show that the ordering $<^*$, defined on $\$^*$ by $\dagger <^* \ddagger$ if and only if $\ddagger < \dagger$ on $\$$, is a well-defined linear ordering.

10. Prove Properties C.3 and C.4.

11. Prove Properties C.10 and C.12.

12. Show that the core of a resolution space is abelian.

13. Show that the commutant of the core of a resolution space is the algebra of memoryless operators.

14. Show that the core of a resolution space is the center of the algebra of memoryless operators.

15. Show that the set of all partitions of a resolution of the identity forms a directed set when ordered by containment.

16. Give precise definitions for the uniform, strong, ultraweak, and weak operator topologies.

17. Give an example of a function F and a resolution of the identity $\mathscr{E}$ such that

$$(M) \int F(\dagger)\, dE(\dagger) \neq (M) \int dE(\dagger)\, F(\dagger)$$

18. Compute $(m) \int E_\dagger U \, dE(\dagger)$ where U is the bilateral shift on $l_2(-\infty, \infty)$ and $\mathscr{E}$ is its usual resolution structure.

19. Prove Corollary E.4.

20. Compute

(i) $$(M) \int I \, dE(\dagger)$$

(ii) $$(m) \int dE(\dagger) \, E_\dagger \, dE(\dagger)$$

(iii) $$s\,(M) \int E^\dagger \, dE(\dagger) \, E_\dagger$$

(Be sure to give a precise definition of the last integral.)

21. Define the various integrals of triangular truncation by taking limits in the weak and ultraweak operator topologies and show that Propositions E.2, E.5, and E.7 remain valid when these topologies are employed.

REFERENCES

1. Kailath, T., and Duttweiler, D., An RKHS approach to detection and estimation problems—Part III: Generalized innovations representations and a likelihood ration formula, *IEEE Trans. Informat. Theory* **IT-18**, 730–745 (1972).
2. Saeks, R., and Goldstein, R. A., Cauchy integrals and spectral measures, *Indiana Math. J.* **22**, 367–378 (1972).
3. Youla, D. C., Carlin, H. J., and Castriota, L. J., Bounded real scattering matrices and the foundations of linear passive network theory. *IRE Trans. Circuit Theory* **CT-6**, 102–124 (1959).

2

Hypercausality

A. STRICT CAUSALITY

1. Remark: Although causality is the fundamental physical constraint underlying system theory, one often requires a stronger concept that in one sense or another implies the present output of the system is not dependent on the present input. Coupled with causality, this means that the present output is dependent only on inputs in the "strict" past. In the present chapter we formulate several such *hypercausality* concepts, each of which reduces in the finite-dimensional case to the requirement that an operator have a *strictly lower triangular matrix* representation.

2. Definition: Let A be a bounded linear operator on $(H, \mathscr{E})$. Then A is *strictly causal* if A is causal and if for each $\varepsilon > 0$ there exists a partition $\mathscr{P}'$ of $\mathscr{E}$ such that for every partition $\mathscr{P} = \{E^i : i = 0, 1, \ldots, n\}$ of $\mathscr{E}$ that refines $\mathscr{P}'$

$$\|\Delta^i A \Delta^i\| < \varepsilon, \qquad i = 1, 2, \ldots, n$$

3. Remark: Since Δ^i is the projection onto that part of H that corresponds to the time interval $(\dagger_{i-1}, \dagger_i]$, Definition 2 implies that the response of A in this time interval, due to an input in this time interval, is small. As such, Definition A.2 is consistent with the intuitive concept of hypercausality. We denote the strictly causal operators on $(H, \mathscr{E})$ by $\mathscr{R}$.

Before proceeding to the characterization of the strictly causal operators, we prove a fundamental lemma to the effect that the projection-valued measure defined by a resolution of the identity has bounded variation.

4. Lemma: Let $\mathscr{E}$ be a resolution of the identity on H, let $\mathscr{P} = \{E^i : i = 0, 1, \ldots, n\}$ be a partition of $\mathscr{E}$, and let A^i, $i = 1, 2, \ldots, n$, be bounded linear operators on H. Then

$$\left\| \sum_{i=1}^{n} \Delta^i A^i \Delta \right\| = \max_i \|\Delta^i A^i \Delta^i\|$$

Proof: Let $x \in H$. Then since the Δ^i have mutually orthogonal ranges,

$$\begin{aligned}\left\| \sum_{i=1}^{n} \Delta^i A^i \Delta^i x \right\|^2 &= \sum_{i=1}^{n} \|\Delta^i A^i \Delta^i x\|^2 = \sum_{i=1}^{n} \|\Delta^i A^i \Delta^i \Delta^i x\|^2 \\ &\le \sum_{i=1}^{n} \|\Delta^i A^i \Delta^i\|^2 \|\Delta^i x\|^2 \le \left(\max_i \|\Delta^i A^i \Delta^i\|^2 \right) \sum_{i=1}^{n} \|\Delta^i x\|^2 \\ &= \left(\max_i \|\Delta^i A^i \Delta^i\|^2 \right) \|x\|^2\end{aligned}$$

Since this holds for all x, it implies that

$$\left\| \sum_{i=1}^{n} \Delta^i A^i \Delta^i \right\| \le \max_i \|\Delta^i A^i \Delta^i\|$$

and the opposite inequality may be obtained by working with any nonzero x in the range of Δ^i, $i = 1, 2, \ldots, n$, to complete the proof. //

5. Property: The *integral of triangular truncation*

$$(M) \int E_\dagger A \, dE(\dagger) = (m) \int dE(\dagger) \, AE^\dagger$$

is a projection onto $\mathscr{R}$.

Proof: By the lemma the strict causality condition is equivalent to the condition that there exist a partition $\mathscr{P}'$ of $\mathscr{E}$ such that for every partition $\mathscr{P} = \{E^i : i = 0, 1, \ldots, n\}$ that refines $\mathscr{P}'$

$$\left\| \sum_{i=1}^{n} \Delta^i A \Delta^i \right\| < \varepsilon$$

These are, however, just the partial sums for $(m) \int dE(\dagger)\, A\, dE(\dagger)$. Hence

$$(m) \int dE(\dagger)\, A\, dE(\dagger) = 0$$

On the other hand, the causality condition for a strictly causal operator is equivalent to the requirement that

$$(m) \int E_\dagger A\, dE(\dagger) = A$$

Now by Property 1.D.15

$$\begin{aligned} A &= (m) \int E_\dagger A\, dE(\dagger) = (m) \int dE(\dagger)\, A\, dE(\dagger) + (M) \int E_\dagger A\, dE(\dagger) \\ &= (M) \int E_\dagger A\, dE(\dagger) \end{aligned}$$

showing that A is in the range of the projection $(M) \int E_\dagger A\, dE(\dagger)$. Conversely, if $A = (M) \int E_\dagger A\, dE(\dagger)$, an argument similar to that used in the proof of Property 1.E.2 will reveal that A is causal and hence the validity of the above equality, which in turn implies that $(m) \int dE(\dagger)\, A\, dE(\dagger) = 0$ and that A is strictly causal. //

Combining the above results with Property 1.D.15 we obtain the following alternative characterizations of strict causality:

6. Corollary: The following are equivalent:

(i) A is strictly causal,
(ii) $(M) \int E_\dagger A\, dE(\dagger) = (m) \int dE(\dagger)\, AE^\dagger = A$,
(iii) A is causal and $(M) \int dE(\dagger)\, A\, dE(\dagger) = (m) \int dE(\dagger)\, A\, dE(\dagger) = 0$, and
(iv) $(M) \int E^\dagger A\, dE(\dagger) = (m) \int dE(\dagger)\, AE_\dagger = 0$.

7. Example: In the finite-dimensional Hilbert resolution space of Example 1.B.8 $\mathscr{E}$ is finite, in which case we may without loss of generality work with the maximal partition $\mathscr{P} = \mathscr{E}$. Thus Δ^i is the projection onto the ith element of the basis constructed in that example, and $\Delta^i A \Delta^i$ has the matrix representation

$$(\Delta^i A \Delta^i)_{ij} = \begin{bmatrix} 0 & 0 & 0 & \cdots & 0 & \cdots & 0 \\ 0 & 0 & 0 & \cdots & 0 & \cdots & 0 \\ \vdots & \vdots & \vdots & & \vdots & & \vdots \\ 0 & 0 & 0 & \cdots & a_{ii} & \cdots & 0 \\ \vdots & \vdots & \vdots & & \vdots & & \vdots \\ 0 & 0 & 0 & \cdots & 0 & \cdots & 0 \end{bmatrix}$$

Hence the strict causality condition requires that $a_{ii} = 0, i = 1, 2, \ldots, n$. Combined with the causality condition an operator on this space is thus strictly causal if and only if $a_{ij} = 0$ for $i \leq j$, i.e., if and only if the matrix representation of A is strictly lower triangular.

8. Remark: The example lends credence to the intuition that the strictly causal operators represent an infinite-dimensional extension of the strictly triangular matrices. Unfortunately, this intuition does not extend to $l_2[0, \infty)$, on which the unilateral shift has a strictly lower triangular (infinite) matrix representation but is not strictly causal since

$$(M) \int E_\dagger V dE(\dagger)$$

does not exist (see Example 1.D.10).

9. Example: Consider the integral operator with kernel in $L_2[(0, 1) \times (0, 1)]$

$$[Af](\dagger) = \int_0^1 G(\dagger, \ddagger) f(\ddagger)\, d\ddagger$$

defined on the space $L_2(0, 1)$ with its usual resolution structure. Here

$$\|A\|^2 = \int_0^1 \int_0^1 |G(\dagger, \ddagger)|^2\, d\dagger\, d\ddagger < \infty$$

and it follows from Example 1.B.10 that A is causal if and only if $G(\dagger, \ddagger) = 0$ for $\dagger < \ddagger$. Now, if $\mathscr{P} = \{E^i : i = 0, 1, \ldots, n\}$ is a partition of the usual resolution structure, Δ^i is the operator that multiplies a function in $L_2(0, 1)$ by the characteristic function of the interval $(\dagger_{i-1}, \dagger_i]$. Therefore $\Delta_i A \Delta^i$ is represented by the kernel

$$\chi_{(\dagger_{i-1}, \dagger_i]}(\dagger) G(\dagger, \ddagger) \chi_{(\dagger_{i-1}, \dagger_i]}(\ddagger)$$

with norm

$$\|\Delta^i A \Delta^i\|^2 = \int_{\dagger_{i-1}}^{\dagger_i} \int_{\dagger_{i-1}}^{\dagger_i} |G(\dagger, \ddagger)|^2\, d\dagger\, d\ddagger$$

which goes to zero as the partition is refined. Hence if $G(\dagger, \ddagger) = 0, \dagger < \ddagger$, the integral operator is actually strictly causal. This property is, however, dependent both on the finite time interval employed and on the fact that

$G(\dagger, \ddagger)$ lies in $L_2[(0, 1) \times (0, 1)]$. Indeed, if $G(\dagger, \ddagger)$ were allowed to be distributional with finite mass on the diagonal (say a $\delta(\dagger - \dagger)$ term) it would not be strictly causal.

10. Property: $\mathscr{R}$ is a uniformly closed two-sided ideal in $\mathscr{C}$.

Proof: Since $(M) \int E_\dagger A \, dE(\dagger)$ is linear, $\mathscr{R}$ is closed under addition. Now let $A \in \mathscr{R}$ and $B \in \mathscr{C}$; we desire to show that $BA \in \mathscr{R}$. Since the causals are closed under multiplication, BA is causal and it suffices to show that $(m) \int dE(\dagger) \, BA \, dE(\dagger) = 0$ given that $(m) \int dE(\dagger) \, A \, dE(\dagger) = 0$. First consider an arbitrary partition $\mathscr{P} = \{E^i : i = 0, 1, \ldots, n\}$ and compute

$$\sum_{i=1}^{n} \Delta^i BA\Delta^i = \sum_{i=1}^{n} \Delta^i B\left[\sum_{j=1}^{n} \Delta^j A\right]\Delta^i = \sum_{i=1}^{n} \Delta^i B\Delta^i A\Delta^i$$
$$= \left[\sum_{i=1}^{n} \Delta^i B\Delta^i\right]\left[\sum_{j=1}^{n} \Delta^j A\Delta^j\right]$$

since $\Delta^i\Delta^j = 0$, $i \neq j$, $\Delta^i\Delta^i = \Delta^i$, and A and B are causal. Now, since A is strictly causal there exists a partition $\mathscr{P}'$ such that for any partition $\mathscr{P}$ that refines $\mathscr{P}'$

$$\left\|\sum_{j=1}^{n} \Delta^j A\Delta^j\right\| < \varepsilon$$

Thus with the aid of Lemma 4 we obtain

$$\left\|\sum_{i=1}^{n} \Delta^i BA\Delta^i\right\| = \left\|\left[\sum_{i=1}^{n} \Delta^i B\Delta^i\right]\left[\sum_{j=1}^{n} \Delta^j A\Delta^j\right]\right\|$$
$$\leq \left\|\sum_{i=1}^{n} \Delta^i B\Delta^i\right\| \left\|\sum_{j=1}^{n} \Delta^j A\Delta^j\right\| = \left(\max_i \|\Delta^i B\Delta^i\|\right)\left\|\sum_{j=1}^{n} \Delta^j A\Delta^j\right\|$$
$$\leq \|B\| \left\|\sum_{j=1}^{n} \Delta^j A\Delta^j\right\| < \|B\|\varepsilon$$

implying that

$$(m) \int dE(\dagger) \, BA \, dE(\dagger) = 0$$

Then $BA \in \mathscr{R}$, while a similar argument yields $AB \in \mathscr{R}$. Of course, if $B \in \mathscr{R}$ the above argument shows that $\mathscr{R}$ is closed under multiplication, thereby making it an algebra.

To complete the proof we must verify that $\mathscr{R}$ is closed in the uniform operator topology. Let A_π be a sequence of strictly causal operators that converges uniformly to A. A is clearly causal owing to Property 1.B.6, and it suffices to show that $(m)\int dE(\dagger)\, A\, dE(\dagger) = 0$ given that

$$(m)\int dE(\dagger)\, A_\pi\, dE(\dagger) = 0$$

For any $\varepsilon > 0$ choose γ' such that

$$\|A - A_\gamma\| < \varepsilon/2$$

whenever $\gamma > \gamma'$ and a partition $\mathscr{P}'$ such that for any partition $\mathscr{P} = \{E^i : i = 0, 1, \ldots, n\}$ that refines $\mathscr{P}'$,

$$\left\|\sum_{i=1}^{n} \Delta^i A_\gamma \Delta^i\right\| < \varepsilon/2$$

Then

$$\begin{aligned}\left\|\sum_{i=1}^{n} \Delta^i A \Delta^i\right\| &= \left\|\sum_{i=1}^{n} \Delta^i[A - A_\gamma]\Delta^i + \sum_{i=1}^{n} \Delta^i A \Delta^i\right\| \\ &\leq \left\|\sum_{i=1}^{n} \Delta^i[A - A_\gamma]\Delta^i\right\| + \left\|\sum_{i=1}^{n} \Delta^i A_\gamma \Delta^i\right\| \\ &= \left(\max_i \|\Delta^i[A - A_\gamma]\Delta^i\|\right) + \left\|\sum_{i=1}^{n} \Delta^i A_\gamma \Delta^i\right\| < \varepsilon/2 + \varepsilon/2 = \varepsilon\end{aligned}$$

Hence

$$(m)\int dE(\dagger)\, A\, dE(\dagger) = \lim_{\mathscr{P}}\left[\sum_{i=1}^{n} \Delta^i A \Delta^i\right] = 0$$

and A is strictly causal, as was to be shown. //

11. Remark: To formulate a concept of strict anticausality we employ the *dual resolution of the identity* $\mathscr{E}^*$, where $\mathscr{E}^* = \{E_\dagger : \dagger \in \$^*\}$ and $\$^*$ is the set $\$$ with the *reverse ordering* $\dagger <^* \ddagger$ if and only if $\ddagger < \dagger$, while $\dagger_0^* = \dagger_\infty$ and $\dagger_\infty^* = \dagger_0$.

12. Definition: An operator A on $(H, \mathscr{E})$ is *strictly anticausal* if it is strictly causal on $(H, \mathscr{E}^*)$.

13. Property: A linear operator A on $(H, \mathscr{E})$ is strictly anticausal if and only if A^* is strictly causal.

Proof: For A to be strictly anticausal A must be causal on $(H, \mathscr{E}^*)$ and for each $\varepsilon > 0$ there must exist a partition $\mathscr{P}^{*\prime}$ of $\mathscr{E}^*$ such that for every partition $\mathscr{P}^* = \{E_{\dagger_i^*} : i = 0, 1, \ldots, n\}$ of $\mathscr{E}^*$ that refines $\mathscr{P}^{*\prime}$

$$\|\Delta_{\dagger_i^*} A \Delta_{\dagger_i^*}\| < \varepsilon, \qquad i = 1, 2, \ldots, n$$

Now with each such partition $\mathscr{P}^*$ of $\mathscr{E}^*$ we associate a partition $\mathscr{P} = \{E^j : j = 0, 1, \ldots, n\}$ where $\dagger_j = \dagger_{n-j}^*$. Of course, since \$ and \$* represent the same set, although with different orderings, this new partition of $\mathscr{E}$ is well defined. Moreover,

$$\begin{aligned} \Delta^{\dagger_j} &= [E^{\dagger_j} - E^{\dagger_{j-1}}] = [I - E_{\dagger_j}] - [I - E_{\dagger_{j-1}}] = [E_{\dagger_{j-1}} - E_{\dagger_j}] \\ &= [E_{\dagger_{n-j+1}^*} - E_{\dagger_{n-j}^*}] = [E_{\dagger_i^*} - E_{\dagger_{i-1}^*}] = \Delta_{\dagger_i^*} \end{aligned}$$

where

$$i = n - j + 1, \qquad j = 1, 2, \ldots, n$$

So if we let $\mathscr{P}'$ be the partition associated with $\mathscr{P}^{*\prime}$ and $\mathscr{P}$ be the partition associated with $\mathscr{P}^*$ as defined above, then if A is strictly anticausal, for any $\varepsilon > 0$ there exists a partition $\mathscr{P}'$ such that for every partition $\mathscr{P}$ that refines $\mathscr{P}'$

$$\begin{aligned} \|\Delta^{\dagger_j} A^* \Delta^{\dagger_j}\| &= \|[\Delta^{\dagger_j} A^* \Delta^{\dagger_j}]^*\| = \|\Delta^{\dagger_j} A \Delta^{\dagger_j}\| \\ &= \|\Delta_{\dagger_i^*} A \Delta_{\dagger_i^*}\| < \varepsilon, \qquad j = 1, 2, \ldots, n \end{aligned}$$

On the other hand, since A is causal on $(H, \mathscr{E}^*)$, it is anticausal on $(H, \mathscr{E})$, and so A^* is causal on $(H, \mathscr{E})$. A^* thus satisfies the requirements for strict causality on $(H, \mathscr{E})$, and a dual argument will verify that A is strictly anticausal if A^* is strictly causal on $(H, \mathscr{E})$. //

14. Remark: Consistent with Property 13 we denote the strictly anticausal operators $\mathscr{R}^*$. The characteristics of the strictly anticausal operators are dual to those of the strictly causals, as follows.

15. Property: The integral of triangular truncation

$$(M) \int dE(\dagger)\, A E_\dagger = (m) \int E^\dagger A \, dE(\dagger)$$

is a projection onto $\mathscr{R}^*$.

16. Corollary: The following are equivalent:

(i) A is strictly anticausal,
(ii) $(M) \int dE(\dagger) AE_\dagger = (m) \int E^\dagger A\, dE(\dagger) = A$,
(iii) A is strictly causal and

$$(M) \int dE(\dagger)\, A\, dE(\dagger) = (M) \int dE(\dagger)\, A\, dE(\dagger) = 0$$

and

(iv) $(M) \int dE(\dagger) AE^\dagger = (m) \int E_\dagger A\, dE(\dagger) = 0$.

17. Property: $\mathscr{R}^*$ is a uniformly closed two-sided ideal in $\mathscr{C}^*$.

B. STRONG STRICT CAUSALITY

1. Remark: The purpose of this section is to characterize the range of the strongly convergent integral of triangular truncation

$$\mathrm{s}\,(M) \int E_\dagger A\, dE(\dagger) = \mathrm{s}\,(m) \int dE(\dagger)\, AE^\dagger$$

Unlike

$$\mathrm{s}\,(m) \int E_\dagger A\, dE(\dagger) = \mathrm{s}\,(M) \int dE(\dagger)\, AE^\dagger$$

whose range is the causal operators independently of topology, the range of the former integral is dependent on the choice of topology. Indeed, in Example 1.D.10 it was shown that $\mathrm{s}\,(M) \int E_\dagger V\, dE(\dagger) = V$, where V is the unilateral shift on $l_2[0, \infty)$ with its usual resolution structure, while $(M) \int E_\dagger V\, dE(\dagger)$ does not exist. V is therefore in the range of the strongly convergent integral but not strictly causal.

2. Definition: Let A be a bounded linear operator on $(H, \mathscr{E})$. Then A is *strongly strictly causal* if it is causal and if for $x \in H$ and $\varepsilon > 0$ there exists a partition $\mathscr{P}'$ of $\mathscr{E}$ such that for every partition $\mathscr{P} = \{E^i : i = 0, 1, \ldots, n\}$ of $\mathscr{E}$ that refines $\mathscr{P}'$

$$\sum_{i=1}^{n} \|\Delta^i A \Delta^i x\|^2 < \varepsilon$$

3. Remark: On first impression one might expect to extend the strict causality concept to the strongly convergent case simply by requiring that

$\|\Delta^i A \Delta^i x\| < \varepsilon$, $i = 1, 2, \ldots, n$. This is, however, too weak a condition. (It is satisfied by every bounded linear operator.) With the aid of the Pythagorean theorem we obtain

$$\sum_{i=1}^{n} \|\Delta^i A \Delta^i x\|^2 = \left\| \left[\sum_{i=1}^{n} \Delta^i A \Delta^i \right] x \right\|^2 < \varepsilon$$

Hence the above strong strict causality condition is equivalent to the requirement that $\mathrm{s}\,(M) \int dE(\dagger)\, A\, dE(\dagger) = \mathrm{s}\,(m) \int dE(\dagger)\, A\, dE(\dagger) = 0$.

The strongly strictly causal operators are denoted $\mathscr{S}$. Clearly $\mathscr{R} \subset \mathscr{S}$ since the strongly convergent integral extends the uniformly convergent integral. Of course, the unilateral shift provides an example showing that the containment is proper.

4. Property: The integral of triangular truncation

$$\mathrm{s}\,(M) \int E_{\dagger}\, A\, dE(\dagger) - \mathrm{s}\,(m) \int dE(\dagger)\, A E^{\dagger}$$

is a projection onto $\mathscr{S}$.

5. Corollary: The following are equivalent.

(i) A is strongly strictly causal,

(ii) $\mathrm{s}\,(M) \int E_{\dagger} A\, dE(\dagger) = \mathrm{s}\,(m) \int dE(\dagger)\, A E^{\dagger} = A$,

(iii) A is causal and

$$\mathrm{s}\,(M) \int dE(\dagger)\, A\, dE(\dagger) = \mathrm{s}\,(m) \int dE(\dagger)\, A\, dE(\dagger) = 0$$

and

(iv) $\mathrm{s}\,(M) \int E^{\dagger} A\, dE(\dagger) = \mathrm{s}\,(\mathrm{m}) \int dE(\dagger)\, A E_{\dagger} = 0$.

6. Example: Let $H = l_2(-\infty, \infty)$ with its usual resolution structure and let A be a bounded linear operator on $l_2(-\infty, \infty)$ with matrix representation (a_{ij}) relative to the standard orthonormal basis $\{e_i\}_{-\infty}^{\infty}$ on H. We have already seen in Example 1.B.10 that A is causal if and only if (a_{ij}) is lower triangular, i.e., $a_{ij} = 0$ for $i < j$. Here, we would like to show that A is strongly strictly causal if and only if $a_{ij} = 0$ for $i \leq j$.

Since every strongly strictly causal operator is causal, $a_{ij} = 0$ for $i < j$, and it suffices to show that a causal A will be strongly strictly causal if and only if $a_{ii} = 0$ and its converse. Suppose that $a_{ii} = 0$ for all i. Then given

$\varepsilon > 0$ and $b = \{b_i\}_{-\infty}^{\infty} \in l_2(-\infty,\infty)$, we construct a partition of $\mathscr{E}$ as follows. First, choose N such that

$$\sum_{i=-\infty}^{-N} |b_i|^2 < \varepsilon/2\|A\|^2, \qquad \sum_{i=N}^{\infty} |b_i|^2 < \varepsilon/2\|A\|^2$$

Now define a partition by letting the ith element of the partition be given by E^{i-N-1}, $1 \leq i \leq 2N$, with the zeroth element equal to the zero operator and the $(2N+1)$th element equal to the identity:

$$\mathscr{P} = \{0, E^{-N}, E^{1-N}, E^{2-N}, \ldots, E^{N-1}, I\}$$

To be compatible with the above notation we index this partition by $\mathscr{P} = \{E^i : i = -N-1, -N, \ldots, N\}$ where $E^{-N-1} = 0$ and $E^N = I$. Since $a_{ii} = 0$,

$$\Delta^i A \Delta^i = 0, \qquad -N < i < N$$

and hence

$$\begin{aligned} \left\| \sum_{i=-N}^{N} \Delta^i A \Delta^i b \right\|^2 &= \|E^{-N}AE^{-N}b\|^2 + \|E_{-N}AE_{-N}b\|^2 \\ &\leq \|A\|^2[\|E^{-N}b\|^2 + \|E_{-N}b\|^2] \\ &= \|A\|^2\left[\sum_{i=N}^{\infty}|b_i|^2 + \sum_{i=-\infty}^{-N}|b_i|^2\right] < \varepsilon \end{aligned}$$

showing that A is strongly strictly causal.

On the other hand, if A is strongly strictly causal, let $\varepsilon > 0$ and $b = e_i$, the ith element of the standard basis. Then there exists a partition for which any refinement thereof satisfies

$$\left\| \sum_{j=1}^{n} \Delta^j A \Delta^j e_i \right\| < \varepsilon$$

verifying that $a_{ii} = 0$ for a strongly strictly causal A.

7. Remark: Consistent with the example our intuition to the effect that hypercausality is an abstraction of the strictly lower triangular matrices is valid for the strongly strictly causal operators. Of course, since $\mathscr{R} \subset \mathscr{S}$ every strictly causal operator on $l_2(-\infty, \infty)$ also admits a strictly lower triangular matrix representation, but not the converse. Indeed, the bilateral shift is a counterexample. Of course, the above characterization of strong strict causality also holds for operators on $l_2[0, \infty)$ with the obvious modifications.

8. Property: $\mathscr{S}$ is a uniformly closed right ideal in $\mathscr{C}$.

Proof: Since $\mathrm{s}\,(M) \int E_\dagger A\, dE(\dagger)$ is linear, $\mathscr{S}$ is closed under addition. Now let $A \in \mathscr{S}$ and $B \in \mathscr{C}$; we desire to show that $BA \in \mathscr{S}$. Since the causals are closed under multiplication, BA is causal and it suffices to show that $\mathrm{s}\,(m) \int dE(\dagger)\, BA\, dE(\dagger) = 0$ given that $\mathrm{s}\,(m) \int dE(\dagger)\, A\, dE(\dagger) = 0$. First, consider an arbitrary partition $\mathscr{P} = \{E^i : i = 0, 1, \ldots, n\}$ and compute

$$\sum_{i=1}^{n} \Delta^i BA\Delta^i = \sum_{i=1}^{n} \Delta^i B\left[\sum_{j=1}^{n} \Delta^j A\right]\Delta^i = \sum_{i=1}^{n} \Delta^i B\Delta^i A\Delta^i$$

$$= \left[\sum_{i=1}^{n} \Delta^i B\Delta^i\right]\left[\sum_{j=1}^{n} \Delta^j A\Delta^j\right]$$

since $\Delta^i\Delta^j = 0$, $i \neq j$, $\Delta^i\Delta^i = \Delta^i$, and A and B are causal. Now, since A is strongly strictly causal for any $x \in H$ and $\varepsilon > 0$, there exists a partition $\mathscr{P}'$ such that for any partition $\mathscr{P}$ that refines $\mathscr{P}'$

$$\left\|\left[\sum_{j=1}^{n} \Delta^j A\Delta^j\right]x\right\| < \varepsilon$$

Thus with the aid of Lemma A.4 we obtain

$$\left\|\left[\sum_{i=1}^{n} \Delta^i BA\Delta^i\right]x\right\| = \left\|\left[\sum_{i=1}^{n} \Delta^i B\Delta^i\right]\left[\sum_{j=1}^{n} \Delta^j A\Delta^j\right]x\right\|$$

$$\leq \left\|\sum_{i=1}^{n} \Delta^i B\Delta^i\right\| \left\|\left[\sum_{j=1}^{n} \Delta^j A\Delta^j\right]x\right\|$$

$$= \left\{\max_i \|\Delta^i B\Delta^i\|\right\} \left\|\left[\sum_{j=1}^{n} \Delta^j A\Delta^j\right]x\right\|$$

$$\leq \|B\| \left\|\left[\sum_{=1}^{n} \Delta^i A\Delta^i\right]x\right\| < \|B\|\varepsilon$$

implying that

$$\mathrm{s}\,(m) \int dE(\dagger)\, BA\, dE(\dagger) = 0$$

Thus $BA \in \mathscr{S}$. Of course, if $B \in \mathscr{S}$ the above argument shows that $\mathscr{S}$ is closed under multiplication, making it an algebra.

To complete the proof we must verify that $\mathscr{S}$ is closed in the uniform operator topology. Let A_π be a sequence of strongly strictly causal operators that converges uniformly to A. A is clearly causal by Propperty 1.B.6 and it suffices to show that s (m) $\int dE(\dagger)\, A\, dE(\dagger) = 0$ given that

$$\mathrm{s}\,(m) \int dE(\dagger)\, A_\pi\, dE(\dagger) = 0$$

For any $\varepsilon > 0$ and $x \in H$ choose γ' such that

$$\|A - A_\gamma\| < \varepsilon/2\|x\|$$

whenever $\gamma > \gamma'$ and a partition $\mathscr{P}'$ such that for any partition $\mathscr{P} = \{E^i : i = 0, 1, \ldots, n\}$ that refines $\mathscr{P}'$

$$\left\| \left[\sum_{i=1}^{n} \Delta^i A_\gamma \Delta^i \right] x \right\| < \varepsilon/2$$

Then

$$\begin{aligned}
\left\| \left[\sum_{i=1}^{n} \Delta^i A \Delta^i \right] x \right\| &= \left\| \left[\sum_{i=1}^{n} \Delta^i [A - A_\gamma] \Delta^i + \sum_{i=1}^{n} \Delta^i A_\gamma \Delta^i \right] x \right\| \\
&\leq \left[\sum_{i=1}^{n} \Delta^i [A - A_\gamma] \Delta^i \right] x \Bigg\| + \left\| \left[\sum_{i=1}^{n} \Delta^i A_\gamma \Delta^i \right] x \right\| \\
&\leq \left\| \sum_{i=1}^{n} \Delta^i [A - A_\gamma] \Delta' \right\| \|x\| + \left\| \left[\sum_{i=1}^{n} \Delta^i A_\gamma \Delta^i \right] x \right\| \\
&= \max_i \|\Delta^i [A - A_\gamma] \Delta^i\| \, \|x\| + \left\| \left[\sum_{i=1}^{n} \Delta^i A_\gamma \Delta^i \right] x \right\| \\
&\leq \|A - A_\gamma\| \, \|x\| + \left\| \sum_{i=1}^{n} \Delta^i A \Delta^i x \right\|
\end{aligned}$$

Thus

$$\mathrm{s}\,(m) \int dE(\dagger)\, A\, dE(\dagger) = \mathrm{s} \lim_i \sum_{i=1}^{n} \Delta^i A \Delta^i = 0$$

and A is strongly strictly causal, as was to be shown. //

9. Remark: Note that even though we are working with strongly strictly causal operators and strongly convergent integrals of triangular truncation we have shown that $\mathscr{S}$ is closed in the uniform operator topology—not in the strong topology.

10. Remark: The development of the strongly strictly anticausal operators closely parallels the development of the strictly anticausal operators. We therefore shall simply state the major results.

11. Definition: An operator on $(H, \mathscr{E})$ is *strongly strictly anticausal* if it is strongly strictly causal on $(H, \mathscr{E}^*)$.

12. Property: A linear operator A on $(H, \mathscr{E})$ is strongly strictly anticausal if and only if A^* is strongly strictly causal.

13. Remark: Consistent with Property 12, the strongly strictly anticausal operators on $(H, \mathscr{E})$ are denoted $\mathscr{S}^*$.

14. Property: The integral of triangular truncation

$$\mathrm{s}\,(M) \int dE(\dagger)\, AE_{\dagger} = \mathrm{s}\,(m) \int E^{\dagger}A\, dE(\dagger)$$

is a projection onto $\mathscr{S}^*$.

15. Corollary: The following are equivalent:

(i) A is strongly strictly anticausal,
(ii) $\mathrm{s}\,(M) \int dE(\dagger)\, AE_{\dagger} = \mathrm{s}\,(m) \int E^{\dagger}A\, dE(\dagger) = A$,
(iii) A is anticausal and

$$\mathrm{s}\,(M) \int dE(\dagger)\, A\, dE(\dagger) = \mathrm{s}\,(M) \int dE(\dagger)\, A\, dE(\dagger) = 0$$

and

(iv) $\mathrm{s}\,(M) \int dE(\dagger)\, AE^{\dagger} = \mathrm{s}\,(m) \int E_{\dagger}A\, dE(\dagger) = 0$.

16. Property: $\mathscr{S}^*$ is a uniformly closed left ideal in $\mathscr{C}^*$.

C. STRONG CAUSALITY

1. Remark: If one carefully analyses those operators, such as the unilateral and bilateral shift, that are intuitively hypercausal but fail to satisfy the axioms for strict causality, one finds that the difficulty resides with the Δ^1 and Δ^n terms, which may correspond to an infinite time interval no matter how much a partition is refined. This difficulty may be alleviated by working with a generalized partition.

2. Definition: Let $\mathscr{E}$ be a resolution of the identity on H. A *generalized partition* of $\mathscr{E}$ is a sequence $\mathscr{P} = \{E^{\dagger_i} : \dagger_i \in \$\}$ such that

(i) $\dagger_i \leq \dagger_j$ for $i < j$, $i, h \in Z$,
(ii) $V\{E^{\dagger_i}(H) : \dagger_i \in \$\} = H$, and
(iii) $\cap\{E^{\dagger_i}(H) : \dagger_i \in \$\} = \{0\}$.

3. Remark: A generalized partition is denoted $\mathscr{P} = \{E^i : i \in Z\}$ where $E^i = E^{\dagger_i}$. We let $\Delta^i = [E^i - E^{i-1}]$. Of course, every (finite) partition is a generalized partition.

4. Definition: Let A be a bounded linear operator on $(H, \mathscr{E})$, Then A is *strongly causal* if A is causal and if for each $\varepsilon > 0$ there exists a generalized partition $\mathscr{P}'$ of $\mathscr{E}$ such that for every generalized partition $\mathscr{P} = \{E^i : i \in Z\}$ of $\mathscr{E}$ that refines $\mathscr{P}'$:

$$\|\Delta^i A \Delta^i\| < \varepsilon, \qquad i \in Z$$

5. Remark: The strongly causal operators are denoted $\mathscr{T}$. Clearly $\mathscr{R} \subset \mathscr{T}$ since every partition is also a generalized partition. In general the relationship between $\mathscr{S}$ and $\mathscr{T}$ is unknown. One exception to this is $l_2(-\infty, \infty)$ with its usual resolution structure as indicated in the following example.

6. Example: Consider the resolution space $l_2(-\infty, \infty)$ with its usual resolution structure $\mathscr{E}$. Since $\mathscr{E}$ is countable we may without loss of generality work with the maximal generalized partition $\mathscr{P} = \mathscr{E}$, in which case Δ^i is the projection onto $V\{e_i\}$ where $\{e_i\}_{-\infty}^{\infty}$ is the standard basis for $l_\infty(-\infty, \infty)$. Hence if an operator A has a matrix representation (a_{ij}) relative to the standard basis, then $\Delta^i A \Delta^i$ is the operator whose matrix representation is zero except for the iith entry, which equals a_{ii}. The strong causality condition therefore requires that $a_{ii} = 0$ for all i, which together with the causality condition to the effect that $a_{ij} = 0$ for $i < j$ implies that A is strongly causal if and only if $a_{ij} = 0$ for $i \leq j$. Of course, a similar argument applies to $l_2[0\ \infty)$. Thus the shifts are strongly causal and, moreover, strong causality and strong strict causality coincide on the l_2-spaces with their usual resolution structure.

7. Remark: Given the definition of the strong causal operators it would presumably be possible to extend the integral theory for strictly causal operators to the strongly causal case by defining integrals of triangular truncation relative to a family of generalized partitions. In fact, however, the significance of the strongly causal operators in our theory is not sufficient to justify such an endeavor. We, therefore, leave it to the reader (see Exercise 7).

8. Property: $\mathscr{T}$ is a uniformly closed two-sided ideal in $\mathscr{C}$.

Proof: If $A \in \mathscr{T}$ and $B \in \mathscr{C}$, then for any generalized partition

$$\|\Delta^i BA\Delta^i\| = \|\Delta^i B\Delta^i A\Delta^i\| < \|B\| \, \|\Delta^i A\Delta^i\|$$

where we have used the fact that both A and B are causal in the above derivation. Hence if A is strongly causal, so is BA, and a similar argument applies to BA. To prove that $\mathscr{T}$ is closed under addition let $\varepsilon > 0$. Then if A and B are strongly causal, there exist generalized partitions $\mathscr{P}'$ and $\mathscr{P}''$ such that if $\mathscr{P}'''$ is any common refinement of both $\mathscr{P}'$ and $\mathscr{P}''$ and $\mathscr{P}$ is a refinement of $\mathscr{P}'''$, then

$$\|\Delta^i A\Delta^i\| < \varepsilon/2 \qquad \text{and} \qquad \|\Delta^i B\Delta^i\| < \varepsilon/2$$

Hence

$$\|\Delta^i[A + B]\Delta^i\| = \|\Delta^i A\Delta^i + \Delta^i B\Delta^i\| \leq \|\Delta^i A\Delta^i\| + \|\Delta^i B\Delta^i\| < \varepsilon$$

and $A + B$ is strongly causal. Finally, if A_π is a net of strongly causal operators that converges to A, fix $\varepsilon > 0$ and choose γ' such that if $\gamma > \gamma'$, then

$$\|A - A_\gamma\| < \varepsilon/2$$

and choose a generalized partition $\mathscr{P}'$ such that if $\mathscr{P} = \{E^i : i \in Z\}$ refines $\mathscr{P}'$:

$$\|\Delta^i A_\gamma \Delta^i\| < \varepsilon/2, \qquad i \in Z$$

Now

$$\begin{aligned}\|\Delta^i A\Delta^i\| &= \|\Delta^i[A - A_\gamma]\Delta^i + \Delta^i A_\gamma \Delta^i\| \leq \|\Delta^i[A - A_\gamma]\Delta^i\| + \|\Delta^i A_\gamma \Delta^i\| \\ &\leq \varepsilon/2 + \varepsilon/2 = \varepsilon\end{aligned}$$

showing that A is strongly causal and completing the proof. //

9. Definition: An operator A on $(H, \mathscr{E})$ is strongly anticausal if it is strongly causal on $(H, \mathscr{E}^*)$.

10. Remark: By invoking an argument similar to that used in the proof of Property A.13 we may verify.

11. Property: A linear operator A on $(H, \mathscr{E})$ is strongly anticausal if and only if A^* is strongly causal.

12. Remark: Consistent with the above we denote the strongly anti-causal operators $\mathscr{T}^*$.

13. Property: $\mathscr{T}^*$ is a uniformly closed two-sided ideal in $\mathscr{C}^*$.

D. THE RADICAL

1. Remark: In the preceeding section we have defined three ideals of causal operators, each of which reduces to the strictly lower triangular matrices in the finite-dimensional case. Although these ideals were defined topologically, one can define a similar ideal, which also reduces to the strictly lower triangular matrices, in purely algebraic terms. This ideal, which plays a significant role in ring theory, is usually termed the *radical* or *Jacobson radical* and is most commonly defined as the intersection of all maximal left (or right) ideals in a ring. For our purposes we adopt an equivalent definition that is formulated in terms of more classical operator theoretic notation.

2. Definition: Let $\mathscr{A}$ be a Banach algebra with identity. Then

$$\mathscr{A}_{\mathscr{R}} = \{A \in \mathscr{A} : \sigma(AB) = \{0\} \text{ for all } B \in \mathscr{A}\}$$

3. Remark: In Definition 2 σ denotes the *spectrum* of an element of $\mathscr{A}$, and we could equally well have defined $\mathscr{A}_{\mathscr{R}}$ in terms of elements for which $\sigma(BA) = \{0\}$. In general, if $\sigma(A) = \{0\}$ for $A \in \mathscr{A}$ we say that A is *quasi-nilpotent*. Equivalently, A is quasinilpotent if and only if

$$\inf_i \|A^i\|^{1/i} = 0$$

4. Example: Consider the lower triangular $n \times n$ matrices (a_{ij}) where $a_{ij} = 0$ for $i < j$. Now

$$\sigma[(a_{ij})] = \{a_{ii} : i = 1, 2, \ldots, n\}$$

and hence (a_{ij}) is quasinilpotent (actually *nilpotent* in this case) if and only if $a_{ii} = 0$, $i = 1, 2, \ldots, n$; i.e., (a_{ij}) is strictly lower triangular. Moreover, a strictly lower triangular matrix multiplied by a lower triangular matrix is strictly lower triangular. Hence for this Banach algebra $\mathscr{A}_{\mathscr{R}}$ is just the strictly lower triangular matrices. Of ocurse, if these matrices are employed as a representation of the causal operators on a finite-dimensional space, as in Example 1.B.8, then an operator will be in $\mathscr{A}_{\mathscr{R}}$ if and only if its matrix respesentation is strictly lower triangular.

5. Remark: Consistent with Example 4, the radical of the causals is a natural candidate for hypercausality. In fact, we shall show that $\mathscr{C}_{\mathscr{R}} = \mathscr{R}$, the strictly causals.[2,3]

6. Lemma: $\mathscr{R}$ is quasinilpotent.

Proof: Let $\lambda \neq 0$ be a complex number. Then if A is strictly causal, there exists a partition $\mathscr{P}'$ of $\mathscr{E}$ such that for any partition

$$\mathscr{P} = \{E^i : i = 0, 1, \ldots, n\}$$

that refines $\mathscr{P}'$

$$\|\Delta^i A \Delta^i\| < |\lambda|$$

The operator $B = [\sum_{i=1}^n \Delta^i A \Delta^i - \lambda I]$ is thus invertible, and its inverse is

$$B^{-1} = \sum_{i=1}^{n} [\Delta^i A \Delta^i - \Delta^i \lambda]^{-1}$$

A simple computation shows that

$$B^{-1}\left[\sum_{j=1}^{n} \sum_{i=j+1}^{n} \Delta^i A \Delta^j\right] = \left[\sum_{j=1}^{n} \sum_{i=j+1}^{n} \Delta^i B^{-1} A \Delta^j\right]$$

and that

$$\left[\sum_{j=1}^{n} \sum_{i=j+1}^{n} \Delta^i B^{-1} A \Delta^j\right]^n = 0$$

As such, $[B^{-1}[\sum_{j=1}^n \sum_{1=j+1}^n \Delta^i A \Delta^j]$ is nilpotent and hence

$$I + B^{-1}\left[\sum_{j=1}^{n} \sum_{i=j+1}^{n} \Delta^i A \Delta^j\right]$$

is invertible. Finally, since A is causal

$$\begin{aligned}[A - \lambda I] &= \left[\sum_{i=1}^{n} \Delta^i A \Delta^i\right] - [\lambda I] + \left[\sum_{j=1}^{n} \sum_{i=j+1}^{n} \Delta^i A \Delta^j\right] \\ &= B\left[I + B^{-1}\left[\sum_{j=1}^{n} \sum_{i=j+1}^{n} \Delta^i A \Delta^j\right]\right]\end{aligned}$$

showing that $\lambda \notin \sigma(A)$ if $A \in \mathscr{R}$ and $\lambda \neq 0$. Of course, since $\mathscr{R}$ is a proper ideal A is not invertible. Hence $0 \in \sigma(A)$, verifying that $\sigma(A) = \{0\}$, as was to be shown. //

Radical Theorem: For a Hilbert resolution space $(H, \mathscr{E})$, $\mathscr{C}_{\mathscr{R}} = \mathscr{R}$.

Proof: Since $\mathscr{R}$ is an ideal of quasinilpotent operators in $\mathscr{C}$, $\mathscr{R} \subset \mathscr{C}_{\mathscr{R}}$ and so suppose $A \notin \mathscr{R}$. Then, via the argument of reference 1, there exists an $\varepsilon > 0$ and an increasing sequence of projectors $E^{\dagger_i} \in \mathscr{E}$ such that

$$\|\Delta^i A \Delta^i\| > \varepsilon, \qquad i = 1, 2, \ldots$$

where $\Delta^i = [E^{\dagger_i} - E^{\dagger_{i-1}}]$. By linearity we can assume without loss of generality that $\varepsilon = 1$. Now, for each i there exists a unit vector $x^i \in \Delta^i(H)$ such that

$$\|\Delta^i A x_i\| > \varepsilon = 1$$

Let $y_i = \Delta^i A x_i$ and define a family of operators by

$$S_i = (1/\|y_i\|^2)[y_i \otimes x_{i+1}]$$

where $[y_i \otimes x_{i+1}](x) = (x, y_i)x_{i+1}$. Now, by the Schwarz inequality

$$\|S_i\| = (1/\|y_i\|^2)\|y_i\|\,\|x_{i+1}\| = 1/\|y_i\| < 1$$

since $\|y_i\| = \|\Delta^i A x_i\| > 1$. Moreover, by construction the response of S_i to an input in $\Delta^i(H)$ lies in $\Delta^{i+1}(H)$, so that S_i is causal (actually strictly causal). Since the Δ^is are projections on mutually orthogonal subspaces,

$$S = \operatorname{s\,lim}_n \left[\sum_{i=1}^{n} S_i\right]$$

converges and $\|S\| \le 1$.

We now claim that for each n there exists a vector z_{n+1} orthogonal to $\sum_{i=1}^{n+1} \otimes \Delta^i(H)$ such that

$$[SA]^n x_1 = x_{n+1} + z_{n+1}$$

If this is indeed the case, then

$$\|[SA]^n x_1\| \ge \|x_{n+1}\| \ge 1$$

showing that SA is not quasinilpotent and that $A \notin \mathscr{C}_{\mathscr{R}}$. We construct such a z_n by induction. Now, $Ax_1 = y_1 + a_1$, where $a_1 = E_1 A x_1$ is orthogonal to $\Delta^1(H)$. Thus

$$Sy_1 = S_1 y_1 = (1/\|y_1\|^2)(y_1 y_1)x_2 = x_2$$

and $Sa_1 = \sum_{i=2}^{\infty} S_i a_i$ is orthogonal to $\Delta^1(H) \oplus \Delta^2(H)$. Accordingly, if $z_2 = Sa_1$, then

$$[SA]x_1 = x_2 + z_2$$

as required.

Now assume that

$$[SA]^{n-1}x_1 = x_n + z_n$$

where z_n is orthogonal to $\sum_{i=1}^{n} \oplus \Delta^i(H)$. Then $Ax_n = y_n + a_n$, where a_n is orthogonal to $\sum_{i=1}^{n} \oplus \Delta^i(H)$ and

$$Sa_n = \left[\sum_{i=n+1}^{\infty} S_i\right]a_n$$

is orthogonal to $\sum_{i=1}^{n+1} \oplus {}^i(H)$ while

$$SAz_n = S\left[I - \sum_{i=1}^{n} \Delta^i\right]A\left[I - \sum_{i=1}^{n} \Delta^i\right]z_n = \left[I - \sum_{i=1}^{n+1} \Delta^i\right]SAz_n$$

by causality. Therefore,

$$\begin{aligned}[SA]^n x_1 &= SAx_n + SAz_n = Sy_n + Sa_n + SAz_n \\ &= x_{n+1} + \left[I - \sum_{i=1}^{n+1} \Delta^i\right][Sa_n + SAz_n] = x_{n+1} + z_{n+1}\end{aligned}$$

where

$$Z_{n+1} = \left[I - \sum_{i=1}^{n+1} \Delta^i\right][Sa_n + SAz_n]$$

is orthogonal to $\sum_{i=1}^{n+1} \oplus \Delta^i(H)$, as required. //

7. Corollary: If A is causal, then $E_\dagger A - AE_\dagger = E_\dagger AE^\dagger$ is strictly causal.

Proof: Since A is causal, $AE_\dagger = E_\dagger AE_\dagger$. Thus if B is causal,

$$[E_\dagger AE^\dagger BE_\dagger AE^\dagger] = [E_\dagger AE^\dagger][E^\dagger BE^\dagger][E_\dagger AE^\dagger] = 0$$

Thus if

$$D = \sum_i B_i E_\dagger AE^\dagger C_i$$

where $B_i, C_i \in \mathscr{C}$, then $D^2 = 0$. The ideal generated by $E_\dagger AE_\dagger$ is therefore nilpotent and is thus contained in $\mathscr{C}_R = \mathscr{R}$. //

8. Corollary: $\mathscr{R}$ is strongly dense in $\mathscr{S}$.

Proof: If $A \in \mathscr{S}$, then

$$A = \mathrm{s}\,(M)\int E_\dagger A\, dE(\dagger) = \mathrm{s}\lim_{\mathscr{P}}\left[\sum_{i=1}^{n} E_i A\Delta^i\right] = \mathrm{s}\lim_{\mathscr{P}}\left[\sum_{i=1}^{n} E_i AE^i\Delta^i\right]$$

since $\Delta^i = E^i\Delta^i$. Now E_iAE^i is strictly causal by Corollary 7, and hence so are $E_iAE^i\Delta^i$ and $\sum_{i=1}^n E_iAE^i\Delta^i = \sum_{i=1}^n E_iA\Delta^i$. It follows that every strongly strictly causal operator is a strong limit of strictly causal operators, as was to be shown. //

9. Remark: The operators

$$\sum_{i=1}^{n} E_i A\Delta^i = \sum_{i=1}^{n} \Delta^i AE^{i-1}$$

are sometimes termed the *pre–strictly causal operators* and denoted $\mathcal{Q}$. By Corollary 7 and the argument used in the proof of Corollary 8, $\mathcal{Q} \subset \mathcal{R}$. Since $\mathcal{R}$ is properly contained in $\mathcal{S}$, Corollary 8 implies that $\mathcal{R}$ is not strongly closed. It is, however, uniformly closed because of Property A.10.

10. Remark: The entire theory developed here can be applied to anticausal operators simply by taking adjoints. Hence $\mathcal{R}^*$ is quasinilpotent, coincides with $\mathcal{C}_{\mathcal{R}}^*$ and is strongly dense in $\mathcal{S}^*$. The *pre–strictly anticausal operators*

$$\sum_{i=1}^{n} \Delta^i AE_i = \sum_{i=1}^{n} E^{i-1} A\Delta^i$$

are denoted by $\mathcal{Q}^* \subset \mathcal{R}^*$.

11. Remark: Consistent with the theorem, the strictly causal operators admit a purely algebraic characterization. Hence the concept can be extended to an algebraic resolution space, wherein we may simply define the strictly causal operators to be the radical of the causal operators (which are always algebraically defined). In fact, however, this proves to be too strong a concept. We thus formulate an alternative concept of algebraic strict causality.[4]

12. Definition: Let $A: H \to H$ where $(H, \mathcal{E})$ is an algebraic resolution space. Then A is *algebraically strictly causal* if for any $u \in H$ the equality

$$E_{\dagger} AE^{\dagger} u = 0, \qquad \dagger \le \ddagger$$

implies that

$$E^{\ddagger} Au = 0$$

13. Remark: The algebraically strictly causal operators are denoted $\mathcal{V}$ as per the following property:

14. Property: $\mathcal{S} \subset \mathcal{V} \subset \mathcal{C}$.

Proof: If $A \in \mathscr{S}$, then it follows from Corollary B.5 that

$$A = \mathrm{s}\,(m) \int dE(\dagger)\, AE^{\dagger}$$

Thus if

$$E_{\dagger} AE^{\dagger}u = 0, \qquad \dagger \leq \ddagger$$

then

$$E^{\ddagger}Au = E^{\ddagger}\left[\mathrm{s}\,(m) \int dE(\dagger)\, AE^{\dagger}\right]u = \mathrm{s}\,(m) \int_{\dagger_0}^{\ddagger} dE(\dagger)\, E_{\dagger} AE^{\dagger}u$$

$$= \mathrm{s}\,(m) \int_{\dagger_0}^{\ddagger} dE(\dagger) 0 = 0$$

showing that $A \in \mathscr{V}$. On the other hand, if $A \in \mathscr{V}$ and $u \in H_{\ddagger}$, then $u = E_{\ddagger}u$ and

$$E_{\dagger} AE^{\dagger}u = E_{\dagger} AE^{\dagger}E_{\ddagger}u = 0, \qquad \dagger \leq \ddagger$$

which implies that $E^{\ddagger}Au = 0$ or, equivalently, that $Au \in H_{\ddagger}$ and that $A \in \mathscr{C}$, as was to be shown. //

15. Property: Let $A \in \mathscr{V}$ and $B \in \mathscr{M}$. Then $AB \in \mathscr{V}$.

Proof: If $A \in \mathscr{V}$ and $E_{\dagger} AE^{\dagger}u = 0$, $\dagger \leq \ddagger$, then $E^{\ddagger}Au = 0$. Hence if

$$0 = E_{\dagger} ABE^{\dagger}u = E_{\dagger} AE^{\dagger}(Bu), \qquad \dagger \leq \ddagger$$

then

$$E^{\ddagger}(AB)u = E^{\ddagger}A(Bu) = 0$$

showing that $AB \in \mathscr{V}$. //

PROBLEMS

1. Show that the integral operator

$$[Af](\dagger) = \int_0^{\infty} G(\dagger, \ddagger) f(\ddagger)\, d\ddagger$$

on $L_2(0, \infty)$ with kernel in $L_2[(0, \infty) \times (0, \infty)]$ and $G(\dagger, \ddagger) = 0$ for $\dagger < \ddagger$ is strongly causal but may not be strictly causal.

2. Consider the convolution operator $h = g * f$ mapping $L_2[0, \infty)$ with its usual resolution structure to itself, where g is a distribution with support in $[0, \infty)$. Show that the operator is strongly causal if and only if g has no mass at zero.

3. Show that $\mathscr{R}$ is not weakly closed.

4. Consider the resolution space $l_2[0, \infty)$ with its usual resolution structure and let a linear operator A have a matrix representation (a_{ij}) relative to the standard basis. Show that A is strongly strictly causal if and only if $a_{ij} = 0$ for $i \leq j$. Show that it is strongly causal if and only if $a_{ij} = 0$ for $i \leq j$.

5. Show that $0 \in \sigma(A)$ for every $A \in \mathscr{S}$ and for every $A \in \mathscr{T}$.

6. Show that $\mathscr{S}$ is not ultraweakly closed (weakly closed).

7. Formulate an integration theory defined in terms of the generalized partitions of a resolution of the identity. Can you formulate an *integral of triangular truncation* that is a projection on $\mathscr{T}$?

8. Show that the radical of a Banach algebra is the intersection of all maximal left (right) ideals in the algebra.

9. An operator is *nilpotent* if $A^n = 0$ for some n. Show that every nilpotent operator is quasinilpotent,

10. Show that A is quasinilpotent if and only if

$$\inf_i \|A^i\|^{1/i} = 0$$

11. Show that the pre–strictly causal operators are strongly dense in $\mathscr{R}$.

12. If $A \in \mathscr{C}$ is nilpotent show that $[I - A]^{-1}$ exists and is causal.

13. Is $\mathscr{V}$ an ideal in $\mathscr{C}$?

REFERENCES

1. Ashton, G., Ph.D. dissertation, Univ. of London (1981).
2. Erdos, J. A., and Longstaff, N. A., The convergence of triangular integrals of operators on Hilbert space, *Indiana Math. J.* **22**, 929–938 (1973).
3. Feintuch, A., Strictly and strongly causal linear operators, *SIAM J. Math. Anal.* **10**, 603–613 (1979).
4. Schumitzky, A., State feedback control for general linear systems, *Proc. Internat. Symp. Math. Networks and Systems* pp. 194–200. T. H. Delft (July 1979).

3

Decomposition

A. DECOMPOSITION IN $\mathscr{B}$

1. Remark: In Chapters 1 and 2 we characterized several classes of causal and anticausal operators on a Hilbert resolution space, each of which is the range of an appropriate integral of triangular truncation. Here we consider the problem of decomposing an arbitrary bounded operator $A \in \mathscr{B}$ into the sum of causal and anticausal operators. In the process, we characterize the domains of the various integrals of triangular truncation.

2. Property: Let $A \in \mathscr{B}$ be a bounded linear operator on $(H, \mathscr{E})$. Then

$$A = [A]_{\mathscr{R}} + [A]_{\mathscr{M}} + [A]_{\mathscr{R}^*}$$

where $[A]_{\mathscr{R}} \in \mathscr{R}$ is strictly causal, $[A]_{\mathscr{M}} \in \mathscr{M}$ is memoryless, and $[A]_{\mathscr{R}^*} \in \mathscr{R}^*$ is strictly anticausal if and only if two of the three integrals of triangular truncation

$$(M)\int E_{\dagger} A\, dE(\dagger) = (m)\int dE(\dagger)\, AE^{\dagger}$$

$$(M)\int dE(\dagger)\, A\, dE(\dagger) = (m)\int dE(\dagger)\, A\, dE(\dagger)$$

and

$$(M)\int dE(\dagger)\, AE_{\dagger} = (m)\int E^{\dagger} A\, dE(\dagger)$$

exist, in which case all three integrals exist and the decomposition is given by

$$[A]_{\mathscr{R}} = (M)\int E_{\dagger} A\, dE(\dagger) = (m)\int dE(\dagger)\, AE^{\dagger}$$

$$[A]_{\mathscr{M}} = (M)\int dE(\dagger)\, A\, dE(\dagger) = (m)\int dE(\dagger)\, A\, dE(\dagger)$$

and

$$[A]_{\mathscr{R}^*} = (M)\int dE(\dagger)\, AE_{\dagger} = (m)\int E^{\dagger} A\, dE(\dagger)$$

Moreover, when the decomposition exists it is unique.

Proof: It follows from Property 1.D.15 that if two of the three integrals exists, then so does the third, in which case the decomposition exists and is given by these formula since the ranges of the three integrals are $\mathscr{R}$, $\mathscr{M}$, and $\mathscr{R}^*$, respectively. It therefore suffices to show that if $A = A_1 + A_2 + A_3$ is an arbitrary decomposition of A, where $A_1 \in \mathscr{R}$, $A_2 \in \mathscr{M}$, and $A_3 \in \mathscr{R}^*$, then

$$A_1 = (M)\int E_{\dagger} A\, dE(\dagger)$$

$$A_2 = (M)\int dE(\dagger)\, A\, dE(\dagger)$$

and

$$A_3 = (m)\int E^{\dagger} A\, dE(\dagger)$$

By Property 2.A.5, since $A_1 \in \mathscr{R}$ and $(M) \int E_\dagger A\, dE(\dagger)$ is a projection onto $\mathscr{R}$,

$$A_1 = (M) \int E_\dagger A_1\, dE(\dagger)$$

Similarly, since $A_2 \in \mathscr{M}$ and $A_3 \in \mathscr{R}^*$, $A_2 + A_3 \in \mathscr{C}^*$, and thus Corollary 1.E.6 implies that

$$(M) \int E_\dagger [A_2 + A_3]\, dE(\dagger) = 0$$

so that

$$[A]_{\mathscr{R}} = (M) \int E_\dagger A\, dE(\dagger) = (M) \int E_\dagger [A_1 + A_2 + A_3]\, dE(\dagger)$$

$$= (M) \int E_\dagger A_1\, dE(\dagger) + (M) \int E_\dagger [A_2 + A_3]\, dE(\dagger) = A_1 + 0 = A_1$$

as was to be shown. A similar argument yields $[A]_{\mathscr{M}} = A_2$ and $[A]_{\mathscr{R}^*} = A_3$, thus completing the proof. //

3. Remark: Since the topology in which the integrals of triangular truncation are defined is not employed in the above argument, a parallel result employing strong limits can be formulated, as follows.

4. Property: Let $A \in \mathscr{B}$ be a bounded linear operator on $(H, \mathscr{E})$. Then

$$A = [A]_{\mathscr{S}} + [A]_{\mathscr{M}} + [A]_{\mathscr{S}^*}$$

where $[A]_{\mathscr{S}} \in \mathscr{S}$ is strongly strictly causal, $[A]_{\mathscr{M}} \in \mathscr{M}$ is memoryless, and $[A]_{\mathscr{S}^*} \in \mathscr{S}^*$ is strongly strictly anticausal if and only if two of the three integrals of triangular truncation

$$\mathrm{s}\,(M) \int E_\dagger A\, dE(\dagger) = \mathrm{s}\,(m) \int dE(\dagger)\, A E^\dagger$$

$$\mathrm{s}\,(M) \int dE(\dagger)\, A\, dE(\dagger) = \mathrm{s}\,(m) \int dE(\dagger)\, A\, dE(\dagger)$$

and

$$\mathrm{s}\,(M) \int dE(\dagger)\, A E_\dagger = \mathrm{s}\,(m) \int E^\dagger A\, dE(\dagger)$$

exist, in which case all three integrals exist and the decomposition is given by

$$[A]_{\mathscr{S}} = \mathrm{s}\,(M) \int E_{\dagger} A \, dE(\dagger) = \mathrm{s}\,(m) \int dE(\dagger)\, AE_{\dagger}$$

$$[A]_{\mathscr{S}} = \mathrm{s}\,(M) \int dE(\dagger)\, A \, dE(\dagger) = \mathrm{s}\,(m) \int dE(\dagger)\, A \, dE(\dagger)$$

and

$$[A]_{\mathscr{S}^*} = \mathrm{s}\,(M) \int dE(\dagger)\, AE_{\dagger} = \mathrm{s}\,(m) \int E^{\dagger} A \, dE(\dagger)$$

Moreover, when the decomposition exists it is unique.

5. Remark: Not that the use of the same notation in Properties 2 and 4 for the memoryless part $[A]_{\mathscr{M}}$ represents an abuse of notation since $[A]_{\mathscr{M}}$ is given by different integrals in the two cases. The strong integral is, however, an extension of the uniform integral and hence no difficulty arises.

6. Remark: Occasionally, rather than decomposing an operator into three parts we prefer to decompose it into only two terms

$$A = [A]_{\mathscr{C}} + [A]_{\mathscr{R}^*}, \qquad A = [A]_{\mathscr{C}} + [A]_{\mathscr{S}^*},$$

$$A = [A]_{\mathscr{S}} + [A]_{\mathscr{C}^*}, \qquad \text{or} \qquad A = [A]_{\mathscr{R}} + [A]_{\mathscr{C}^*}$$

where $[A]_{\mathscr{C}} \in \mathscr{C}$ and $[A]_{\mathscr{C}^*} \in \mathscr{C}^*$. One can then formulate a decomposition theory paralleling the preceding: $[A]_{\mathscr{C}}$ and $[A]_{\mathscr{C}^*}$ are given by the appropriate integrals of triangular truncation that define projections onto $\mathscr{C}$ and $\mathscr{C}^*$ as per 1.E.2 and 1.E.5.

7. Remark: Although Properties 2 and 4 represent a definitive starting point for our decomposition theory, they in effect transform the existence criteria for the operator decompositions into the characterization of the domains of the five integrals of triangular truncations (actually three owing to Property 1.D.15). However, no complete characterization of these domains, unlike their ranges, is known. Rather, we consider two special cases in the following sections, causal and compact operators. The difficulty encountered with the general decomposition theory is illustrated by the following example.

8. Example: Consider the Hilbert resolution space $l_2[0, \infty)$ with its usual resolution structure and the operator A on $l_2[0, \infty)$ characterized by the semi-infinite *Toeplitz matrix*

$$A = \begin{bmatrix} 0 & -1 & -\frac{1}{2} & -\frac{1}{3} & -\frac{1}{4} & -\frac{1}{5} & -\frac{1}{6} & \cdots \\ 0 & 0 & -1 & -\frac{1}{2} & -\frac{1}{3} & -\frac{1}{4} & -\frac{1}{5} & \cdots \\ \frac{1}{2} & 1 & 0 & -1 & -\frac{1}{2} & -\frac{1}{3} & -\frac{1}{4} & \cdots \\ \frac{1}{3} & \frac{1}{2} & 1 & 0 & -1 & -\frac{1}{2} & -\frac{1}{3} & \cdots \\ \frac{1}{4} & \frac{1}{3} & \frac{1}{2} & 1 & 0 & -1 & -\frac{1}{2} & \cdots \\ \frac{1}{5} & \frac{1}{4} & \frac{1}{3} & \frac{1}{2} & 1 & 0 & -1 & \cdots \\ \frac{1}{6} & \frac{1}{5} & \frac{1}{4} & \frac{1}{3} & \frac{1}{2} & 1 & 0 & \cdots \\ \vdots & \vdots & \vdots & \vdots & \vdots & \vdots & \vdots & \end{bmatrix}$$

which is defined by the sequence

$$a_k = a_{i(i-k)} = \begin{cases} 1/k, & k \neq 0 \\ 0, & k = 0 \end{cases}$$

For such an operator

$$\|A\| = \operatorname*{ess\,sup}_{\theta} |\hat{a}(\theta)|$$

where $\hat{a}$ is the Fourier transform of the defining sequence a_k,

$$\hat{a}(\theta) = \sum_{k=-\infty}^{\infty} a_k e^{-ik\theta}, \qquad 0 \leq \theta < 2\pi$$

In particular, for this A

$$\hat{a}(\theta) = (\pi - \theta)/i$$

and hence $\|A\| = \pi$.

Now, it follows from Example 2.B.6 and Remark 2.B.7 that an operator on $l_2[0, \infty)$ is strongly strictly causal if and only if its matrix representation relative to the standard basis is strictly lower triangular, memoryless if and only if its matrix representation is diagonal, and strongly strictly anticausal if and only if its matrix representation is strictly upper triangular. Hence the only possible candidate for an additive decomposition of A is

$$A = \begin{bmatrix} 0 & 0 & 0 & 0 & \cdots \\ 1 & 0 & 0 & 0 & \cdots \\ \frac{1}{2} & 1 & 0 & 0 & \cdots \\ \frac{1}{3} & \frac{1}{2} & 1 & 0 & \cdots \\ \vdots & \vdots & \vdots & \vdots & \end{bmatrix} + \begin{bmatrix} 0 & 0 & 0 & 0 & \cdots \\ 0 & 0 & 0 & 0 & \cdots \\ 0 & 0 & 0 & 0 & \cdots \\ 0 & 0 & 0 & 0 & \cdots \\ \vdots & \vdots & \vdots & \vdots & \end{bmatrix} + \begin{bmatrix} 0 & -1 & -\frac{1}{2} & -\frac{1}{3} & \cdots \\ 0 & 0 & -1 & \frac{1}{2} & \cdots \\ 0 & 0 & 0 & -1 & \cdots \\ 0 & 0 & 0 & 0 & \cdots \\ \vdots & \vdots & \vdots & \vdots & \end{bmatrix}$$

$$= [A]_{\mathscr{S}} + [A]_{\mathscr{M}} + [A]_{\mathscr{S}^*}$$

where $[A]_{\mathscr{S}}$ is defined by the sequence

$$(a_{\mathscr{S}})_k = \begin{cases} 1/k, & k > 0 \\ 0, & k \leq 0 \end{cases}$$

$[A]_{\mathscr{M}} = 0$, and $[A]_{\mathscr{S}*}$ is defined by the sequence

$$(a_{\mathscr{S}*})_k = \begin{cases} 0, & k \geq 0 \\ 1/k, & k < 0 \end{cases}$$

Now a little algebra will reveal that

$$\hat{a}_{\mathscr{S}}(\theta) = -\ln(1 - e^{-i\theta})$$

while

$$\hat{a}_{\mathscr{S}*}(\theta) = \ln(1 - e^{i\theta})$$

As such,

$$\|[A]_{\mathscr{S}}\| = \operatorname*{ess\,sup}_{\theta} |-\ln(1 - e^{-i\theta})| = \infty$$

and

$$\|[A]_{\mathscr{S}*}\| = \operatorname*{ess\,sup}_{\theta} |\ln(1 - e^{i\theta})| = \infty$$

Thus even though our original matrix represented a bounded operator on $l_2[0, \infty)$, its "obvious decomposition" into strictly upper and lower triangular matrices defines a "decomposition" where both terms represent unbounded operators. Since this is the only potential decomposition for A, we thus conclude that A does not admit a decomposition as per Property 4. Of course, since $[A]_{\mathscr{R}}$ fails to exist whenever $[A]_{\mathscr{S}}$ fails to exist (but not conversely), a decomposition as per Property 2 also fails.

9. Remark: The example illustrates the inherent difficulty encountered in decomposition theory. Indeed, the operator A defined by a Toeplitz matrix is of the simplest possible type and admits an "obvious decomposition." Unfortunately, this "decomposition" does not define a bounded operator, and hence no true decomposition exists. Since such phenomena are typical when dealing with general bounded linear operators, a general theory of decomposition in $\mathscr{B}$ has yet to be formulated. Significant progress has, however, been achieved for the cases of causal and compact operators. The remainder of the chapter is devoted to the derivation of a decomposition theory for these two classes.

B. DECOMPOSITION IN $\mathscr{C}$

1. Remark: In this section we consider the problem of decomposing a causal operator A into the sum of a strictly causal and a memoryless operator:[2,3]

$$A = [A]_{\mathscr{R}} + [A]_{\mathscr{M}}$$

For $A \in \mathscr{C}$ Corollary 1.E.4 implies that

$$(M) \int dE(\dagger)\, AE_{\dagger} = (m) \int E^{\dagger} A\, dE(\dagger) = 0$$

and similarly for the strongly convergent integral. Hence Properties A.2 and A.4 imply that the above decomposition exist if and only if

$$(M) \int dE(\dagger)\, A\, dE(\dagger) = (m) \int dE(\dagger)\, A\, dE(\dagger)$$

exists. For an arbitrary partition $\mathscr{P} = \{E^i : i = 0, 1, \ldots, n\}$ of $\mathscr{E}$ the partial sums for these integrals take the form $\sum_{i=1}^{n} \Delta^i A \Delta^i$, and hence Lemma 2.A.4 implies that

$$\left\| \sum_{i=1}^{n} \Delta^i A \Delta^i \right\| = \max_i \|\Delta^i A \Delta^i\| \leq \|A\|$$

Thus $[A]_{\mathscr{M}}$ cannot become unbounded even though it may not converge. Moreover, since $A \in \mathscr{C}$ this implies that $[A]_{\mathscr{R}}$ cannot become unbounded either. Hence the pathological phenomena encountered in Example A.8 cannot occur when $A \in \mathscr{C}$, greatly simplifying the process of characterizing those causal operators that admit a decomposition into a strictly causal operator and a memoryless operator. We begin this process by formulating an *extension* of $[A]_{\mathscr{M}}$ which preserves many of its properties and exists for all $A \in \mathscr{B}$. Although this extension cannot be employed to construct a decomposition by means of Properties A.2 and A.4, it proves to be a useful tool in our decomposition theory.

2. Definition: An *invariant mean* on a group G is a uniformly continuous linear functional $M(\;)$ defined on the space of all bounded complex functions on G satisfying the following:

(i) For real-valued functions $f: G \to R$,

$$\inf_G \{f(g) : g \in G\} \leq M(f) \leq \sup_G \{f(g) : g \in G\}$$

(ii) If $h \in G$ and $F: G \to C$ and $f_h: G \to C$ is defined by

$$f_h(g) = f(hg), \qquad g \in G$$

then $M(f_h) = M(f)$.

3. Remark: Although the existence of an invariant mean is not ensured for all groups, every abelian group G admits at least one invariant mean. For the present purposes the group of interest is the abelian group of *unitary operators* $\mathscr{U}$ *contained in* $\mathscr{K}_{\mathscr{E}}$, the core of a resolution of the identity $\mathscr{E}$. We denote such an invariant mean $M_{\mathscr{U}}(\ )$. Although this invariant mean may not be unique, the properties of the extension of $[\ ]_{\mathscr{M}}$ derived therefrom are independent of the choice of invariant mean.

4. Remark: Let $\mathscr{K}_1$ denote the *nuclear* (or *trace class*) operators on a Hilbert space H. Recall that the trace is a continuous linear function defined on $\mathscr{K}_1$ by

$$\mathrm{tr}[A] = \sum_i \lambda_i$$

where λ_i is the ith eigenvalue of the compact operator $A \in \mathscr{K}_1$ repeated according to multiplicity. Furthermore, $\mathscr{K}_1$ is an ideal in $\mathscr{B}$ and its dual is $\mathscr{B} = \mathscr{K}_1^*$. In particular, every bounded linear functional on $\mathscr{K}_1$ takes the form $f: \mathscr{K}_1 \to \mathscr{C}$ where

$$f(A) = \mathrm{tr}[BA]$$

for some $B \in \mathscr{B}$. Note that since $\mathscr{K}_1$ is an ideal in $\mathscr{B}$, $BA \in \mathscr{K}_1$ and hence $\mathrm{tr}[BA]$ is well defined.

5. Definition: For any bounded linear operator $A \in \mathscr{B}$, $[A]^{\mathscr{M}}$ is the operator on $\mathscr{B}$ defined by the equality

$$\mathrm{tr}[[A]^{\mathscr{M}}K] = M_{\mathscr{U}}(\mathrm{tr}[U^*AUK])$$

where $M_{\mathscr{U}}(\ )$ is an invariant mean on $\mathscr{U}$ and K is an arbitrary nuclear operator.

6. Remark: Since $K \in \mathscr{K}_1$ so is U^*AUK, and hence $\mathrm{tr}[U^*AUK]$ is a well-defined complex-valued function on $\mathscr{U}$ that is bounded, since $\mathrm{tr}[\ ]$ is a bounded linear functional and $U^*AUK \in \mathscr{K}_1$. Now, $M_{\mathscr{U}}(\ )$ is a bounded linear functional on such complex-valued functions, and hence $M_{\mathscr{U}}(\mathrm{tr}[U^*AUK]) \in \mathscr{C}$ is independent of U and defines a bounded linear functional on $\mathscr{K}_1$. Since every such functional on $\mathscr{K}_1$ is of the form $\mathrm{tr}[BK]$ for some uniquely defined $B \in \mathscr{B}$, the equality defining $[A]^{\mathscr{M}}$ is assured of admitting a unique solution $[A]^{\mathscr{M}}$. As such, for a given $M_{\mathscr{U}}(\ )$ $[A]^{\mathscr{M}}$ is uniquely defined. Note, however, that $[A]^{\mathscr{M}}$ is dependent on the choice of $M_{\mathscr{U}}(\ )$. Fortunately, the fundamental properties of $[A]^{\mathscr{M}}$ are independent of the choice of $M_{\mathscr{U}}(\ )$, as shown by the following property. For reasons that will become apparent $[\ ]^{\mathscr{M}}$ is termed the *expectation operator*.

7. Property: Let $[\ ]^{\mathscr{M}}: \mathscr{B} \to \mathscr{B}$ be defined as above; then the following hold:

(i) $[\ ]^{\mathscr{M}}$ is linear.
(ii) $[A]^{\mathscr{M}} \in \mathscr{M}$ for all $A \in \mathscr{B}$.
(iii) If $B \in \mathscr{M}$ and $A \in \mathscr{B}$, then $[BA]^{\mathscr{M}} = B[A]^{\mathscr{M}}$ and $[AB]^{\mathscr{M}} = [A]^{\mathscr{M}}B$.
(iv) $[\ ]^{\mathscr{M}}$ is a projection of $\mathscr{B}$ onto $\mathscr{M}$.
(v) $[\ ]^{\mathscr{M}}$ is uniformly continuous and $\|[\ ]^{\mathscr{M}}\| = 1$.
(vi) If $A \in \mathscr{L}$ or $A \in \mathscr{L}^*$, then $[A]^{\mathscr{M}} = 0$.
(vii) If $A \in \mathscr{R}$ or $A \in \mathscr{R}^*$, then $[A]^{\mathscr{M}} = 0$.
(viii) If $A \in \mathscr{T}$ or $A \in \mathscr{T}^*$, then $[A]^{\mathscr{M}} = 0$.
(ix) If A is hermitian, then $[A]^{\mathscr{M}}$ is hermitian.
(x) If A is positive, then $[A]^{\mathscr{M}}$ is positive.
(xi) If A is positive definite, then $[A]^{\mathscr{M}}$ is positive definite.
(xii) If $[A]_{\mathscr{M}}$ exists in the uniform topology, then $[A]^{\mathscr{M}} = [A]_{\mathscr{M}}$.

Proof: The proof is greatly simplified by the observation that the nuclear operators are generated by the *rank*-1 *operators* $x \otimes y$ so that without loss of generality we may take K in the defining equality for $[A]^{\mathscr{M}}$ to be $x \otimes y$. Moreover, for any $A \in \mathscr{B}$

$$\mathrm{tr}[A(x \otimes y)] = (y, Ax)$$

Hence with $K = x \otimes y$ the defining equality for $[A]^{\mathscr{M}}$ reduces to

$$(y, [A]^{\mathscr{M}}x) = M_{\mathscr{U}}((Uy, AUx))$$

(i) The verification that $[\ ]^{\mathscr{M}}$ is linear follows trivially from the linearity of $M_{\mathscr{U}}(\ )$ and the bilinearity of the inner product.

(ii) Let $V \in \mathscr{U}$ and consider

$$(y, V[A]^{\mathscr{M}}x) = (V^*y, [A]^{\mathscr{M}}x) = M_{\mathscr{U}}((UV^*y, AUx)) = M_{\mathscr{U}}((UV^{-1}y, AUx))$$

since $V^{-1} = V^*$ for a unitary operator. Now, by the definition of the invariant mean $M_{\mathscr{U}}(f_V) = M_{\mathscr{U}}(f)$. In particular, if we define f by $f(U) = (UV^{-1}y, AUx)$, then $f_V(U) = (Uy, AUVx)$ since V and U commute. Thus it follows from the above equality and the invariance property that

$$(y, V[A]^{\mathscr{M}}x) = M_{\mathscr{U}}((UV^{-1}y, AUx)) = M_{\mathscr{U}}((Uy, AUVx)) = (y, [A]^{\mathscr{M}}Vx)$$

Since this equality holds for all $x, y \in H$ we have verified that $V[A]^{\mathscr{M}} = [A]^{\mathscr{M}}V$ for all $V \in \mathscr{U}$. Finally, since $E^{\dagger} \in \mathscr{K}_{\mathscr{E}}$, which is generated by $\mathscr{U}$,

$$E^{\dagger}[A]^{\mathscr{M}} = [A]^{\mathscr{M}}E^{\dagger}, \qquad \dagger \in \$$$

showing that $[A]^{\mathscr{M}} \in \mathscr{M}$.

(iii) If $B \in \mathscr{M}$, it commutes with $\mathscr{U}$; hence for $A \in \mathscr{B}$

$$\begin{aligned}(y, [BA]^{\mathscr{M}}x) &= M_{\mathscr{U}}((Uy, BAUx)) = M_{\mathscr{U}}((B^*Uy, AUx)) \\ &= M_{\mathscr{U}}((UB^*y, AUx)) \\ &= (B^*y, [A]^{\mathscr{M}}x) = (y, B[A]^{\mathscr{M}}x)\end{aligned}$$

Since this holds for all $x, y \in H$, it implies that $[BA]^{\mathscr{M}} = B[A]^{\mathscr{M}}$, and similarly for the second equality.

(iv) If $A \in \mathscr{M}$, it commutes with $\mathscr{U}$, and hence

$$\begin{aligned}(y, [A]^{\mathscr{M}}x) &= M_{\mathscr{U}}((Uy, AUx)) = M_{\mathscr{U}}((y, U^*AUx)) = (M_{\mathscr{U}}((y, AU^*Ux)) \\ &= M_{\mathscr{U}}((y, Ax)) = (y, Ax)\end{aligned}$$

where we have used the facts that $U^*U = I$ and that $M_{\mathscr{U}}(f) = f$ if f is a constant function (i.e., independent of U). Since the above equality holds for all $x, y \in H$, it implies that $[A]^{\mathscr{M}} = A$ if A is memoryless, which together with the fact that the range of $[\ ]^{\mathscr{M}}$ is contained in $\mathscr{M}$ implies that $[\ ]^{\mathscr{M}}$ is a projection onto $\mathscr{M}$.

(v) To verify uniform continuity we observe that

$$\begin{aligned}(y, [A]^{\mathscr{M}}x) &= M_{\mathscr{U}}((Uy, AUx)) \le \sup_{\mathscr{U}} (Uy, AUx) \\ &\le \|Uy\|\,\|A\|\,\|Ux\| = \|A\|\,\|y\|\,\|x\|\end{aligned}$$

showing that $\|[A]^{\mathscr{M}}\| \le \|A\|$. The opposite inequality follows from the idempotence of $[\ ]^{\mathscr{M}}$. Hence $\|[A]^{\mathscr{M}}\| = 1$ and $[\ ]^{\mathscr{M}}$ is uniformly continuous.

(vi) If A is pre-strictly causal, then there exists a partition $\mathscr{P}'$ such that for any partition $\mathscr{P} = \{E^i : i = 0, 1, \ldots, n\}$ that refines $\mathscr{P}'$

$$\Delta^i A \Delta^i = 0, \qquad i = 1, 2, \ldots, n$$

Now for such an operator

$$[A]^{\mathscr{M}} = \sum_{i=1}^{n} \Delta^i[A]^{\mathscr{M}} = \sum_{i=1}^{n} \Delta^i[A]^{\mathscr{M}}\Delta^i = \sum_{i=1}^{n} [\Delta^i A \Delta^i]^{\mathscr{M}} = \sum_{i=1}^{n} [0]^{\mathscr{M}} = 0$$

while a similar argument holds for pre-strictly anticausal operators.

(vii) By Property 2.A.5 every strictly causal operator is the uniform limit of pre–strictly causal operators. Accordingly, if $A \in \mathscr{R}$ there exists a net $A_\pi \in \mathscr{Q}$ such that A_π converges uniformly to A. Thus, (v) implies that $[A_\pi]^{\mathscr{M}} = 0$ converges uniformly to $[A]^{\mathscr{M}}$. As such $[A]^{\mathscr{M}} = 0$, as was to be shown. Of course, a similar argument can be applied to $A \in \mathscr{R}^*$.

(viii) If $A \in \mathscr{T}$, then for any $\varepsilon > 0$ there exists a generalized partition $\mathscr{P}'$ such that, if $\mathscr{P} = \{E^i : i \in Z\}$ refines $\mathscr{P}'$

$$\|\Delta^i A \Delta^i\| < \varepsilon, \qquad i \in Z$$

Now since $[\ ]^{\mathscr{M}}$ has norm 1,

$$\|[\Delta^i A \Delta^i]^{\mathscr{M}}\| < \varepsilon, \qquad i \in Z$$

but since $\Delta^i \in \mathscr{M}$, (iii) implies that

$$[\Delta^i A \Delta^i]^{\mathscr{M}} = \Delta^i [A]^{\mathscr{M}} \Delta^i = [A]^{\mathscr{M}} \Delta^i$$

Thus

$$[A]^{\mathscr{M}} = \sum_i [A]^{\mathscr{M}} \Delta^i = \sum_i [\Delta^i A \Delta^i]^{\mathscr{M}}$$

Now, invoking a generalized version of Lemma 2.A.4 in which the finite partition of that lemma is replaced by a generalized partition (with the proof remaining unchanged), we have

$$\|[A]^{\mathscr{M}}\| = \left\| \sum_i [\Delta^i A \Delta^i] \right\| = \sup_i \|\Delta^i A \Delta^i\| \leq \varepsilon$$

Since this holds for all $\varepsilon > 0$ it implies that $[A]^{\mathscr{M}} = 0$, as was to be shown. Of course, a similar argument holds for $A \in \mathscr{T}^*$.

(ix) If $A = A^*$ we have

$$(x, [A]^{\mathscr{M}} x) = M_{\mathscr{U}}((Ux, AUx))$$

which is real because of condition (i) of Definition 2 and because (Ux, AUx) is real for hermitian A. Since this holds for all x, it implies that $[A]^{\mathscr{M}}$ is also hermitian, as was to be shown.

(x) If $A \geq 0$, then $(Ux, Aux) \geq 0$. Hence

$$(x, [A]^{\mathscr{M}} x) = M_{\mathscr{U}}((Ux, AUx)) \geq \inf_{\mathscr{U}} (Ux, AUx) \geq 0$$

implying that $[A]^{\mathscr{M}} \geq 0$. A similar argument holds for $A > 0$, verifying (xi).

(xii) If $[A]_{\mathscr{M}}$ exists in the uniform topology, then

$$[A]_{\mathscr{M}} = \lim_{\mathscr{P}} \left[\sum_{i-1}^{n} \Delta^i A \Delta^i \right]$$

exists and is memoryless. Thus

$$[A]_{\mathscr{M}} = [[A]_{\mathscr{M}}]^{\mathscr{M}} = \left[\lim_{\mathscr{P}} \sum_{i=1}^{n} \Delta^i A \Delta^i \right]^{\mathscr{M}} = \lim_{\mathscr{P}} \left[\sum_{i=1}^{n} \Delta^i A \Delta^i \right]^{\mathscr{M}}$$

$$= \lim_{\mathscr{P}} \left[\sum_{i=1}^{n} \Delta^i [A]^{\mathscr{M}} \Delta^i \right] = \lim_{\mathscr{P}} \left[\sum_{i=1}^{n} \Delta^i [A]^{\mathscr{M}} \right] = [A]^{\mathscr{M}}$$

where we have invoked (i)–(iii) and (v). //

8. Remark: By (xii) the expectation operator $[A]^{\mathscr{M}}$ is indeed an extension of $[A]_{\mathscr{M}}$ defined for all $A \in \mathscr{B}$ and preserving most of the properties of $[A]_{\mathscr{M}}$. Note that even though $[A]^{\mathscr{M}}$ is not unique its fundamental properties are independent of the choice of invariant mean.

9. Remark: Note that if A is strongly causal, then $[A]^{\mathscr{M}} = 0$. Moreover, we could define a new class of *hypercausal operators* by requiring that A be causal and satisfy $[A]^{\mathscr{M}} = 0$. Indeed any causal A may then be decomposed into the sum of such a hypercausal operator and a memoryless operator:

$$A = [A]^{\mathscr{M}} + (A - [A]^{\mathscr{M}})$$

yielding a decomposition for an arbitrary causal operator. Unfortunately, this class of hypercausal operators does not appear to be sufficiently structured to make such a decomposition useful and it will not be considered further.

10. Definition: Let $\mathscr{A}$ be a subalgebra of $\mathscr{B}$. Then a *derivation* from $\mathscr{A}$ into $\mathscr{B}$ is a linear map $\partial: \mathscr{A} \to \mathscr{B}$ such that $\partial[AB] = A\,\partial[B] + \partial[A]\,B$ for all $A, B \in \mathscr{A}$.

11. Lemma: If ∂ is a derivation from $\mathscr{K}_{\mathscr{E}}$ into $\mathscr{B}$ and $M_{\mathscr{U}}$ is an invariant mean on $\mathscr{U}$, then there exists a linear operator $T \in \mathscr{B}$ such that

$$\partial[A] = AT - TA$$

for all $A \in \mathscr{K}_{\mathscr{E}}$. Furthermore, one such T is defined by the equality

$$\operatorname{tr}[TK] = M_{\mathscr{U}}(\operatorname{tr}[U^* \partial[U]\,K]), \qquad K \in \mathscr{K}_1$$

and if $T' \in \mathscr{B}$ is any other operator satisfying $\partial[A] = AT' - T'A$, then T' is related to T as

$$T = T' - [T']^{\mathscr{M}}$$

In particular, $[T]^{\mathscr{M}} = 0$.

Proof: Since every derivation on a von Neumann algebra is bounded, $U^* \partial[U] K \in \mathscr{K}_1$ and hence its trace is a bounded complex function of $\mathscr{U}$. Thus

$$M_{\mathscr{U}}(\mathrm{tr}[U^* \partial[U] K]) \in C$$

defines a bounded linear functional on $\mathscr{K}_1$ and must be of the form $\mathrm{tr}[TK]$ for some $T \in \mathscr{B} = \mathscr{K}_1^*$. Accordingly, T is well defined. Now given this $T \in \mathscr{B}$, $A \in \mathscr{U}$, and $K \in \mathscr{K}_1$ and invoking the equality $\mathrm{tr}[AB] = \mathrm{tr}[BA]$ when both AB and BA are in $\mathscr{K}_1$, we have

$$\begin{aligned}
\mathrm{tr}[(AT - TA)K] &= \mathrm{tr}[ATK] - \mathrm{tr}[TAK] = \mathrm{tr}[TKA] - \mathrm{tr}[TAK] \\
&= \mathrm{tr}[T(KA - AK)] = M_{\mathscr{U}}(\mathrm{tr}[U^* \partial[U] (KA - AK)]) \\
&= M_{\mathscr{U}}(\mathrm{tr}[AU^* \partial[U] K]) - M_{\mathscr{U}}(\mathrm{tr}[U^* \partial[U] AK]) \\
&= M_{\mathscr{U}}(\mathrm{tr}[AU^* \partial[U] K]) - M_{\mathscr{U}}(\mathrm{tr}[U^* \partial[UA] K]) \\
&\quad + M_{\mathscr{U}}(\mathrm{tr}[U^*U \partial[A] K]) \\
&= M_{\mathscr{U}}(\mathrm{tr}[AU^* \partial[U] K]) - M_{\mathscr{U}}(\mathrm{tr}[A(UA)^* \partial[UA] K]) \\
&\quad + M_{\mathscr{U}}(\mathrm{tr}[\partial[A] K]) \\
&= M_{\mathscr{U}}(\mathrm{tr}[AU^* \partial[U] K]) - M_{\mathscr{U}}(\mathrm{tr}[AU^* \partial[U]K]) \\
&\quad + \mathrm{tr}[\partial[A] K] \\
&= \mathrm{tr}[\partial[A] K]
\end{aligned}$$

Here we have invoked the definition of $\partial[\ ]$, the invariance of $M_{\mathscr{U}}$, and the fact that $M_{\mathscr{U}}(f) = f$ when f is constant. Since this equality holds for all $K \in \mathscr{K}_1$, we have verified that $\partial[A] = AT - TA$ whenever $A \in \mathscr{U}$. However, since $\mathscr{U}$ generates $\mathscr{K}_{\mathscr{E}}$ the equality extends to all $A \in \mathscr{K}_{\mathscr{E}}$ as was to be shown.

If one is also given the equality $\partial[A] = AT' - T'A$, then for $K \in \mathscr{K}_1$

$$\begin{aligned}
\mathrm{tr}[TK] &= M_{\mathscr{U}}(\mathrm{tr}[U^* \partial[U] K]) = M_{\mathscr{U}}(\mathrm{tr}[U^*(UT' - T'U)K]) \\
&= M_{\mathscr{U}}(\mathrm{tr}[(T' - U^*T'U)K]) \\
&= \mathrm{tr}[T'K] - \mathrm{tr}[[T']^{\mathscr{M}}K]
\end{aligned}$$

showing that

$$T = T' - [T']^{\mathscr{M}}$$

Finally, since $[\ ]^{\mathscr{M}}$ is a projection we have

$$[T]^{\mathscr{M}} = [T']^{\mathscr{M}} - [[T']^{\mathscr{M}}]^{\mathscr{M}} = [T']^{\mathscr{M}} - [T']^{\mathscr{M}} = 0$$

completing the proof. //

12. Remark: Since $\mathscr{R}$ is an ideal in $\mathscr{C}$ one can define the *quotient space* of equivalence classes $\mathscr{C}/\mathscr{R} = \{[A] : A \in \mathscr{C}\}$ where $[A] \simeq [B]$ if $A - B \in \mathscr{R}$. Moreover, this space has a natural norm

$$\|[A]\|_{\mathrm{q}} = \inf_{\mathscr{R}} \|A + B\|, \qquad B \in \mathscr{R}$$

For our purposes, rather than working in the quotient space $\mathscr{C}/\mathscr{R}$, we shall work with the *quotient seminorm* induced on $\mathscr{C}$ by $\mathscr{R}$ by

$$\|A\|_{\mathrm{q}} = \|[A]\|_{\mathrm{q}}, \qquad A \in \mathscr{C}$$

Note that the use of a common notation for the quotient seminorm on $\mathscr{C}$ and the norm on the quotient space causes no difficulty since the elements on which they act are always denoted differently, i.e., $[A]$ versus A.

13. Lemma: Let $\mathscr{E}$ be a resolution of the identity on H, let $\mathscr{P} = \{E^i : i = 0, 1, \ldots, n\}$ be a partition of $\mathscr{E}$, and let $A^i \in \mathscr{C}$, $i = 1, 2, \ldots, n$. Then

$$\left\| \sum_{i=1}^{n} \Delta^i A^i \Delta^i \right\|_{\mathrm{q}} = \max_i \|\Delta^i A^i \Delta^i\|_{\mathrm{q}}$$

Proof: If $B^i \in \mathscr{R}$, $i = 1, 2, \ldots, n$, then

$$\left\| \sum_{i=1}^{n} \Delta^i A^i \Delta^i \right\|_{\mathrm{q}} \leq \left\| \sum_{i=1}^{n} \Delta^i (A^i + B^i) \Delta^i \right\| = \max_i \|\Delta^i (A^i + B^i) \Delta^i\|$$

$$= \max_i \|\Delta^i (\Delta^i A^i \Delta^i + B^i) \Delta^i\| \leq \max_i \|\Delta^i A^i \Delta^i + B^i\|$$

where we have employed Lemma 2.A.4, the multiplicative triangle inequality, and the definition of $\|\ \|_{\mathrm{q}}$. Since this inequality holds for all $B^i \in \mathscr{R}$, we may take the infimum over $B^i \in \mathscr{R}$, obtaining

$$\left\| \sum_{i=1}^{n} \Delta^i A^i \Delta^i \right\|_{\mathrm{q}} \leq \max_i \|\Delta^i A^i \Delta^i\|_{\mathrm{q}}$$

On the other hand,

$$\|\Delta^j A^j \Delta^j\|_q = \left\|\Delta^j\left[\sum_{i=1}^{n} \Delta^i A^i \Delta^i\right]\Delta^j\right\|_q \leq \left\|\sum_{i=1}^{n} \Delta^i A^i \Delta^i\right\|_q$$

implying that

$$\max_j \|\Delta^j A^j \Delta^j\|_q \leq \left\|\sum_{i=1}^{n} \Delta^i A^i \Delta^i\right\|_q$$

and therefore completing the proof. //

14. Remark: Although for most purposes a linear ordering proves to be sufficient structure for our time set $, one can naturally define an *order topology* on $ by letting the intervals $(\dagger, \ddagger)$, $\dagger < \ddagger$, $[-\infty, \ddagger)$, and $(\dagger, \infty]$ be the basic open sets. Since $\mathscr{E}$ is order complete and order isomorphic to $, $ is also order complete, and hence $ is complete in the order topology. Moreover, since $ includes its minimum and maximum elements $\dagger_0$ and $\dagger_\infty$, $ is compact in its order topology. This fact proves to be useful in the derivation of the following lemma.

15. Lemma: For each $A \in \mathscr{C}$ there exists a $\dagger \in \$$ such that one of the following holds:

(i) Let $\Delta = E^\dagger - E^\ddagger$ where $\dagger > \ddagger \geq \dagger_0$. Then

$$\|\Delta A \Delta\|_q = \|\Delta A\|_q = \|A\Delta\|_q = \|A\|_q$$

(ii) Let $\Delta = E_\dagger - E_\ddagger$ where $\dagger < \ddagger \leq \dagger_\infty$. Then

$$\|\Delta A \Delta\|_q = \|\Delta A\|_q = \|A\Delta\|_q = \|A\|_q$$

Proof: By the argument employed in the proof of Corollary 2.D.7 one may verify that $\Delta A \Delta - \Delta A$ and $\Delta A \Delta - A\Delta$ are strictly causal, indeed pre-strictly causal. Hence

$$[\Delta A \Delta \simeq [\Delta A] \simeq [A\Delta]$$

and

$$\|\Delta A \Delta\|_q = \|\Delta A\|_q = \|A\Delta\|_q$$

Moreover, the triangle inequality implies that

$$\|\Delta A\|_q \leq \|\Delta\|_q \|A\|_q \leq \|A\|_q$$

Thus all that need be verified is that $\|\Delta A\|_q \geq \|A\|_q$ when the hypotheses of the lemma are satisfied. If this is not the case, then

(i) there exists a $\dagger'' \in \$$, $\dagger'' > \dagger_0$, such that

$$\|E_{\dagger''}A\|_q < \|A\|_q$$

(ii) there exists a $\dagger' \in \$$, $\dagger' < \dagger_\infty$, such that

$$\|E^{\dagger'}A\|_q < \|A\|_q$$

and

(iii) for each $\dagger \in \$$, $\dagger_0 < \dagger < \dagger_\infty$, there exists $\ddagger_1 > \dagger > \ddagger_2$ such that

$$\|(E_{\ddagger_1} - E_\dagger)A\|_q < \|A\|_q, \qquad \|(E_\dagger - E_{\ddagger_2})A\|_q < \|A\|_q$$

Indeed, if any of these conditions fails, the conditions of the lemma are satisfied.

Now consider the set of intervals in $\$$ taking the form $(\dagger_1, \dagger_2)$, $\dagger \in \$$, $[\dagger_0, \dagger')$, and $(\dagger'', \dagger_\infty]$. Each of these intervals is open in the order topology of $\$$ and they cover $\$$. Thus, since $\$$ is compact in its order topology, there exists a finite subcover of intervals I_i, $i = 1, 2, \ldots, n$, where $I_1 = [\dagger_0, \dagger')$, $I_i = (\dagger_{1_i}, \dagger_{2_i})$, $i = 2, 3, \ldots, n-1$, and $I_n = (\dagger'', \dagger_\infty]$. Since these intervals cover $\$$, one can choose a partition $\mathscr{P} = \{E^i : i = 0, 1, \ldots, n\}$ such that

$$\Delta^1 < E^{\dagger'}, \qquad \Delta^i < [E^{\dagger_{2i}} - E^{\dagger_{1i}}], \qquad i = 2, 3, \ldots, n-1$$

and

$$\Delta^n < E_{\dagger''}$$

Hence $\|\Delta^i A\|_q < \|A\|_q$, $i = 1, 2, \ldots, n$, and

$$\|A\|_q = \left\| \left[\sum_{i=1}^n \Delta^i \right] A \right\|_q = \left\| \left[\sum_{i=1}^n \Delta^i A \right] \right\|_q = \left\| \left[\sum_{i=1}^n \Delta^i A \Delta^i \right] \right\|_q$$
$$= \max_i \|\Delta^i A \Delta^i\|_q = \max_i \|\Delta^i A\|_q < \max_i \|A\|_q = \|A\|_q$$

which is a contradiction. Here we have used $[\Delta^i A] \simeq [\Delta^i A \Delta^i]$ at two points in the derivation. The assumption that the conditions of the lemma are not satisfied yields a contradiction, thus verifying the lemma. //

Causal Decomposition Theorem: Let A be causal. Then A admits a decomposition into the sum of a strictly causal and a memoryless operator

$$A = [A]_{\mathscr{R}} + [A]_{\mathscr{M}}$$

if and only if $AK - KA \in \mathscr{R}$ for all $K \in \mathscr{K}_{\mathscr{E}}$.

Proof: By Property A.2 it suffices to show that

$$[A]_{\mathscr{M}} = (M)\int dE(\dagger)\,A\,dE(\dagger) = (m)\int dE(\dagger)\,A\,dE(\dagger)$$

exists if and only if the hypotheses of the theorem are satisfied. If this is the case, then

$$[A]_{\mathscr{R}} = (M)\int E_{\dagger}A\,dE(\dagger) = (m)\int dE(\dagger)\,AE^{\dagger}$$

also exists, and for $K \in \mathscr{K}_{\mathscr{E}}$,

$$\begin{aligned} AK - KA &= ([A]_{\mathscr{R}} + [A]_{\mathscr{M}})K - K([A]_{\mathscr{R}} + [A]_{\mathscr{M}}) \\ &= [A]_{\mathscr{R}}K + [A]_{\mathscr{M}}K - K[A]_{\mathscr{R}} - K[A]_{\mathscr{M}} \\ &= [A]_{\mathscr{R}}K + [A]_{\mathscr{M}}K - K[A]_{\mathscr{R}} - [A]_{\mathscr{M}}K \\ &= [A]_{\mathscr{R}}K - K[A]_{\mathscr{R}} \in \mathscr{R} \end{aligned}$$

as was to be shown. Now suppose that $A \in \mathscr{C}$ and $AK - KA \in \mathscr{R}$ for all $K \in \mathscr{K}_{\mathscr{E}}$. We shall show that A has the desired decomposition. Indeed, we shall show that $[A]_{\mathscr{M}} = [A]^{\mathscr{M}}$ and $[A]_{\mathscr{R}} = A - [A]^{\mathscr{M}}$. Since $[A]^{\mathscr{M}}$ always exists and is memoryless, it suffices to show that $[A]_{\mathscr{R}} = A - [A]^{\mathscr{M}} \in \mathscr{R}$. To this end let $\partial: \mathscr{K}_{\mathscr{E}} \to \mathscr{B}$ be the derivation induced by $[A]_{\mathscr{R}}$, i.e.,

$$\partial[K] = K[A]_{\mathscr{R}} - [A]_{\mathscr{R}}K$$

and note that $[[A]_{\mathscr{R}}]^{\mathscr{M}} = 0$ since $[\ ]^{\mathscr{M}}$ is a projection. By Lemma 15 there exists a $\dagger \in \$$ such that one of the following holds:

(i) Let $\Delta = E^{\dagger} - E^{\ddagger}$ where $\dagger > \ddagger \geq \dagger_0$. Then

$$\|\Delta[A]_{\mathscr{R}}\Delta\|_{\mathrm{q}} = \|A\|_{\mathrm{q}}$$

(ii) Let $\Delta = E_{\dagger} - E_{\ddagger}$ where $\dagger < \ddagger \leq \dagger_{\infty}$. Then

$$\|\Delta[A]_{\mathscr{R}}\Delta\|_{\mathrm{q}} = \|A\|_{\mathrm{q}}$$

Assume that $\dagger$ satisfies (i). The proof where (ii) holds is similar.

First assume that $E_{\dagger}$ has an immediate predecessor $E_{\dagger_-}$. Then $\Delta = E^{\dagger} - E^{\dagger-}$ is a minimal projection in $\mathscr{K}_{\mathscr{E}}$ (i.e., there is no projection P such that $P < \Delta$), and so ΔK, $K \in \mathscr{K}_{\mathscr{E}}$, consists of scalar multiples of Δ (since $\mathscr{K}_{\mathscr{E}}$ is abelian). Thus if $K \in \mathscr{K}_{\mathscr{E}}$ there exists a scalar c such that

$$K(\Delta[A]_{\mathscr{R}}\Delta) = c(\Delta[A]_{\mathscr{R}}\Delta) = (\Delta[A]_{\mathscr{R}}\Delta)K$$

Since $\mathscr{M}$ is the commutant of $\mathscr{K}_{\mathscr{E}}$, this implies that $(\Delta[A]_{\mathscr{R}}\Delta) \in \mathscr{M}$. Hence

$$(\Delta[A]_{\mathscr{R}}\Delta) = [(\Delta[A]_{\mathscr{R}}\Delta)]^{\mathscr{M}} = \Delta[[A]_{\mathscr{R}}]^{\mathscr{M}}\Delta = \Delta(0)\Delta = 0$$

while

$$0 = \|(\Delta[A]_{\mathscr{R}}\Delta)\|_{\mathrm{q}} = \|[A]_{\mathscr{R}}\|_{\mathrm{q}}$$

which implies that $[A]_{\mathscr{R}} \in \mathscr{R}$, as was to be shown.

If $E_{\dagger}$ does not have an immediate predecessor, then there exists a net of projection $E^{\dagger_\pi}$ converging strongly to $E_{\dagger}$ from below. Then $\Delta_\pi = E^{\dagger} - E^{\dagger_\pi}$ converges strongly to 0, and Lemma 15 implies that

$$\|\Delta_\pi[A]_{\mathscr{R}}\Delta_\pi\|_{\mathrm{q}} = \|[A]_{\mathscr{R}}\|_{\mathrm{q}}$$

for all π. Thus

$$\|\Delta_\pi[A]_{\mathscr{R}}\Delta_\pi\| \geq \|\Delta_\pi[A]_{\mathscr{R}}\Delta_\pi\|_{\mathrm{q}} = \|A\|_{\mathrm{q}}$$

for all π. Now if Δ_π is fixed, the net

$$[\Delta_\pi - \Delta_\gamma][A]_{\mathscr{R}}[\Delta_\pi - \Delta_\gamma]$$

converges strongly to $\Delta_\pi[A]_{\mathscr{R}}\Delta_\pi$. Furthermore, for a fixed $x \in H$,

$$[\Delta_\pi[A]_{\mathscr{R}}\Delta_\gamma]x \to 0 \qquad \text{and} \qquad [\Delta_\gamma[A]_{\mathscr{R}}\Delta_\pi] \to 0$$

while

$$\|[\Delta_\gamma[A]_{\mathscr{R}}\Delta_\gamma]x\| \leq \|[A]_{\mathscr{R}}\| \, \|\Delta_\gamma x\| \to 0$$

Thus by the lower semicontinuity of the norm in the strong operator topology, for each π there exists γ such that $\Delta_\gamma < \Delta_\pi$ and

$$\|[\Delta_\pi - \Delta_\gamma][A]_{\mathscr{R}}[\Delta_\pi - \Delta_\gamma]\| > \tfrac{1}{2}\|[A]_{\mathscr{R}}\|_{\mathrm{q}}$$

since $\|\Delta_\pi[A]_{\mathscr{R}}\Delta_\pi\| \geq \|[A]_{\mathscr{R}}\|_{\mathrm{q}}$, while the remaining terms on the left side of the above inequality converge strongly to zero. We may now construct a decreasing sequence of projections Δ_n by setting $\Delta_1 = \Delta_\pi$, $\Delta_2 = \Delta_\gamma$. We then repeat the process with Δ_π replaced by Δ_γ and construct a new $\Delta_{\gamma'}$ satisfying the inequality that we take for Δ_3. Inductively, we may thus construct a decreasing sequence of projections Δ_i, $i = 1, 2, \ldots,$ such that

$$\|\Delta^i[A]_{\mathscr{R}}\Delta^i\| > \tfrac{1}{2}\|[A]_{\mathscr{R}}\|_{\mathrm{q}}$$

where $\Delta^i = \Delta_i - \Delta_{i+1}$, $i = 1, 2, \ldots,$. Note further that Δ^i is of the form

$$\Delta^i = \Delta_i - \Delta_{i+1} = [E^{\dagger} - E^{\dagger_i}] - [E^{\dagger} - E^{\dagger_{i+1}}] = [E^{\dagger_{i+1}} - E^{\dagger_i}]$$

where $\dagger_i$ is increasing to $\dagger$, and that the Δ^is are mutually orthogonal. Therefore $E_{\dagger_i}$ is a decreasing sequence of projections bounded below by $E_\dagger$ and we may let

$$E_\ddagger = \operatorname{s}\lim_i\, [E_{\dagger_i}]$$

which exists by the completeness of $\mathscr{E}$. Then $E_\dagger \le E_\ddagger$ and $E_\ddagger < E_{\dagger_i}$, $i = 1, 2, \ldots$.

If $\dagger' < \ddagger$, then since $\dagger_i$ converges to $\ddagger$ from below, $E_\ddagger < E_{\dagger_i} < E_{\dagger'}$ for some i and $[E^\ddagger - E^{\dagger'}] > \Delta^k$, $k \ge i$. Since $[[A]_\mathscr{R}] = 0$ and $\partial[K] = K[A]_\mathscr{R} - [A]_\mathscr{R} K$, Lemma 11 implies that

$$\operatorname{tr}[[A]_\mathscr{R} L] = M_\mathscr{U}(\operatorname{tr}[U^* \partial[U]\, L]), \qquad L \in \mathscr{K}_1$$

Now, if for some i $\|\Delta^i \partial[K]\Delta^i\| \le \frac{1}{2}\|[A]_\mathscr{R}\|_q$ for all $K \in \mathscr{K}_\mathscr{E}$ with $\|K\| \le 1$, then

$$\begin{aligned}\operatorname{tr}[\Delta^i[A]_\mathscr{R}\Delta^i L] &= \operatorname{tr}[[A]_\mathscr{R}\Delta^i L \Delta^i] = M_\mathscr{U}(\operatorname{tr}[U^* \partial[U]\Delta^i L \Delta^i]) \\ &= M_\mathscr{U}(\operatorname{tr}[\Delta^i U^* \partial[U]\Delta^i L]) = M_\mathscr{U}(\operatorname{tr}[U^*\Delta^i \partial[U]\Delta^i L])\end{aligned}$$

so

$$\begin{aligned}|\operatorname{tr}[\Delta^i[A]_\mathscr{R}\Delta^i L]| &\le \left|\sup_\mathscr{U} \operatorname{tr}[U^*\Delta^i \partial[U]\Delta^i L]\right| = \left|\sup_\mathscr{U} \operatorname{tr}[\Delta^i \partial[U]\Delta^i L U^*]\right| \\ &\le \|\Delta^i \partial[U]\, \partial^i\|\, \|LU^*\|_1 \le \|\Delta^i \partial[U]\Delta^i\|\, \|U^*\|\, \|L\|_1 \\ &\le \tfrac{1}{2}\|[A]_\mathscr{R}\|_q\, \|L\|_1\end{aligned}$$

where $\|L\|_1$ is the norm of L in $\mathscr{K}_1$. This, however, contradicts the fact that $\|\Delta^i[A]_\mathscr{R}\Delta^i\| > \frac{1}{2}\|[A]_\mathscr{R}\|_q$. Hence for each i there exists a $K^i \in \mathscr{K}_\mathscr{E}$ with $\|K^i\| \le 1$ such that

$$\|\Delta^i \partial[K^i]\Delta^i\| > \tfrac{1}{2}\|[A]_\mathscr{R}\|_q$$

Let $K = \sum_i K^i\Delta^i$, which converges strongly since the Δ^i are mutually orthogonal and the K^i and Δ^i commute. Then

$$\begin{aligned}\Delta^i \partial[K]\Delta^i &= \Delta^i[K[A]_\mathscr{R} - [A]_\mathscr{R} K]\Delta^i \\ &= \Delta^i[K^i[A]_\mathscr{R} - [A]_\mathscr{R} K^i]\Delta^i \\ &= \Delta^i \partial[K^i]\Delta^i\end{aligned}$$

Thus

$$\|\Delta^i \partial[K]\Delta^i\| > \tfrac{1}{2}\|[A]_\mathscr{R}\|_q, \qquad i = 1, 2, \ldots$$

Finally, by assumption $\partial[K] = K[A]_{\mathscr{R}} - [A]_{\mathscr{R}}K$ is strictly causal. Thus if $\|[A]_{\mathscr{R}}\|_q > 0$, there exists $\dagger' < \ddagger$ such that

$$\|[E\ddagger - E^{\dagger'}]\,\partial[K][E^{\ddagger} - E^{\dagger'}]\| < \tfrac{1}{4}\|[A]_{\mathscr{R}}\|_q$$

However, $[E^{\ddagger} - E^{\dagger'}] > \Delta^i$ for some i, implying that

$$\|\Delta^i\,\partial[K]\,\partial^i\| < \tfrac{1}{4}\|[A]_{\mathscr{R}}\|_q$$

which is a contradiction. Therefore $\|[A]_{\mathscr{R}}\|_q = 0$, showing that $[A]_{\mathscr{R}} \in \mathscr{R}$ and, at long last, completing the proof. //

16. Remark: A careful inspection of the proof and its preliminary lemmas will reveal that the above derivation also characterizes those causal operators that admit a decomposition into a memoryless operator and a strongly causal operator. Indeed, one need only define

$$A = [A]_{\mathscr{T}} + [A]_{\mathscr{M}}$$

where $[A]_{\mathscr{M}} = [A]^{\mathscr{M}}$ and $[A]_{\mathscr{T}} = A - [A]^{\mathscr{M}}$ and follow the same derivation to verify $[A]_{\mathscr{T}}$ will be strongly causal if $[A]_{\mathscr{T}}K - K[A]_{\mathscr{T}} \in \mathscr{T}$ for all $K \in \mathscr{K}_{\mathscr{E}}$. Of course, in the modified derivation $\|\ \|_q$ must be taken to be the quotient seminorm induced by $\mathscr{C}/\mathscr{T}$. Then if $A \in \mathscr{C}$, A admits a decomposition into the sum of a strongly causal and memoryless operator,

$$A = [A]_{\mathscr{T}} + [A]_{\mathscr{M}}$$

if and only if $AK - KA \in \mathscr{T}$ for all $K \in \mathscr{K}_{\mathscr{E}}$.

17. Remark: Unlike in the strongly causal case, this derivation fails in the strongly strictly causal case. The necessity proof, however, remains valid, and indeed the sufficiency proof fails only in the final paragraph. It is an open question whether the above theorem can be extended to characterize the decomposition of a causal operator into the sum of a strongly strictly causal operator and a memoryless operator. We do, however, have the following:

18. Property: Let A be causal and assume that A admits a decomposition into the sum of a strongly strictly causal operator and a memoryless operator. Then $AK - KA \in \mathscr{S}$ for all $K \in \mathscr{K}_{\mathscr{E}}$.

19. Remark: Of course, the obvious dual theorems hold to the effect that $A \in \mathscr{C}^*$ can be decomposed as

$$A = [A]_{\mathscr{M}} + [A]_{\mathscr{R}^*}$$

where $[A]_{\mathscr{M}} \in \mathscr{M}$ and $[A]_{\mathscr{R}^*} \in \mathscr{R}^*$ if and only if $AK - KA \in \mathscr{R}^*$ for all $K \in \mathscr{K}_{\mathscr{E}}$, with a similar result for the decomposition of A into a memoryless and a strongly anticausal operator.

20. Remark: Note that for any causal A and $E^\dagger \in \mathscr{E}$, $AE^\dagger - E^\dagger A \in \mathscr{R}$. Moreover, since $\mathscr{R}$ is a uniformly closed linear space, $AK - KA$ will be strictly causal for any K that can be expressed as the uniform limit of linear combinations of $\mathscr{E}$. $\mathscr{R}$ is, however, not weakly closed, and hence if one allows K to be the weak limit of linear combinations of $\mathscr{E}$, this may no longer be the case. Since such weak limits are included in the core $\mathscr{K}_{\mathscr{E}}$, the condition that $AK - KA \in \mathscr{R}$ for all $K \in \mathscr{K}_{\mathscr{E}}$ defines a proper subset of the causal operators, the operators that admit a decomposition into the sum of a strictly causal operator and a memoryless operator by the terms of the theorem.

21. Example: Consider the operator $I + V$ where V is the unilateral shift on $l_2[0, \infty)$, with its usual resolution structure. Here $I + V$ has the semi-infinite matrix representation

$$I + V = \begin{bmatrix} 1 & 0 & 0 & 0 & 0 & \cdots \\ 1 & 1 & 0 & 0 & 0 & \cdots \\ 0 & 1 & 1 & 0 & 0 & \cdots \\ 0 & 0 & 1 & 1 & 0 & \cdots \\ 0 & 0 & 0 & 1 & 1 & \cdots \\ \vdots & \vdots & \vdots & \vdots & \vdots & \end{bmatrix}$$

relative to the standard basis $\{e_i\}_0^\infty$ on $l_2[0, \infty)$. Now, if we let E^i be the projection onto the closed linear space of $\{e_0, e_1, e_2, \ldots, e_{i-1}\}$, then $[I + V]E^3$ has the matrix representation

$$[I + V]E^3 = \begin{bmatrix} 1 & 0 & 0 & 0 & 0 & \cdots \\ 1 & 1 & 0 & 0 & 0 & \cdots \\ 0 & 1 & 1 & 0 & 0 & \cdots \\ 0 & 0 & 1 & 0 & 0 & \cdots \\ 0 & 0 & 0 & 0 & 0 & \cdots \\ \vdots & \vdots & \vdots & \vdots & \vdots & \end{bmatrix}$$

while $E^3[I + V]$ has the matrix representation

$$E^3[I + V] = \begin{bmatrix} 1 & 0 & 0 & 0 & 0 & \cdots \\ 1 & 1 & 0 & 0 & 0 & \cdots \\ 0 & 1 & 1 & 0 & 0 & \cdots \\ 0 & 0 & 0 & 0 & 0 & \cdots \\ 0 & 0 & 0 & 0 & 0 & \cdots \\ \vdots & \vdots & \vdots & \vdots & \vdots & \end{bmatrix}$$

Thus $[I + V]E^3 - E^3[I + V]$ has the matrix representation

$$[I + V]E^3 - E^3[I + V] = \begin{bmatrix} 0 & 0 & 0 & 0 & 0 & \cdots \\ 0 & 0 & 0 & 0 & 0 & \cdots \\ 0 & 0 & 0 & 0 & 0 & \cdots \\ 0 & 0 & 1 & 0 & 0 & \cdots \\ 0 & 0 & 0 & 0 & 0 & \cdots \\ \vdots & \vdots & \vdots & \vdots & \vdots & \end{bmatrix}$$

which is strictly causal as expected. Indeed, it is pre–strictly causal (use the partition $\mathscr{P}' = \{0, E^3, I\}$ or any refinement thereof). In general, if one replaces E^3 by E^i, $[I + V]E^i - E^i[I + V]$ will have a matrix representation that is zero except for a 1 in the $(i, i - 1)$th entry and will also be pre–strictly causal.

Now let $\Delta^i = E^i - E^{i-1}$, $i = 1, 2, \ldots,$ and let

$$K = \sum_j \Delta^{2j}$$

which is strongly (and hence weakly) convergent since the Δ^{2j} have mutually orthogonal ranges. $[I + V]K$ thus has the matrix representation

$$[I + V]K = \begin{bmatrix} 0 & 0 & 0 & 0 & 0 & 0 & 0 & \cdots \\ 0 & 1 & 0 & 0 & 0 & 0 & 0 & \cdots \\ 0 & 1 & 0 & 0 & 0 & 0 & 0 & \cdots \\ 0 & 0 & 0 & 1 & 0 & 0 & 0 & \cdots \\ 0 & 0 & 0 & 1 & 0 & 0 & 0 & \cdots \\ 0 & 0 & 0 & 0 & 0 & 1 & 0 & \cdots \\ 0 & 0 & 0 & 0 & 0 & 1 & 0 & \cdots \\ \vdots & \vdots & \vdots & \vdots & \vdots & \vdots & \vdots & \end{bmatrix}$$

while $K[I + V]$ has the matrix representation

$$K[I + V] = \begin{bmatrix} 0 & 0 & 0 & 0 & 0 & 0 & 0 & \cdots \\ 1 & 1 & 0 & 0 & 0 & 0 & 0 & \cdots \\ 0 & 0 & 0 & 0 & 0 & 0 & 0 & \cdots \\ 0 & 0 & 1 & 1 & 0 & 0 & 0 & \cdots \\ 0 & 0 & 0 & 0 & 0 & 0 & 0 & \cdots \\ 0 & 0 & 0 & 0 & 1 & 1 & 0 & \cdots \\ 0 & 0 & 0 & 0 & 0 & 0 & 0 & \cdots \\ \vdots & \vdots & \vdots & \vdots & \vdots & \vdots & \vdots & \end{bmatrix}$$

and $[I + V]K - K[I + V]$ has the matrix representation

$$[I + K]K - K[I + V] = \begin{bmatrix} 0 & 0 & 0 & 0 & 0 & 0 & 0 & \cdots \\ -1 & 0 & 0 & 0 & 0 & 0 & 0 & \cdots \\ 0 & 1 & 0 & 0 & 0 & 0 & 0 & \cdots \\ 0 & 0 & -1 & 0 & 0 & 0 & 0 & \cdots \\ 0 & 0 & 0 & 1 & 0 & 0 & 0 & \cdots \\ 0 & 0 & 0 & 0 & -1 & 0 & 0 & \cdots \\ 0 & 0 & 0 & 0 & 0 & 1 & 0 & \cdots \\ \vdots & \vdots & \vdots & \vdots & \vdots & \vdots & \vdots & \end{bmatrix}$$

Thus

$$[I + V]K - K[I + V] = AV$$

where A is the memoryless operator mapping the sequence $\{b_i\}_0^\infty$ into the $\{(-1)^i b_i\}_0^\infty$; i.e., A has a diagonal matrix representations with $a_{ii} = (-1)^i$. Now if $[I + V]K - K[I + V]$ were strictly causal, then since $A^{-1} = A$ is memoryless and the strictly causals form an ideal,

$$V = A^{-1}([I + V]K - K[I + V]) = A([I + V]K - K[I + V])$$

would also be strictly causal. Since this is not the case, $[I + V]K - K[I + V] \notin \mathscr{R}$ showing that $[I + V]$ cannot be decomposed into a memoryless operator and a strictly causal operator. Of course, it has the obvious decomposition into a memoryless operator and a strongly causal operator consistent with the fact that $[I + V]K - K[I + V] \in \mathscr{T}$.

22. Remark: It follows from the proof of the theorem that if $A \in \mathscr{C}$ can be decomposed into the sum of a memoryless operator and a strictly causal operator, then $[A]_{\mathscr{M}} = [A]^{\mathscr{M}}$ and $[A]_{\mathscr{R}} = A - [A]^{\mathscr{M}}$ with the decomposition existing if and only if $A - [A]^{\mathscr{M}} \in \mathscr{R}$. Applying this to the operator $[I + V]$ from the above example, we obtain

$$[A]_{\mathscr{R}} = [I + V] - [I + V]^{\mathscr{M}} = [I+V] - I = V$$

since V is strongly causal. On the other hand since $V \notin \mathscr{R}$ we are assured that no decomposition into a memoryless and a strictly causal operator can exist. The above criterion for the existence of a decomposition for $A \in \mathscr{C}$ often proves to be a useful tool and is formalized as follows.

23. Corollary: Let $A \in \mathscr{C}$. Then A admits a decomposition $A = [A]_{\mathscr{M}} + [A]_{\mathscr{H}}$, where $[A]_{\mathscr{M}} \in \mathscr{M}$ and $[A]_{\mathscr{H}} \in \mathscr{H}$, $\mathscr{H} = \mathscr{R}$ (or $\mathscr{T}$) if and only if $(A - [A]^{\mathscr{M}}) \in \mathscr{H}$, in which case $[A]_{\mathscr{M}} = [A]^{\mathscr{M}}$ and $[A]_{\mathscr{H}} = (A - [A]^{\mathscr{M}})$.

24. Remark: In essence the corollary implies that $[A]^{\mathscr{M}}$ and $(A - [A]^{\mathscr{M}})$ are the only candidates for the decomposition of a causal operator into the sum of a memoryless operator and a hypercausal operator (in $\mathscr{R}$ or $\mathscr{T}$).

C. DECOMPOSITION IN $\mathscr{K}$

1. Remark: The one class of noncausal operators for which a viable decomposition theory can be formalized are the compact operators $\mathscr{K}$. Recall that the compact operators form an ideal in $\mathscr{B}$ and contain a number of additional ideals. These include the nuclear operators $\mathscr{K}_1$, which we have already encountered, and the *Hilbert–Schmidt operators* $\mathscr{K}_2$. Recall further that if $A, B \in \mathscr{K}_2$, then $AB^* \in \mathscr{K}_1$ and $\mathrm{tr}[AB^*]$ defines an inner product on $\mathscr{K}_2$, which makes it a Hilbert space with norm $\|A\|_2^2 = \mathrm{tr}[AA^*]$.

2. Lemma: Let $\mathscr{P} = \{E^i : i = 0, 1, \ldots, n\}$ be a partition of $\mathscr{E}$. Then for $A \in \mathscr{K}_2$

$$\|A\|_2^2 = \sum_{i=1}^{n} \sum_{j=1}^{n} \|\Delta^i A \Delta^j\|_2^2$$

Proof: Consider the operators $\Delta^i A \Delta^j$ and $\Delta^p A \Delta^q$ with either $j \neq q$ or $i \neq p$. In the first instance

$$\mathrm{tr}[(\Delta^i A \Delta^j)(\Delta^p A \Delta^q)^*] = \mathrm{tr}\, \Delta^i A \Delta^j \Delta^q A^* \Delta^p = \mathrm{tr}[0] = 0$$

while in the latter case

$$\begin{aligned}\mathrm{tr}[(\Delta^i \Delta A^j)(\Delta^p A \Delta^q)^*] &= \mathrm{tr}[\Delta^i A \Delta^j \Delta^q A^* \Delta^p] \\ &= \mathrm{tr}[A \Delta^j \Delta^q \Delta^* \Delta^p \Delta^i] = \mathrm{tr}[0] = 0\end{aligned}$$

As such, the terms $\Delta^i A \Delta^j$, $i, j = 1, 2, \ldots, n$ are mutually orthogonal in $\mathscr{K}_2$. Hence the lemma follows from the pythagorean theorem for $\mathscr{K}_2$. //

Hilbert–Schmidt Decomposition Theorem: Let $A \in \mathscr{K}_2$. Then A admits a decomposition into the sum of a strictly causal, a memoryless, and a strictly anticausal operator.

Proof: Given Property A.2 it suffices to verify the existence of $(m)\int dE(\dagger)\, AE^\dagger$ and $(m)\int E\dagger A\, dE(\dagger)$. Moreover, since $\|A\| \le \|A\|_2$ it suffices to verify convergence in the *Hilbert–Schmidt norm*, $\|\ \|_2$. Let $\mathscr{P} = \{E^i : i = 0, 1, \ldots, n\}$ be a partition of $\mathscr{E}$ and consider the partial sum

$$\Sigma_{\mathscr{P}} = \sum_{i=1}^{n} E^{i-1} A \Delta^i = \sum_{i=1}^{n} \sum_{j=1}^{i-1} \Delta^j A \Delta^i$$

By the lemma,

$$\|\Sigma_{\mathscr{P}}\|_2^2 = \sum_{i=1}^{n} \sum_{j=1}^{i-1} \|\Delta^j A \Delta^i\|_2^2 \le \sum_{i=1}^{n} \sum_{j=1}^{n} \|\Delta^j A \Delta^i\|_2^2 = \|A\|_2^2$$

Moreover, if $\mathscr{P}'$ refines $\mathscr{P}$, it follows from the lemma that

$$\|\Sigma_{\mathscr{P}'} - \Sigma_{\mathscr{P}}\|_2^2 = \|\Sigma_{\mathscr{P}'}\|_2^2 - \|\Sigma_{\mathscr{P}}\|_2^2$$

Hence

$$\|\Sigma_{\mathscr{P}'}\|_2 \ge \|\Sigma_{\mathscr{P}}\|_2$$

As such, $\|\Sigma_{\mathscr{P}}\|_2$ is a bounded (by $\|A\|_2$) increasing function defined on the directed set of partitions with

$$\sup_{\mathscr{P}} \|\Sigma_{\mathscr{P}}\|_2 = M \le \|A\|_2$$

For any $\varepsilon > 0$ choose a partition $\mathscr{P}$ such that

$$M^2 \ge \|\Sigma_{\mathscr{P}}\|_2^2 > M^2 - \varepsilon^2/4$$

Then if $\mathscr{P}'$ and $\mathscr{P}''$ are any two partitions that refine $\mathscr{P}$,

$$\begin{aligned} \|\Sigma_{\mathscr{P}'} - \Sigma_{\mathscr{P}''}\|_2^2 &\le 2[\|\Sigma_{\mathscr{P}'} - \Sigma_{\mathscr{P}}]\|_2^2 + \|\Sigma_{\mathscr{P}''} - \Sigma_{\mathscr{P}}\|_2^2 \\ &= 2[\|\Sigma_{\mathscr{P}'}\|_2^2 + \|\Sigma_{\mathscr{P}''}\|_2^2 - 2\|\Sigma_{\mathscr{P}}\|_2^2] \end{aligned}$$

and hence

$$\|\Sigma_{\mathscr{P}'} - \Sigma_{\mathscr{P}''}\|_2^2 \le 4[M^2 - \|\Sigma_{\mathscr{P}}\|_2^2] < \varepsilon^2$$

verifying that the partial sums for $(m)\int E^{\dagger}A\,dE(\dagger)$ are Cauchy and hence that the integral converges in the Hilbert–Schmidt norm and thus also the uniform norm. Of course, the proof for the other integral is similar. //

3. Remark: In the above we have actually shown that the integral converges in the Hilbert–Schmidt norm and hence that the terms in the decomposition of a Hilbert–Schmidt operator are in fact Hilbert–Schmidt. Indeed, the integrals of triangular truncation are *orthogonal projections* of $\mathscr{K}_2$ onto their ranges.

4. Corollary: The integrals of triangular truncation restricted to $\mathscr{K}_2$ are orthogonal projections.

Proof: Consistent with the theorem and Property 1.D.15, the integrals of triangular truncation are well defined on $\mathscr{K}_2$; we already know that they are projections. Therefore it suffices to show that they are hermitian operators on $\mathscr{K}_2$. To this end consider the partial sum

$$\Sigma_{\mathscr{P}}(A) = \sum_{i=1}^{n} E^{i-1}A\Delta^{i}$$

where $A \in \mathscr{K}_2$ and let $B \in \mathscr{K}_2$. Then

$$\begin{aligned}\operatorname{tr}[(\Sigma_{\mathscr{P}}(A))B^*] &= \sum_{i=1}^{n} \operatorname{tr}[E^{i-1}A\Delta^{i}B^*] = \sum_{i=1}^{n} \operatorname{tr}[A\Delta^{i}B^*E^{i-1}] \\ &= \sum_{i=1}^{n} \operatorname{tr}[A(E^{i-1}B\Delta^{i})^*] = \operatorname{tr}[A(\Sigma_{\mathscr{P}}(B))^*]\end{aligned}$$

As such, $\Sigma_{\mathscr{P}}$ is hermitian, and hence so is its limit $(m)\int E^{\dagger}A\,dE(\dagger)$. A similar argument applies to the other integrals. //

5. Remark: Although $\mathscr{K}_2$ is the most commonly used ideal of compact operators in our theory, it is not the largest ideal of causal operators for which a decomposition exists. Indeed, one can formulate an ideal of compact operators $\mathscr{K}_{\omega}$ that contains $\mathscr{K}_2$ as a proper subset on which a decomposition is well defined. Moreover, this is the largest ideal of compact operators on which a decomposition is well defined in the sense that if $(H, \mathscr{E})$ is an arbitrary resolution space then there exists $A \in \mathscr{K} \backslash \mathscr{K}_{\omega}$ that does not admit a decomposition relative to $(H, \mathscr{E})$. The details of this theory are beyond the scope of the present text and will not be considered further.[1] We shall, however, give an example of a compact operator that does not admit a decomposition.

6. Example: Let

$$A_n = \begin{bmatrix} 0 & -1 & -\frac{1}{2} & \cdots & -1/n \\ 1 & 0 & -1 & \cdots & -1/(n-1) \\ \frac{1}{2} & 1 & 0 & \cdots & -1/(n-2) \\ \vdots & \vdots & \vdots & & \vdots \\ 1/n & 1/(n-1) & 1/(n-2) & \cdots & 0 \end{bmatrix}$$

be the $n \times n$ matrix corresponding to the first n rows and n columns of the semi-infinite matrix A used in Example A.8. Now, since $\|A\| = \pi$, $\|A_n\| \leq \pi$. On the other hand, since the lower triangular part of A represented an unbounded operator, $\|[A_n]_{\mathscr{R}}\| \to \infty$. Now

$$\sigma_n = 1/\sqrt{\|[A_n]_{\mathscr{R}}\|} \to 0$$

and the semiinfinite matrix

$$B = \operatorname{diag}[\sigma_i A_i] = \begin{bmatrix} 0 & & & & & \\ & 0 & -\sigma_1 & & & \\ & -\sigma_1 & 0 & & & \\ & & & 0 & -\sigma_2 & -\sigma_2/2 \\ & & & \sigma_2 & 0 & -\sigma_2 \\ & & & \sigma_2/2 & -\rho_2 & 0 \end{bmatrix}$$

defines an operator on $l_2[0, \infty)$. Indeed, since $\|\sigma_i A_i\| \to 0$, B is compact. Now, if we let $[A_n]_{\mathscr{R}}$ denote the strictly lower triangular submatrix associated with A_n, it follows that $[B]_{\mathscr{R}}$ must be characterized by the semi-infinite matrix

$$[B]_{\mathscr{R}} = \operatorname{diag}[\sigma_i[A_i]_{\mathscr{R}}]$$

if it exists. This matrix, however, fails to define a bounded operator since

$$\|\sigma_i[A_i]_{\mathscr{R}}\| = \sqrt{\|[A_n]_{\mathscr{R}}\|} \to \infty$$

Hence B is a compact operator that does not admit a decomposition.

7. Remark: Even though $[K]_{\mathscr{R}}$ and $[K]_{\mathscr{R}^*}$ may fail to exist for a compact K, $[K]_{\mathscr{M}}$ will always exist:

8. Corollary: Let $A \in \mathscr{K}$. Then

$$(M) \int dE(\dagger)\, A\, dE(\dagger) = (m) \int dE(\dagger)\, A\, dE(\dagger)$$

exists.

Proof: Let $\mathscr{P} = \{E^i : i = 0, 1, \ldots, n\}$ and recall from Remark B.1 that

$$\|\Sigma_{\mathscr{P}}(A)\| = \left\| \sum_{i=1}^{n} \Delta^i A \Delta^i \right\| \leq \|A\|$$

for $A \in \mathscr{K}$. Moreover, since $\mathscr{K}_2$ is dense in K, for any $\varepsilon > 0$ there exist $B \in \mathscr{K}_2$ such that $\|A - B\| < \varepsilon$ and partitions $\mathscr{P}'$ and $\mathscr{P}''$ such that

$$\begin{aligned} \|\Sigma_{\mathscr{P}'}(A) - \Sigma_{\mathscr{P}''}(A)\| &\leq \|\Sigma_{\mathscr{P}'}(A - B)\| + \|\Sigma_{\mathscr{P}'}(B) - \Sigma_{\mathscr{P}''}(B)\| \\ &\quad + \|\Sigma_{\mathscr{P}''}(B - A)\| \\ &\leq 2\|A - B\| + \|\Sigma_{\mathscr{P}'}(B) - \Sigma_{\mathscr{P}''}(B)\| \end{aligned}$$

showing that $\Sigma_{\mathscr{P}}(A)$ is Cauchy since $\|A - B\| < \varepsilon$ and $\Sigma_{\mathscr{P}}(B)$ is Cauchy by the theorem. As such, the required integrals of triangular truncation exist. //

9. Corollary: Let $A \in \mathscr{C} \cap \mathscr{K}$. Then A admits a decomposition into a memoryless operator and a strictly causal operator. Let $A \in \mathscr{C}^* \cap \mathscr{K}$. Then A admits a decomposition into a memoryless operator and a strictly anticausal operator.

10. Corollary: Let $A \in \mathscr{K}$. Then the following are equivalent:

(i) $A \in \mathscr{R}$,
(ii) $A \in \mathscr{T}$, and
(iii) A is causal and $[A]^{\mathscr{M}} = 0$.

Proof: To show the equivalence of (i) and (ii) we first observe that $\mathscr{R} \subset \mathscr{T}$. Now, since $A \in \mathscr{K}$ Corollary 9 implies that A has the decomposition

$$A = [A]_{\mathscr{M}} + [A]_{\mathscr{R}}$$

whereas if $A \in \mathscr{T}$, $[A]^{\mathscr{M}} = 0$ and hence

$$0 = [A]^{\mathscr{M}} = [[A]_{\mathscr{M}}]^{\mathscr{M}} + [[A]_{\mathscr{R}}]^{\mathscr{M}} = [A]_{\mathscr{M}} + 0 = [A]_{\mathscr{M}}$$

showing that $A = [A]_{\mathscr{R}} \in \mathscr{R}$. Note in the last equality that we have used the fact that $[\ ]^{\mathscr{M}}$ is a projection onto $\mathscr{M}$ and that $[[A]_{\mathscr{R}}]^{\mathscr{M}} = 0$ since $[A]_{\mathscr{R}} \in \mathscr{R}$ (derived in Property B.7). Since these properties also hold for operators for which $[A]^{\mathscr{M}} = 0$ a similar proof will verify the equivalence of (i) and (iii). //

11. Corollary: Let $A \in \mathscr{K}$. Then the following are equivalent:

(i) $A \in \mathscr{R}^*$,
(ii) $A \in \mathscr{T}^*$, and
(iii) A is anticausal and $[A]^{\mathscr{M}} = 0$.

12. Remark: Given the above corollaries, one need not make any distinction in the decomposition theory for compact operators among the various classes of hypercausal operator.

PROBLEMS

1. Show that *if* $A \in \mathscr{B}$ admits the decomposition $A = [A]_{\mathscr{C}} + [A]_{\mathscr{R}^*}$ where $[A]_{\mathscr{C}} \in \mathscr{C}$ and $[A]_{\mathscr{R}^*} \in \mathscr{R}^*$, then the decomposition is unique.
2. Show that every operator defined on a finite-dimensional resolution space admits a decomposition.
3. Give an example of an operator defined on $L_2(-\infty, \infty)$ that does not admit a decomposition.
4. Let $A \in \mathscr{C}$ be a bounded linear operator and show that the partial sums for the integral $(m) \int dE(\dagger)\, AE^{\dagger}$ are uniformly bounded.
5. Show that every finite group admits a unique invariant mean.
6. Construct an invariant mean for the group Z_p.
7. Show that the formula $\mathrm{tr}[BA]$ defines a continuous linear functional on $\mathscr{K}_1$ when B is a fixed bounded linear operator and $A \in \mathscr{K}_1$.
8. Determine whether $[\ ]^{\mathscr{M}}$ is a strongly continuous mapping from $\mathscr{B}$ to $\mathscr{M}$.
9. Give an example of a nonzero positive hermitian $A \in \mathscr{B}$ such that $[A]^{\mathscr{M}} = 0$.
10. Show that Lemma 2.A.4 remains valid when the finite partition is replaced by a generalized partition.
11. Show that the class of causal operators for which $[A]^{\mathscr{M}} = 0$ is uniformly closed. Does this class form an ideal in $\mathscr{C}$?
12. Show that the derivative is a derivation on $C[0, 1]$.
13. Show that every derivation on a von Neumann algebra is bounded.
14. Show that $\mathrm{tr}[AB] = \mathrm{tr}[BA]$ whenever both AB and BA are in $\mathscr{K}_1$.

15. Show that $\| \ \|_q$ is a well-defined norm on $\mathscr{C}/\mathscr{R}$.
16. Show that the time set $ is compact under its order topology.
17. Show that the bilateral shift U on $l_2(-\infty, \infty)$ with its usual resolution structure cannot be decomposed into the sum of a strictly causal and a memoryless operator.
18. Show that for any $A \in \mathscr{K}_2$ $\|A\| \leq \|A\|_2$.

REFERENCES

1. Gohberg, I. C., and Krein, M. G., "Theory and Application of Volterra Operators in Hilbert Space." American Mathematical Society, Providence, Rhode Island, 1970.
2. Johnson, B., and Parott, S., Operators commuting with a von Neumann algebra modulo the set of compact operators, *J. Funct. Anal.* **11**, 39–61 (1972).
3. Larson, D. R., Ph. D. Dissertation, Univ. of California at Berkeley, Berkeley, California (1976).

4

Factorization

A. REGULAR FACTORIZATION

1. Remark: In Chapter 3 we considered the problem of decomposing an operator into the sum of operators, each with specified causality characteristics. Here we consider the parallel problem of factoring a bounded operator into the product of causal and anticausal operators. Specifically, we work with a bounded operator that admits a bounded inverse $A \in \mathscr{B} \cap \mathscr{B}^{-1}$ that we desire to factor as $A = BC$ where $B \in \mathscr{C} \cap \mathscr{C}^{-1}$ and $C \in \mathscr{C}^* \cap \mathscr{C}^{*-1}$.

2. Definition: Let $A \in \mathscr{B} \cap \mathscr{B}^{-1}$. Then a representation of A in the form $A = BC$ where $B \in \mathscr{C} \cap \mathscr{C}^{-1}$ and $C \in \mathscr{C}^* \cap \mathscr{C}^{*-1}$ is termed a regular factorization on $(H, \mathscr{E})$.

3. Definition: If $A \in \mathscr{B} \cap \mathscr{B}^{-1}$ is positive and hermitian, then a representation of A in the form $A = BB^*$ where $B \in \mathscr{C} \cap \mathscr{C}^{-1}$ is termed a *spectral factorization* on $(H, \mathscr{E})$.

4. Remark: If $B \in \mathscr{C} \cap \mathscr{C}^{-1}$, then $B^* \in \mathscr{C}^* \cap \mathscr{C}^{*-1}$; hence a spectral factorization is a regular factorization. To define a regular factorization in a finite-dimensional space it suffices that $B \in \mathscr{C}$ and $C \in \mathscr{C}^*$, since in finite dimensions the inverse of a causal operator is causal (because the inverse of a triangular matrix is triangular). In general, however, this is not the case. For instance, on $l_2(-\infty, \infty)$ with its usual resolution structure $I = UU^*$ where U is the bilateral shift, $U \in \mathscr{C}$ but $U^{-1} = U^* \in \mathscr{S}^*$; hence $I = UU^*$ is not a spectral factorization for the identity.

5. Remark: We shall continue to restrict ourselves to regular factorizations of the form $A = BC$ where $B \in \mathscr{C} \cap \mathscr{C}^{-1}$ and $C \in \mathscr{C}^* \cap \mathscr{C}^{*-1}$. Of course, a parallel theory can be formulated for *reverse-order factorizations* of the form $A = CB$ where $C \in \mathscr{C}^* \cap \mathscr{C}^{*-1}$ and $B \in \mathscr{C} \cap \mathscr{C}^{-1}$ simply by working with the dual resolution structure. For the sake of brevity, however, we shall leave this latter case to the reader (see Exercises 11–13).

6. Example: In the finite-dimensional case the causal operators are represented by the lower triangular matrices defined with respect to an appropriate basis, as in Example 1.B.8, while the anticausal operators are represented by upper triangular matrices with respect to the same basis. In the finite-dimensional case the regular factorization problem thus reduces to the factorization of an arbitrary matrix into the product of an upper triangular and a lower triangular matrix. Interestingly, such factorizations are rather common in numerical linear algebra, where they are employed as an aid to matrix inversion and to the solution of the eigenvalue problem.[7] In that application a regular factorization is termed an *$\mathscr{L}\mathscr{U}$ factorization.* There are a number of standard algorithms for implementing this factorization (which always exists).

7. Example: Consider the *discrete convolution* operator defined on $l_2(-\infty, \infty)$:

$$a_i = [Gb]_i = \sum_{j=-\infty}^{\infty} G_{i-k} b_j$$

Following Example 1.B.12 this operator is bounded if and only if $\hat{G} \in L_\infty(T)$ where $\hat{G}$ is the discrete Fourier transform of the sequence G_j. Moreover, the operator is causal if and only if $\hat{G} \in H_\infty(T)$. Now, the inverse of such a convolution operator has discrete Fourier transform

$$[\hat{G}^{-1}](e^{i\theta}) = 1/\hat{G}(e^{i\theta}), \qquad 0 \leq \theta < 2\pi$$

Hence $G \in \mathscr{C} \cap \mathscr{C}^{-1}$ if and only if both $\hat{G}$, $1/\hat{G} \in H_{\infty}(T)$, i.e., $\hat{G}$ is an *outer function.* Now, if $A = A^* > 0$ is a discrete convolution operator represented by $\hat{A} > 0$, the spectral factorization problem reduces to the factorization of $\hat{A}$ in the form

$$\hat{A} = \hat{B}\hat{B}^*$$

where $\hat{B}$ is outer since the composition of two convolution operators is represented by the multiplication of their discrete Fourier transforms. Indeed, if $A \in L_1(T)$, it is a standard result of complex function theory that such a factorization exists, and hence any positive hermitian convolution operator with discrete Fourier transforms $\hat{A} \in L_1(T)$ admits a spectral factorization.[2]

8. Remark: An interesting special case of Example 7 is encountered when $\hat{A}$ is a real rational function in $z = e^{i\theta}$. In this case the positivity of $\hat{A}(e^{i\theta})$ implies that the poles and zeros of this rational function have an appropriate quadrilateral symmetry, which can be used to construct the outer factor $\hat{B}$ simply by identifying the poles and zeros of $\hat{A}$ outside the unit circle with $\hat{B}$ and those inside the unit circle with $\hat{B}^*$.[6]

9. Property: Let $A = BC = B'C'$ be two regular factorizations of $A \in \mathscr{B} \cap \mathscr{B}^{-1}$. Then there exists $M \in \mathscr{M} \cap \mathscr{M}^{-1}$ such that $B' = BM$ and $C' = M^{-1}C$. Moreover, if the two factorizations are spectral factorizations, M is unitary.

Proof: Since $BC = B'C'$ we may let

$$M = B^{-1}B' = CC'^{-1} \in \mathscr{M}$$

which satisfies the required equalities and is memoryless since $B^{-1}B' \in \mathscr{C}$ and $CC'^{-1} \in \mathscr{C}^*$. Moreover,

$$M^{-1} = B'^{-1}B = C'C^{-1} \in \mathscr{M}$$

since $B'^{-1}B \in \mathscr{C}$ and $C'C^{-1} \in \mathscr{C}^*$. If the factorizations are spectral, then $C = B^*$ and $C' = B'^*$, and hence

$$M^{-1} = (B'^*)(B^*)^{-1} = (B)^{-1}(B')^* = M^*$$

showing that M is unitary and completing the proof. //

10. Lemma: Let $A \in \mathscr{C} \cap \mathscr{C}^{-1}$ or $\mathscr{C}^* \cap \mathscr{C}^{*-1}$ then $E^\dagger A E^\dagger$ is an invertible mapping on $H^\dagger = E^\dagger(H)$ and $E_\dagger A E_\dagger$ is an invertible mapping on $H_\dagger = E_\dagger(H)$.

Proof: Let H be represented by the product space $H = H^\dagger \otimes H_\dagger$ and let $A \in \mathscr{C} \cap \mathscr{C}^{-1}$ be represented by the 2×2 lower triangular matrix on this product space

$$A = \begin{bmatrix} E^\dagger A E^\dagger & 0 \\ E_\dagger A E^\dagger & E_\dagger A E_\dagger \end{bmatrix}$$

where the causality of A implies that $E^\dagger A E_\dagger = 0$.

Now, since $A^{-1} \in \mathscr{C}$, it has the matrix representation

$$A^{-1} = \begin{bmatrix} E^\dagger(A^{-1})E^\dagger & 0 \\ E_\dagger(A^{-1})E^\dagger & E_\dagger(A^{-1})E_\dagger \end{bmatrix}$$

The formula for the computation of the inverse of a triangular matrix implies that

$$[E^\dagger A E^\dagger]^{-1} = E^\dagger(A^{-1})E^\dagger$$

maps $H^\dagger$ onto $H^\dagger$ while

$$[E_\dagger A E_\dagger]^{-1} = E_\dagger(A^{-1})E_\dagger$$

maps $H_\dagger$ onto $H_\dagger$. Of course a similar argument will yield the same result for $A \in \mathscr{C}^* \cap \mathscr{C}^{*-1}$ with the lower triangular matrices replaced by upper triangular matrices. //

11. Property: If A admits a regular factorization, then $E_\dagger + E^\dagger A E^\dagger$ and $E^\dagger + E_\dagger A^{-1} E_\dagger$ are invertible on B.

Proof: If $A = BC$ is a regular factorization, then

$$E_\dagger + E^\dagger A E^\dagger = E_\dagger + [E^\dagger B E^\dagger][E^\dagger C E^\dagger]$$

Now $[E^\dagger B E^\dagger]$ and $[E^\dagger C E^\dagger]$ are invertible on $H^\dagger$, and hence so is their product while $E_\dagger$ is clearly invertible on $H_\dagger$. Thus the sum is invertible on $H = H^\dagger \otimes H_\dagger$. Moreover, $A^{-1} = C^{-1}B^{-1}$ is a regular factorization of A^{-1} in reverse order, and a similar argument will verify that $E^\dagger + E_\dagger A^{-1} E_\dagger$ is invertible. //

12. Example: Consider the bilateral shift U on $l_2(-\infty, \infty)$ with its usual resolution structure and the infinite matrix representation

$$U = \begin{bmatrix} & \vdots & \vdots & \vdots & \vdots & \vdots & \vdots & \vdots & \vdots & \vdots & \\ \cdots & 0 & 0 & 0 & 0 & 0 & 0 & 0 & 0 & 0 & \cdots \\ \cdots & 1 & 0 & 0 & 0 & 0 & 0 & 0 & 0 & 0 & \cdots \\ \cdots & 0 & 1 & 0 & 0 & 0 & 0 & 0 & 0 & 0 & \cdots \\ \cdots & 0 & 0 & 1 & 0 & 0 & 0 & 0 & 0 & 0 & \cdots \\ \cdots & 0 & 0 & 0 & 1 & 0 & 0 & 0 & 0 & 0 & \cdots \\ \cdots & 0 & 0 & 0 & 0 & 1 & 0 & 0 & 0 & 0 & \cdots \\ \cdots & 0 & 0 & 0 & 0 & 0 & 1 & 0 & 0 & 0 & \cdots \\ \cdots & 0 & 0 & 0 & 0 & 0 & 0 & 1 & 0 & 0 & \cdots \\ \cdots & 0 & 0 & 0 & 0 & 0 & 0 & 0 & 1 & 0 & \cdots \\ & \vdots & \vdots & \vdots & \vdots & \vdots & \vdots & \vdots & \vdots & \vdots & \end{bmatrix}$$

Now the matrix representation for $E_0 + E^0 U E^0$ takes the form

$$E_0 + E^0 U E^0 = \begin{bmatrix} & \vdots & \vdots & \vdots & \vdots & \vdots & \vdots & \vdots & \vdots & \vdots & \\ \cdots & 0 & 0 & 0 & 0 & 0 & 0 & 0 & 0 & 0 & \cdots \\ \cdots & 1 & 0 & 0 & 0 & 0 & 0 & 0 & 0 & 0 & \cdots \\ \cdots & 0 & 1 & 0 & 0 & 0 & 0 & 0 & 0 & 0 & \cdots \\ \cdots & 0 & 0 & 1 & 0 & 0 & 0 & 0 & 0 & 0 & \cdots \\ \cdots & 0 & 0 & 0 & 0 & 1 & 0 & 0 & 0 & 0 & \cdots \\ \cdots & 0 & 0 & 0 & 0 & 0 & 1 & 0 & 0 & 0 & \cdots \\ \cdots & 0 & 0 & 0 & 0 & 0 & 0 & 1 & 0 & 0 & \cdots \\ \cdots & 0 & 0 & 0 & 0 & 0 & 0 & 0 & 1 & 0 & \cdots \\ \cdots & 0 & 0 & 0 & 0 & 0 & 0 & 0 & 0 & 1 & \cdots \\ & \vdots & \vdots & \vdots & \vdots & \vdots & \vdots & \vdots & \vdots & \vdots & \end{bmatrix}$$

which is not invertible since $[E_0 + E^0 A E^0](e_{-1}) = 0$. Therefore, the bilateral shift does not admit a resgular factorization.

13. Remark: As Example 12 indicates, relatively simple operators may fail to admit a regular factorization. Moreover, a definitive characterization of those operators on a Hilbert resolution space that do admit a regular factorization has yet to be obtained. The one class of operators that appears to admit a viable factorization theory consists of the operators on a discrete resolution space, which are considered in the following.[1] Recall from Definition 1.B.14 that a resolution space $(H, \mathscr{E})$ is said to be discrete if each $\dagger \in \$$ admits an immediate predecessor $\ddagger \in \$$ and $H_{\ddagger} \backslash H_{\dagger}$ is finite dimensional.

14. Lemma: Let A be an invertible operator on a discrete resolution space $(H, \mathscr{E})$. Then A may be factored into the form UB where U is unitary and $B \in \mathscr{C} \cap \mathscr{C}^{-1}$.

Proof: Let $H_{\dagger} = E_{\dagger}(H)$. Then for $\dagger < \ddagger$ $H_{\ddagger} \subset H_{\dagger}$ and we may let $G_{\dagger} = A(H_{\dagger})$, where $G_{\ddagger} \subset G_{\dagger}$ if $\dagger < \ddagger$. Moreover, since A is invertible, $G_{\dagger}$ is closed and

$$\mathrm{cls}\{G_{\dagger}\} = H$$

Furthermore, since $\bigcap_{\dagger} \{H_{\dagger}\} = 0$,

$$\bigcap_{\dagger} \{G_{\dagger}\} = 0$$

For any $\dagger \in \$$ let $\dagger_{-}$ denote its immediate predecessor. Then since A is invertible, A maps $[H_{\dagger_{-}} \backslash H_{\dagger}]$ onto $[G_{\dagger_{-}} \backslash G_{\dagger}]$, and since these are finite-dimensional spaces

$$\dim[H_{\dagger_{-}} \backslash H_{\dagger}] = \dim[G_{\dagger_{-}} \backslash G_{\dagger}]$$

while

$$\sum_{\dagger} \oplus [H_{\dagger_{-}} \backslash H] = \sum_{\dagger} \oplus [G_{\dagger_{-}} \backslash G_{\dagger}] = H$$

Thus there exists a unitary operator $V_{\dagger}$ mapping $[G_{\dagger_{-}} \backslash G_{\dagger}]$ onto $[H_{\dagger_{-}} \backslash H_{\dagger}]$ for each $\dagger \in \$$, while

$$V = \sum_{\dagger} \oplus V_{\dagger}$$

maps H onto H, taking $[G_{\dagger_{-}} \backslash G_{\dagger}]$ onto $[H_{\dagger_{-}} \backslash H_{\dagger}]$ for each $\dagger \in \$$.

Since

$$G_{\dagger} = \sum_{\ddagger > \dagger} \oplus [G_{\ddagger} \backslash G_{\ddagger}] \qquad \text{and} \qquad H_{\dagger} = \sum_{\ddagger > \dagger} \oplus [H_{\ddagger} \backslash H_{\ddagger}]$$

we have $V(G_{\dagger}) = H_{\dagger}$. Hence if we set $B = VA$, we have $B(H_{\dagger}) = H_{\dagger}$, showing that $B \in \mathscr{C} \cap \mathscr{C}^{-1}$. Finally, if we let $U = V^*$, we have

$$A = V^*VA = UB$$

as required. //

15. Remark: Note that since $B \in \mathscr{C} \cap \mathscr{C}^{-1}$ if $A = UB$ by the conditions of Lemma 14, then $A \in \mathscr{C}$ if and only if $U \in \mathscr{C}$, and $A \in \mathscr{C}^{-1}$ if and only if $U \in \mathscr{C}^{-1}$. Then $A \in \mathscr{C} \cap \mathscr{C}^{-1}$ if and only if $U \in \mathscr{C} \cap \mathscr{C}^{-1}$. Since $U^{-1} = U^*$, this, however, implies that $U \in \mathscr{M}$. Thus $A \in \mathscr{C} \cap \mathscr{C}^{-1}$ if and only if $U \in \mathscr{M}$.

Discrete Spectral Factorization Theorem: Let A be a positive definite hermitian operator on a discrete resolution space. Then A admits a spectral factorization.

Proof: Since $A = A^* > 0$, $A^{-1} = A^{*-1} > 0$. Hence A^{-1} admits a unique positive hermitian square root $\sqrt{A}^{-1}$, which we may factor according to the lemma to obtain $\sqrt{A}^{-1} = UB$, where U is unitary and $B \in \mathscr{C} \cap \mathscr{C}^{-1}$. Moreover, since $\sqrt{A}^{-1}$ is hermitian,

$$\sqrt{A}^{-1} = [\sqrt{A}]^{-1*} = [UB] = B^*U^* = B^*U^{-1}$$

Hence

$$A^{-1} = [\sqrt{A}^{-1}][\sqrt{A}^{-1}] = [B^*U^{-1}][UB] = B^*B$$

Finally

$$A = [B^{-1}][B^{*-1}]$$

where $B^{-1} \in \mathscr{C} \cap \mathscr{C}^{-1}$, as required. //

16. Remark: The theorem implies that every positive hermitian operator on a discrete resolution space admits a spectral factorization. In general, however, an arbitrary operator on a discrete resolution space may fail to admit a regular factorization. Indeed, the bilateral shift on $l^2(-\infty, \infty)$ is an appropriate counter example, as shown in Example 12.

17. Remark: Unlike the case of a discrete resolution space, the problem of constructing a spectral factorization for a continuous resolution space is far from resolved. Indeed, at the time of this writing a long standing conjecture of Ringrose to the effect that every positive definite hermitian operator admits a spectral factorization has been proven false,[5] though a constructive counterexample has yet to be exhibited. Moreover, the problem of classifying those operators that do admit a spectral factorization remains open.

B. RADICAL FACTORIZATION

1. Remark: If, rather than working with a factorization $A = BC$ where $B \in \mathscr{C} \cap \mathscr{C}^{-1}$ and $C \in \mathscr{C}^* \cap \mathscr{C}^{*-1}$, one requires that $B = [M + D]$ and $C = [N + F]$, $D \in \mathscr{R}$, $F \in \mathscr{R}^*$, and $M, N \in \mathscr{M}$, enough additional structure is obtained to allow one to formulate a viable existence criterion for the factorization.

2. Definition: Let $A \in \mathscr{B} \cap \mathscr{B}^{-1}$. Then a representation of A in the form $A = [I + D]M[I + F]$, $D \in \mathscr{R}$, $M \in \mathscr{M}$, and $F \in \mathscr{R}^*$ is termed a *radical factorization* of A.

3. Remark: Note that since $\mathscr{R}$ and $\mathscr{R}^*$ are *quasinilpotent*, $(I + D)^{-1} \in \mathscr{C}$ and $(l + F)^{-1} \in \mathscr{C}^*$, so that every radical factorization is also a regular factorization. On the other hand, since $\mathscr{S}$ and $\mathscr{S}^*$ (or $\mathscr{T}$ and $\mathscr{T}^*$) are not quasi-nilpotent the concept of a radical factorization does not generalize to the case where $D \in \mathscr{S}$ and $F \in \mathscr{S}^*$ (or $D \in \mathscr{T}$ and $F \in \mathscr{T}^*$).

4. Example: Let U be the bilateral shift on $l_2(-\infty, \infty)$ with its usual resolution structure and consider the operator $e^{[U+U^*]}$. Since $UU^* = U^*U$,

$$e^{[U+U^*]} = [e^U][e^{U^*}] = [e^U][e^U]^*$$

where $[e^U] \in \mathscr{C}$ and $[e^U]^{-1} = [e^{-U}] \in \mathscr{C}$ since $\mathscr{C}$ is closed. As such, e^{U+U^*} admits a regular, indeed a spectral, factorization. Now, since every radical factorization is a regular factorization and any two regular factorizations differ by a memoryless factor if e^{U+U^*} also admits a radical factorization, it must be of the form

$$e^{U+U^*} = [I + D]M[I + F] = [I + D](M[I + F])$$

where

$$[I + D] = [e^U]N$$

for $N \in \mathscr{M} \cap \mathscr{M}^{-1}$. Moreover, since $U \in \mathscr{S}$

$$[[e^U]]^{\mathscr{M}} = [[I + U + U^2/2 + U^3/6 + \cdots]]^{\mathscr{M}} = I$$

while the fact that $D \in \mathscr{R}$ and $N \in \mathscr{M}$ implies that

$$\begin{aligned} I = [I]^{\mathscr{M}} &= [I]^{\mathscr{M}} + [D]^{\mathscr{M}} = [I + D]^{\mathscr{M}} \\ &= [[e^U]N]^{\mathscr{M}} = [[e^U]]^{\mathscr{M}}N = IN = N \end{aligned}$$

Therefore if e^{U+U^*} admits a radical factorization, then $[e^U] - I \in \mathscr{R}$. A little algebra with the infinite matrix representation for this operator will, however, reveal that this is not the case. (See Example 1.D.10 for a similar argument using the unilateral shift on $l_2[0, \infty)$.) We conclude that e^{U+U^*} does not admit a radical factorization, and we have thus constructed an example of an operator on a discrete resolution space that admits a regular factorization (and a spectral factorization) but not a radical factorization.

5. Property: If $A \in \mathscr{B} \cap \mathscr{B}^{-1}$ admits a radical factorization, then it is unique.

Proof: Assume that

$$A = [I + D](M[I + F]) = [I + D'](M'[I + F'])$$

$D, D' \in \mathscr{R}$, $M, M' \in \mathscr{M}$, and $F, F' \in \mathscr{R}^*$. Since every radical factorization is a regular factorization, there exists an $N \in \mathscr{M} \cap \mathscr{M}^{-1}$ such that

$$[I + D] = [I + D']N$$

Hence

$$\begin{aligned} I = [I]^{\mathscr{M}} &= [I]^{\mathscr{M}} + [D]^{\mathscr{M}} = [I + D]^{\mathscr{M}} = [[I + D']N]^{\mathscr{M}} = [I + D']^{\mathscr{M}}N \\ &= [[I]^{\mathscr{M}} + [D']^{\mathscr{M}}]N = IN = N \end{aligned}$$

It follows that $D = D'$. and a similar argument implies that $M = M'$ and $F = F'$, thereby verifying the uniqueness of the radical factorization. //

6. Property: Assume that $A = A^* > 0$ admits a radical factorization. Then the radical factorization of A is a spectral factorization.

Proof: If $A = [I + D]M[I + F]$ where $A = A^*$,

$$A = A^* = ([I + D]M[I + F])^* = [I + F^*]M^*[I + D^*]$$

is a radical factorization of A. However, since A admits a unique radical factorization we have $D = F^*$ and $M = M^*$, implying that

$$\begin{aligned} A &= [I + D]M[I + D^*] = ([I + D]\sqrt{M})(\sqrt{M}[I + D^*]) \\ &= ([I + D]\sqrt{M})([I + D]\sqrt{M})^* \end{aligned}$$

admits a spectral factorization. Here $M > 0$ since $A > 0$, and hence— admits the hermitian square foot $\sqrt{M}$. //

7. Lemma: For any $A \in \mathscr{R}$, $(I + A)^{-1}$ exists and takes the form $(I + A)^{-1} = I + B$ where $B \in \mathscr{R}$. For any $F \in \mathscr{R}^*$, $(I + F)^{-1}$ exists and takes the form $(I + F)^{-1} = I + G$ where $G \in \mathscr{R}^*$.

Proof: Since A is quasinilpotent, the Neumann series

$$(I + A)^{-1} = \sum_{i=0}^{\infty} A^i = I + \left[\sum_{i=1}^{\infty} A^i\right]$$

converges uniformly. Moreover, since $A^i \in \mathscr{R}$ and $\mathscr{R}$ is closed in the uniform operator topology,

$$B = \left[\sum_{i=1}^{\infty} A^i\right] \in \mathscr{R}$$

Hence $(I + A)^{-1} = I + B$, $B \in \mathscr{R}$, and a similar argument applies to the dual problem. //

Radical Factorization Theorem: Let $A \in \mathscr{B} \cap \mathscr{B}^{-1}$. Then A admits a radical factorization if and only if $[E^\dagger + E_\dagger A^{-1} E_\dagger]^{-1}$ exists for all $\dagger \in \$$ and the integrals

$$(M) \int [E^\dagger + E_\dagger A^{-1} E_\dagger]^{-1} E_\dagger A^{-1}\, dE(\dagger)$$

$$(M) \int dE(\dagger)\, A^{-1} E_\dagger [E^\dagger + E_\dagger A^{-1} E_\dagger]^{-1}$$

exist, in which case the integral

$$(m) \int dE(\dagger)\, [E^\dagger + E_\dagger A^{-1} E_\dagger]^{-1}\, dE(\dagger)$$

also exists and A admits the radical factorization $A = [I + D]M[I + F]$ where

$$D = -(M) \int [E^\dagger + E_\dagger A^{-1} E_\dagger]^{-1} E_\dagger A^{-1}\, dE(\dagger)$$

$$F = -(M) \int dE(\dagger)\, A^{-1} E_\dagger [E^\dagger + E_\dagger A^{-1} E_\dagger]^{-1}$$

and

$$M = -(m) \int dE(\dagger)\, [E^\dagger + E_\dagger A^{-1} E_\dagger]^{-1}\, dE(\dagger)$$

Proof: Assume that A admits radical factorization

$$A = [I + D]M[I + F] = [I + D](M[I + F])$$

$D \in \mathscr{R}$, $M \in \mathscr{M}$, and $F \in \mathscr{R}^*$. Since this factorization is also regular, Property A.11 implies that $[E^\dagger + E_\dagger A^{-1} E_\dagger]^{-1}$ exists for all $\dagger \in \$$. Indeed,

$$\begin{aligned}
&[E^\dagger + E_\dagger A^{-1} E_\dagger][E^\dagger + [I + D]E_\dagger M[I + F]] \\
&\quad = E^\dagger + E_\dagger A^{-1} E_\dagger [I + D] E_\dagger M[I + F] \\
&\quad = E^\dagger + E_\dagger A^{-1} [I + D] E_\dagger M[I + F] \\
&\quad = E^\dagger + E_\dagger (M[I + F])^{-1} E_\dagger (M[I + F]) \\
&E^\dagger + E_\dagger (M[I + F])^{-1} (M[I + F]) = E^\dagger + E_\dagger = I
\end{aligned}$$

where we have used Lemma 7 to justify the equality $[I + F]^{-1} = [I + Y]$, $Y \in \mathscr{R}^*$. Then

$$[E^\dagger + E_\dagger A^{-1} E_\dagger]^{-1} = [E^\dagger + [I + D]E_\dagger M[I + F]]$$

Now, since $D \in \mathscr{R}$: Lemma 7 implies that $[I + D]^{-1} = [I + X]$, $X \in \mathscr{R}$, so that

$$\begin{aligned}
[E^\dagger + E_\dagger A^{-1} E_\dagger]^{-1} E_\dagger A^{-1} &= [E^\dagger + [I + D]E_\dagger M[I + F]]E_\dagger A^{-1} \\
&= [I + D]E_\dagger M[I + F]E_\dagger A^{-1} \\
&= [I + D]E_\dagger M[I + F]A^{-1} \\
&= [I + D]E_\dagger [I + X]
\end{aligned}$$

Accordingly,

$$\begin{aligned}
(M)\int [E^\dagger + E_\dagger A^{-1} E_\dagger]^{-1} E_\dagger A^{-1}\, dE(\dagger) &= (M)\int [I + D]E_\dagger [I + X]\, dE(\dagger) \\
&= [I + D]\left[(M)\int E_\dagger [I + X]\, dE(\dagger)\right] \\
&= [I + D]X = [I + D]([I + D]^{-1} - 1) = -D
\end{aligned}$$

where we have used the fact that $(M)\int E_\dagger[I + X]\, dE(\dagger) = X$ since $I \in \mathscr{M}$ and $X \in \mathscr{R}$. By a similar argument one may verify that

$$(M)\int dE(\dagger)\, A^{-1} E_\dagger [E + E_\dagger A^{-1} E_\dagger]^{-1} = -F$$

showing that the two integrals exist and completing the necessity proof. Conversely, assume that $[E^\dagger + E_\dagger A^{-1} E_\dagger]^{-1}$ exists for all $\dagger \in \$$ and that the two integrals exist. Now let $\mathscr{P} = \{E^i : i = 0, 1, \ldots, n\}$ be a partition of $\mathscr{E}$ and consider the partial sum

$$\Sigma_{\mathscr{P}} = \sum_{i=1}^{n} [E^i + E_i A^{-1} E_i]^{-1} E_i A^{-1} \Delta^i$$

Also

$$\begin{aligned}
&[I - A^{-1}[E^i + E_i A^{-1} E_i]^{-1} E_i] \\
&\quad = E_i[I - A^{-1}[E^i + E_i A^{-1} E_i]^{-1} E_i] \\
&\qquad + E^i[I - A^{-1}[E^i + E_i A^{-1} E_i]^{-1} E_i] \\
&\quad = [E_i - E_i] + E^i[I - A^{-1}[E^i + E_i A^{-1} E_i]^{-1} E_i] \\
&\quad = E^i[I - A^{-1}[E^i + E_i A^{-1} E_i] E_i]
\end{aligned}$$

while

$$A^{-1} = \sum_{i=1}^{n} A^{-1} \Delta^i$$

Hence

$$\begin{aligned}
A^{-1}\left[I - \sum_{\mathscr{P}}\right] &= A^{-1} - A^{-1}\left[\sum_{i=1}^{n} [E^i + E_i A^{-1} E_i]^{-1} E_i A^{-1} \Delta^i\right] \\
&= \sum_{i=1}^{n} [I - A^{-1}[E^i + E_i A^{-1} E_i]^{-1} E_i] A^{-1} \Delta^i \\
&= \sum_{i=1}^{n} E^i[I - A^{-1}[E^i + E_i A^{-1} E_i]^{-1} E_i] A^{-1} \Delta^i
\end{aligned}$$

By hypothesis $\sum_{\mathscr{P}}$ converges uniformly to

$$-D = (M) \int [E^\dagger + E_\dagger A^{-1} E_\dagger]^{-1} E_\dagger A^{-1}\, dE(\dagger)$$

and thus the left side of the above equality is convergent to the invertible operator $A^{-1}[I + D]$. Hence the right side is also convergent to an invertible operator:

$$\begin{aligned}
A^{-1}[I + D] &= C^{-1} \\
&= (M) \int E^\dagger[I - A^{-1}[E^\dagger + E_\dagger A^{-1} E_\dagger]^{-1} E_\dagger] A^{-1}\, dE(\dagger)
\end{aligned}$$

Moreover, $C^{-1} \in \mathscr{C}^*$ since each partial sum in the defining integral is in $\mathscr{C}^*$. In fact, C is also anticausal. To see this compute

$$E_\dagger C^{-1} E_\dagger = E_\dagger A^{-1}[I + D] E_\dagger = E_\dagger A^{-1} E_\dagger E_\dagger [I + D] E_\dagger$$

where we have used the fact that D is strictly causal (since the partial sums in its defining integral are pre-strictly causal). Now, since $[I + D] \in \mathscr{C} \cap \mathscr{C}^{-1}$ ($[I + D]^{-1} = [I + X]$, $X \in \mathscr{R}$), $E_\dagger[I + D]E_\dagger$ is invertible on $H_\dagger = E_\dagger(H)$. Also, since $[E^\dagger + E_\dagger A^{-1} E_\dagger]$ is invertible on H, $E_\dagger A^{-1} E_\dagger$ is invertible on $H_\dagger$. The above equality thus implies that $E_\dagger C^{-1} E_\dagger$ is invertible on $H_\dagger$. Now, express the equality $C^{-1}C = I$ as a 2×2 operator matrix equality on $H = H^\dagger \oplus H_\dagger$:

$$C^{-1}C = \begin{bmatrix} E^\dagger C^{-1} E^\dagger & E^\dagger C^{-1} E_\dagger \\ 0 & E_\dagger C^{-1} E_\dagger \end{bmatrix} \begin{bmatrix} E^\dagger C E^\dagger & E^\dagger C E_\dagger \\ E_\dagger C E^\dagger & E_\dagger C E_\dagger \end{bmatrix} = \begin{bmatrix} E^\dagger & 0 \\ 0 & E_\dagger \end{bmatrix} = 1$$

where $E_\dagger C^{-1} E^\dagger = 0$ since $C^{-1} \in \mathscr{C}^*$. Now, the $(2-1)$th component of the above equation takes the form

$$[E_\dagger C^{-1} E_\dagger][E_\dagger C E^\dagger] = 0$$

which implies that $E_\dagger C E^\dagger = 0$ since $E_\dagger C^{-1} E_\dagger$ is invertible on $H_\dagger$, and $C \in \mathscr{C}^*$ as claimed.

The existence of the integral and the invertibility of $[E^\dagger + E_\dagger A^{-1} E^\dagger]$ for all $\dagger \in \$$ thus yields a factorization for A in the form

$$A = [I + D]C$$

where $D \in \mathscr{R}$ and $C \in \mathscr{C}^* \cap \mathscr{C}^{*-1}$. A parallel argument with the other integral yields a corresponding factorization

$$A = B[I + F]$$

where $F \in \mathscr{R}^*$ and $B \in \mathscr{C} \cap \mathscr{C}^{-1}$. Although these are not the required radical factorizations, they are both regular factorizations for A, and hence Property A.9 implies that there exists $M \in \mathscr{M} \cap \mathscr{M}^{-1}$ such that $B = [I + D]M$. We thus obtain the required radical factorization in the form

$$A = B[I + F] = [I + D]M[I + F]$$

The existence of both integrals together with the invertibility of $[E^\dagger + E_\dagger A^{-1} E_\dagger]$ for all $\dagger \in \$$ thus implies the existence of the required radical factorization, with $-D$ and $-F$ being given by the specified integrals. It remains only to verify the claimed integral representation for M. Since the existence of the radical factorization has been verified, we may invoke the equality

$$[E^\dagger + E_\dagger A^{-1} E_\dagger]^{-1} = E^\dagger + [I + D]E_\dagger M[I + F]$$

derived in the necessity proof. Now for a partition $\mathscr{P} = \{E^i : i = 0, 1, \ldots, n\}$ we have

$$\sum_{i=1}^{n} \Delta^i [E^{i-1} + E_{i-1} A^{-1} E_{i-1}]^{-1} \Delta^i$$

$$= \sum_{i=1}^{n} \Delta^i [E^{i-1} + [I + D] E_{i-1} [I + F]] \Delta^i$$

$$= \sum_{i=1}^{n} \Delta^i E^i [I + D] E_{i-1} M [I + F] \Delta^i$$

$$= \sum_{i-1}^{n} \Delta^i E^i [I + D] E^i E_{i-1} M [I + F] \Delta^i$$

$$= \sum_{i=1}^{n} \Delta^i [I + D] \Delta^i M \Delta^i [I + F] \Delta^i$$

$$= \left[\sum_{i=1}^{n} \Delta^i [I + D] \Delta^i\right] \left[\sum_{j=1}^{n} \Delta^j \Delta M^j\right] \left[\sum_{k=1}^{n} \Delta^k M \Delta^k\right]$$

Since $D \in \mathscr{R}$, $F \in \mathscr{R}^*$, and $M \in \mathscr{M}$, the three sets of partial sums on the right-hand side of this equality converge to

$$(m) \int dE(\dagger) [I + D] \, dE(\dagger) = I$$

$$(m) \int dE(\dagger) \, M \, dE(\dagger) = M$$

and

$$(m) \int dE(\dagger) [I + F] \, dE(\dagger) = \mathrm{I}$$

Hence the partial sums on the left side of the equation converge to

$$(m) \int dE(\dagger) [E^\dagger + E_\dagger A^{-1} E_\dagger]^{-1} \, dE(\dagger) = IMI = M$$

as claimed. //

8. Remark: In the literature these integrals are often formulated in terms of $T = I - A^{-1}$ rather than A^{-1}. Indeed, simply by substituting $I - T = A^{-1}$ into the above integrals and carrying out the indicated algebraic operations, one that finds that

$$[E^\dagger + E_\dagger A^{-1} E_\dagger] = [I - E_\dagger T E_\dagger]$$

while

$$D = (M)\int [I - E_\dagger T E_\dagger]^{-1} E_\dagger T\, dE(\dagger)$$

$$F = (M)\int dE(\dagger)\, T[I - E_\dagger T E_\dagger]^{-1}$$

and

$$M = (m)\int dE(\dagger)\,[I - E_\dagger T E_\dagger]^{-1}\, dE(\dagger)$$

9. Remark: Occasionally, rather than working with a radical factorization in the form

$$A = [I + D]M[I + F]$$

we desire to work with a *special radical factorization* in which $M = I$, i.e.,

$$A = [I + D][I + F]$$

$D \in \mathscr{R}$ and $F \in \mathscr{R}^*$. Of course, since the radical factorization is already unique, the existence of a special radical factorization is also characterized by the theorem simply by requiring that

$$M = -(m)\int dE(\dagger)\,[E^\dagger + E_\dagger A^{-1} E_\dagger]^{-1}\, dE(\dagger) = I$$

C. COMPACT PERTURBATIONS OF THE IDENTITY

1. Remark: Although the radical factorization theorem is both necessary and sufficient, it yields little insight into the class of operators that admit a radical factorization. The present section is devoted to the formulation of sufficient conditions for the existence of a radical factorization in the class of operators that may be modeled as *compact perturbations of the identity*. We begin with a study of the operators of the form $I + K$ where K is of rank 1.

2. Lemma: Assume that $(I + K)$ is an invertible operator and $K = x \otimes y$ is of rank one. Then $(I + K)^{-1} = (I + L)$ where $L = w \otimes z$ is also of rank 1.

Proof: To compute $(I + K)^{-1}$ we must solve the equation

$$g = (I + K)f = f + Kf = f + (f, x)y$$

for f given g. Equivalently, by letting $\alpha = (f, x)$ we may solve the pair of simultaneous equations

$$f = g - \alpha y$$

and

$$\alpha = (f, x) = (g - \alpha y, x) = (g, x) - \alpha(y, x)$$

From the latter equation,

$$\alpha = (g, x)/[1 + (y, x)]$$

where $1 + (y, x) \neq 0$ since $(I + K)^{-1}$ is assumed to exist. Thus

$$f = g - (g, x)y/[1 + (y, x)] = (I + L)g$$

where

$$L = x \otimes [y/(1 + (y, x))] = w \otimes z$$

as required. //

3. Lemma: Let $K = x \otimes y$ be of rank 1 and assume that $[E^\dagger + E_\dagger(I + K)^{-1}E_\dagger]^{-1}$ exists for all $\dagger \in \$$. Then $(I + K)$ admits a radical factorization.

Proof: Invoking Lemma 2 we can represent $(I + K)^{-1}$ in the form

$$(I + K)^{-1} = (I + L)$$

where L is the rank-1 operator $w \otimes z$, in which case the integrals of the radical factorization theorem reduce to

$$D = -(M) \int [E^\dagger + E_\dagger(I + L)E_\dagger]^{-1}E_\dagger(I + L)\, dE(\dagger)$$

$$= -(M) \int [I + E_\dagger LE_\dagger]^{-1}E_\dagger L\, dE(\dagger)$$

and

$$F = -(M) \int dE(\dagger)(I + L)E_\dagger[E^\dagger + E_\dagger(I + L)E_\dagger]^{-1}$$

$$= -(M) \int dE(\dagger)\, LE_\dagger[I + E_\dagger LE_\dagger]^{-1}$$

We therefore begin the proof by computing $[I + E_\dagger LE_\dagger]^{-1}$. To this end we must solve the equation

$$\begin{aligned} g &= [I + E_\dagger LE_\dagger]f = f + E_\dagger LE_\dagger f = f + (E_\dagger f, w)E_\dagger z \\ &= f + (I, E_\dagger w)E_\dagger z \end{aligned}$$

for f given g at each $\dagger \in \$$. With $\beta = (f, E_\dagger w)$, this reduces to solving the pair of simultaneous equations

$$f = g - \beta E_\dagger z$$

and

$$\beta = (f, E_\dagger w) = (g - \beta E_\dagger z, w) = (g, w) - \beta(E_\dagger z, w)$$

Clearly

$$\beta = (g, w)/(1 + (E_\dagger z, w))$$

where $1 + (E_\dagger z, w) \neq 0$ since $[I + E_\dagger LE_\dagger]^{-1} = [E^\dagger + E_\dagger(I + K)^{-1}E_\dagger]^{-1}$ exists. Hence

$$f = g - (g, w)E_\dagger z/(1 + (E_\dagger z, w)) = [I + E_\dagger LE_\dagger]^{-1}g$$

Now if we let $\mathscr{P}$ be an arbitrary partition of $\mathscr{E}$, a typical term in the partial sum for the integral

$$Dh = -(M) \int [I + E_\dagger LE_\dagger]^{-1} E_\dagger L \, dE(\dagger) \, h$$

takes the form

$$\begin{aligned} [I + E_i LE_i]^{-1} E_i L\Delta^i h &= [I + E_i LE_i]^{-1}[(\Delta^i h, w)E_i z] \\ &= (\Delta^i h, w)E_i z - ((\Delta^i h, w)E_i z, w)E_i z/(1 + (E_i z, w)) \\ &= (\Delta^i h, w)E_i z/(1 + (E_i z, w)) \end{aligned}$$

for an arbitrary $h \in H$.

Finally, since the complex-valued set function $(\Delta^i h, w)/(I + (E_i z, w))$ is of bounded variation and continuous and $1 + (E_i z, w) \neq 0$,

$$\begin{aligned} D &= -(M) \int [I + E_\dagger LE_\dagger]^{-1} E_\dagger L \, dE(\dagger) h \\ &= (M) \int (dE(\dagger) \, h, w)E_\dagger z/(1 + (E_\dagger z, w)) \end{aligned}$$

exists by classical measure theoretic arguments, and hence D is assured to exist. Of course, a similar argument applies to the integral representation for F, thereby completing the proof. //

4. Lemma: Assume that $(I + L)$ admits a radical factorization and that $[E^\dagger + E_\dagger(I + L + K)^{-1}E_\dagger]^{-1}$ exists for all $\dagger \in \$$, where K is a finite-rank operator. Then $(I + L + K)$ admits a radical factorization.

Proof: Initially, consider the case where K is a rank-1 operator. Now, since $(I + L)$ admits a radical factorization, we may express it in the form

$$(I + L) = (I + D)M(I + F)$$

where $D \in \mathscr{R}$, $M \in \mathscr{M}$, and $F \in \mathscr{R}^*$ with all three factors invertible. Then

$$\begin{aligned}(I + L + K) &= (I + D)M(I + F) + K \\ &= (I + D)M[I + M^{-1}(I + D)^{-1}K(I + F)^{-1}](I + F) \\ &= (I + D)M(I + K')(I + F)\end{aligned}$$

where

$$K' = M^{-1}(I + D)^{-1}K(I + F)^{-1}$$

is of rank 1 since K is of rank 1. Moreover, since

$$[E^\dagger + E_\dagger(I + L + K)^{-1}E_\dagger]^{-1}$$

exists for all $\dagger \in \$$, $E_\dagger(I + L + K)^{-1}E_\dagger$ is invertible on $H_\dagger$. Therefore

$$\begin{aligned}E_\dagger(I + K')^{-1}E_\dagger &= E_\dagger(I + F)(I + L + K)^{-1}(I + D)ME_\dagger \\ &= E_\dagger(I + F)E_\dagger(I + L + K)^{-1}E_\dagger(I + D)ME_\dagger\end{aligned}$$

is invertible on $H_\dagger$, showing that $[E^\dagger + E_\dagger(I + K')^{-1}E_\dagger]^{-1}$ exists for all $\dagger \in \$$. Here we have invoked the facts that $(I + F)$ is anticausal and invertible while $(I + D)$ is causal and invertible. Since K' is also of rank 1, Lemma 3 implies that $(I + K')$ admits a radical factorization

$$(I + K') = (I + E)N(I + G)$$

with $E \in \mathscr{R}$, $N \in \mathscr{M}$, and $G \in \mathscr{R}^*$. Substituting this radical factorization into the expression for $(I + L + K)$ now yields

$$\begin{aligned}(I + L + K) &= (I + D)M(I + K')(I + F) \\ &= (I + D)M(I + E)N(I + G)(+F) \\ &= [I + (D + MEM^{-1} + DMEM^{-1})] \\ &\quad \times [MN][I + (G + F + GF)] \\ &= (I + D')M'(I + F')\end{aligned}$$

which is the desired radical factorization with $D' = (D + MEM^{-1} + DMEM^{-1})$, $M' = (MN)$, and $F' = (G + F + GF)$.

Finally, representing a finite-rank K as the sum of rank-1 operators and repeating the above argument for each term, the result can be extended to the case of a finite-rank K. //

5. Property: Let K be a finite-rank operator and assume that $[E^{\dagger} + E_{\dagger}(I + K)^{-1}E_{\dagger}]^{-1}$ exists for all $\dagger \in \$$. Then $(I + K)$ admits a radical factorization.

Proof: The results follows from Lemma 4 with $L = 0$ since the identity has a trivial radical factorization. //

6. Lemma: Let $K \in \mathscr{K}_2$ be a Hilbert–Schmidt operator with Hilbert–Schmidt norm $\|K\|_2 < 1$. Then $(I + K)$ admits a radical factorization.

Proof: First define a Banach algebra $\mathscr{A}$ whose elements take the form $(cI + L)$, where $L \in \mathscr{K}_2$ and c is a scalar, and let $\mathscr{A}$ be normed by

$$\|cI + L\|_{\mathscr{A}} = |c| + \|L\|_2$$

Now, consider the equation

$$G + [KG]_{\mathscr{R}^*} = -[K]_{\mathscr{R}^*}$$

where $G \in \mathscr{A}$. This is well defined by virtue of the Hilbert–Schmidt decomposition theorem since both K and KG are Hilbert–Schmidt operators. Moreover, the operator that takes $G \in \mathscr{A}$ to $[KG]_{\mathscr{R}^*} \in \mathscr{K}_2 \subset \mathscr{A}$ is contractive since

$$\begin{aligned} \|[KG]_{\mathscr{R}^*}\|_{\mathscr{A}} &= \|[KG]_{\mathscr{R}^*}\|_2 \leq \|KG\|_2 = \|KG\|_{\mathscr{A}} \leq \|K\|_{\mathscr{A}}\|G\|_{\mathscr{A}} \\ &= \|K\|_2\|G\|_{\mathscr{A}} < \|G\|_{\mathscr{A}} \end{aligned}$$

Here, we have used the facts that $[\ ]_{\mathscr{R}^*}$ is a projection on $\mathscr{K}_2$ and that $\|\ \|_{\mathscr{A}}$ coincides with $\|\ \|_2$ on $\mathscr{K}_2$. Hence the equation admits a solution G in $\mathscr{A}$. Indeed, G is also in $\mathscr{R}^*$ since $G = [KG]_{\mathscr{R}^*} - [K]_{\mathscr{R}^*}$ so that with the aid of Lemma B.7 we may express $(I + G)^{-1}$ in the form $(I + F)$, $F \in \mathscr{R}^*$. Finally, we define C by the equality

$$C = K + G + KG$$

Now,

$$[C]_{\mathscr{R}^*} = [K]_{\mathscr{R}^*} + [G]_{\mathscr{R}^*} + [KG]_{\mathscr{R}^*} = 0$$

showing that $C \in \mathscr{C}$. We thus have

$$I + C = [I + K + G + KG] = (I + K)(I + G)$$

and we have constructed a regular factorization for $(I + K)$ in the form

$$(I + K) = (I + C)(I + G)^{-1} = (I + C)(I + F)$$

To convert this into a radical factorization we observe that $C \in \mathscr{A}$ since $K \in \mathscr{K}_2$ and $G \in \mathscr{A}$, and hence so is $(I + C)$. We may therefore express $(I + C)$ in the form

$$(I + C) = cI + L$$

where $L \in \mathscr{C} \cap \mathscr{K}_2$ and c is a nonzero scalar. Moreover, since $L \in \mathscr{C} \cap \mathscr{K}_2$ the Hilbert–Schmidt decomposition theorem implies that

$$\begin{aligned}(I + C) &= (cI + L) = (cI + [L]_{\mathscr{M}} + [L]_{\mathscr{R}}) \\ &= (I + [L]_{\mathscr{R}}(cI + [L]_{\mathscr{R}})^{-1})(cI + [L]_{\mathscr{M}}) = (I + D)M\end{aligned}$$

where

$$D = [L]_{\mathscr{R}}(cI + [L]_{\mathscr{M}})^{-1} \qquad \text{and} \qquad M = (cI + [L]_{\mathscr{M}})$$

This, in turn, yields the desired radical decomposition in the form

$$(I + K) = (I + D)M(I + F)$$

7. Remark: Lemma 6 is a specialization to our Hilbert–Schmidt setting of a general Banach space factorization theorem due to Wiener.[2,3] In the general setting one works with a Banach algebra $\mathscr{A}$ that admits a direct sum representation $A = \mathscr{A}_+ \oplus \mathscr{A}_-$ and factors an element $(I + A)$ in the form

$$(I + A) = (I + A_+)(I + A_-)$$

where $A_+ \in \mathscr{A}_+$ and $A_- \in \mathscr{A}_-$ when $\mathscr{A}$ is sufficiently close to the identity.

In our lemma the only property of $\mathscr{K}_2$ that has been used is the Hilbert–Schmidt decomposition theorem, and so the result can be extended to any space of operators in which a decomposition into causal and strictly anticausal parts is assured to exist, say $\mathscr{K}_\omega$.

Hilbert–Schmidt Factorization Theorem: Let $K \in \mathscr{K}_2$ and assume that $[E^\dagger + E_\dagger(I + K)^{-1}E_\dagger]^{-1}$ exists for all $\dagger \in \$$. Then $(I + K)$ admits a radical factorization.

Proof: Since every Hilbert–Schmidt operator can be approximated in the Hilbert–Schmidt norm by a finite-rank operator, let J be a finite-rank operator such that $\|K - J\|_2 < 1$. Then Lemma 6 implies that $(I + [K - J])$ admits a radical factorization, and hence Lemma 4 implies that

$$(I + K) = (I + [K - J] + J)$$

admits a radical factorization, as required. //

8. Remark: Like the Hilbert–Schmidt decomposition theorem, the preceding theorem can be extended beyond the class of Hilbert–Schmidt operators, though at a considerable cost in complexity. In particular, the proof requires compactness and the existence of a causal–strictly anti-causal decomposition. Hence the theorem can be extended to $\mathscr{K}_{\omega}$. A general derivation of the theorem in terms of ideals of compact operators appears elsewhere.[3]

PROBLEMS

1. Assume that A admits a factorization of the form $A = CB$, $C \in \mathscr{C}^* \cap \mathscr{C}^{*-1}$ and $B \in \mathscr{C} \cap \mathscr{C}^{-1}$, and show that $[E_\dagger + E^\dagger A E^\dagger]$ and $[E^\dagger + E^\dagger A^{-1} E^\dagger]$ are invertible on H.

2. Let an operator be factored as $A = UB$ where U is unitary and $B \in \mathscr{C} \cap \mathscr{C}^{-1}$ and show that

(i) $A \in \mathscr{C}$ if and only if $U \in \mathscr{C}$,
(ii) $A \in \mathscr{C}^{-1}$ if and only if $U \in \mathscr{C}^{-1}$, and
(iii) $A \in \mathscr{C} \cap \mathscr{C}^{-1}$ if and only if $U \in \mathscr{M}$.

3. Let $D \in \mathscr{R}$ and show that $[I + D]^{-1} = [I + X]$ where $X \in \mathscr{R}$.

4. Let U be the bilateral shift on $l_2(-\infty,\infty)$ and show that e^U and e^{-U} are causal.

5. For the bilateral shift U on $l_2(-\infty, \infty)$ show that $e^U - I$ is not strictly causal. Show that it is strongly causal.

6. Show that every positive hermitian operator on H admits a unique positive hermitian square root.

7. Derive necessary and sufficient conditions for an operator on $(H, \mathscr{E})$ to admit a factorization in the form $A = [I + F]M[I + D]$, $F \in \mathscr{R}^*$, $M \in \mathscr{M}$, and $D \in \mathscr{R}$. Formulate integrals for computing F, M, and D when they exist.

8. Assume that A admits the radical factorization $A = [I + D]M[I + F]$ and show that $[E^\dagger + E_\dagger A^{-1} E_\dagger]^{-1} E_\dagger A^{-1}$ is causal.

9. Show that $[E^\dagger + E_\dagger A^{-1} E_\dagger] = [I - E_\dagger T E_\dagger]$ if $T = I - A^{-1}$.

10. Let $A \in \mathscr{B}$ and assume that AA^* admits a radical factorization. Then show that $A = BU$ where $B \in \mathscr{C} \cap \mathscr{C}^{-1}$ and U is unitary.

11. Show that every positive hermitian operator on a discrete resolution space admits a *reverse order spectral factorization* $A = B^*B$ where $B \in \mathscr{C} \cap \mathscr{C}^{-1}$.

12. Show that an operator A on $(H, \mathscr{E})$ admits a *reverse-order regular factorization* $A = CB$, $C \in \mathscr{C}^* \cap \mathscr{C}^{*-1}$ and $B \in \mathscr{C} \cap \mathscr{C}^{-1}$, if and only if A^{-1} admits a regular factorization.

13. Show at an operator A on $(H, \mathscr{E})$ admits a *reverse-order radical factorization* if and only if A^{-1} admits a radical factorization.

14. Let $(H, \mathscr{E})$ be a resolution space for which $\mathscr{E}$ is a countable set. Then show that every positive definite hermitian operator on H admits a spectral factorization.

REFERENCES

1. Arveson, W. A., Interpolation in nest algebras, *J. Funct. Anal.* **20**, 208–233 (1975).
2. Douglas, R. G., "Banach Algebra Techniques in Operator Theory." Academic Press, New York, 1972.
3. Erdos, J. A., "The triangular factorization of operators in Hilbert space," *Indiana Univ. Math. J.* **22**, 939–950 (1973).
4. Gohberg, I. C., and Krein, M. G., "Theory and Application of Volterra Operators in Hilbert Space." American Mathematical Society, Providence, Rhode Island, 1970.
5. Larsen, D., unpublished notes, Univ. of Nebraska (1980).
6. Saeks, R., "Generalized Networks." Holt, New York, 1972.
7. Tewarson, R. P., "Sparse Matrices." Academic Press, New York, 1973.

5

Causal Invertibility

A. SUFFICIENT CONDITIONS

1. Remark: Although the causal operators are closed under the usual algebraic operations and weak limits, they are not closed under adjoints or inversion. Fortunately, the adjoint of a causal operator is anticausal and vice versa. In contrast, there exists no such simple characterization for the inverse of a causal operator. Indeed, the inverse of a causal operator may be causal, anticausal, or neither.

2. Example: If A is a causal operator on a finite-dimensional space it is represented by a lower triangular matrix relative to an appropriate basis. Hence if A^{-1} exists, it is causal owing to the usual back substitution algorithm for inverting a triangular matrix.

Similarly, if $A \in \mathscr{M}$, then $E^{\dagger}A = AE^{\dagger}$; hence

$$A^{-1}E^{\dagger} - E^{\dagger}A^{-1} = A^{-1}[E^{\dagger}A - AE^{\dagger}]A^{-1} = A^{-1}[0]A^{-1} = 0$$

showing that every memoryless invertible operator has a memoryless inverse.

3. Example: Consider the bilateral shift on $l_2(-\infty, \infty)$ with its usual resolution structure. Since U is unitary, $U^{-1} = U^* \in \mathscr{C}^*$. As such, the inverse of the bilateral shift is anticausal (actually strongly anticausal). Of course, a similar argument applies to any causal unitary operator; for instance, the *ideal delay* $\tilde{D}$ defined on $L_2(-\infty, \infty)$ with its usual resolution structure by

$$[\tilde{D}f](\dagger) = F(\dagger - 1), \qquad \dagger \in \mathscr{R}$$

4. Property: Let $A \in \mathscr{C}$ with $\|A\| < 1$. Then $(I - A)^{-1} \in \mathscr{C}$.

Proof: Since $\|A\| < 1$,

$$(I - A)^{-1} = \sum_{i=0}^{\infty} A^i$$

exists, and it is causal since $\mathscr{C}$ is closed. //

5. Remark: Note that the above theorem can be generalized to cover the case where the spectral radius of A is less than 1. Indeed, in that case one can renorm $\mathscr{B}$ with a topologically equivalent norm $\| \ \|_A$ for which $\|A\|_A < 1$ and then repeat the preceding argument with the Neumann series convergent in $\| \ \|_A$.

6. Remark: For an operator $A \in \mathscr{B}$ we denote its *spectrum* $\sigma(A)$ and its *resolvant* $\rho(A)$. These are both subsets of the complex plane, and, in general, they may both be made up of several disconnected components. However, since $|\sigma(A)| \leq \|A\|$, $\rho(A)$ always contains a distinguished component $\rho_\infty(A)$ that includes those complex numbers λ for which $|\lambda| > \|A\|$ and, depending on A, may also include complex numbers with $|\lambda| < \|A\|$.

7. Property: Let $A \in \mathscr{C}$ and assume that $0 \in \rho_\infty(A)$. Then $A^{-1} \in \mathscr{C}$.

Proof: Since $0 \in \rho_\infty(A) \subset \rho(A)$, A^{-1} exists, and it suffices to verify that $A^{-1} \in \mathscr{C}$. To this end write

$$(\lambda I - A)^{-1} = (I - A/\lambda)^{-1}/\lambda$$

which is causal when $|\lambda| > \|A\|$ by Property 3. Now for any $\dagger \in \$$, let $x \in H_\dagger$ and $y \in H^\dagger$ and compute

$$f(\lambda) = ((\lambda I - A)^{-1}x, y)$$

which is zero for $|\lambda| > \|A\|$ since $(\lambda I - A)^{-1}x \in H_\dagger$ when $(\lambda I - A)^{-1} \in \mathscr{C}$. The resolvant operator $(\lambda I - A)^{-1}$ is, however, analytic on the connected components of $\rho(A)$, and hence so is $f(\lambda)$. Now, since $f(\lambda)$ is identically zero on an open set ($|\lambda| > \|A\|$) in $\rho_\infty(A)$ it must be identically zero on all of $\rho_\infty(A)$. In particular, if $0 \in \rho_\infty(A)$ then $f(0) = (A^{-1}x, y) = 0$, which implies that $A^{-1} \in \mathscr{C}$ since the equality holds for all $x \in H_\dagger$ and $y \in H^\dagger$. //

8. Property: Assume that $(I + A)^{-1}$ exists where $A \in \mathscr{C} \cap \mathscr{K}$. Then $(I + A)^{-1} \in \mathscr{C}$.

Proof: Since a compact operator has discrete spectrum, so does $(I + A)$, and hence $\rho(A) = \rho_\infty(A)$. Thus, if $(I + A)^{-1}$ exists, then $0 \in \rho(A) = \rho_\infty(A)$, implying that $(I + A)^{-1} \in \mathscr{C}$ owing to Property 7. //

9. Remark: Recall that the *numerical range* of an operator $A \in \mathscr{B}$ is a set of complex numbers $w(A)$ defined by

$$w(A) = \{(Ax, x) : \|x\| = 1\}$$

10. Property: Assume that $A \in \mathscr{C}$ and A^{-1} exists. Then $A^{-1} \in \mathscr{C}$ if $0 \notin w(A)$.

Proof: We show that $A^{-1}(H_\dagger) \subset H_\dagger$. If this is not the case, then $A(H_\dagger)$ is properly contained in $H_\dagger$, and since A^{-1} exists it is a closed subspace. Thus

$$H_\dagger = [A(H_\dagger)] \oplus [H_\dagger \backslash A(H_\dagger)]$$

Now, if $x \in [H_\dagger \backslash A(H)]$ with $\|x\| = 1$, then $Ax \in A(H_\dagger)$, and hence $(Ax, x) = 0$, which contradicts the assumption that $0 \notin w(A)$. As such $A(H_\dagger) = H_\dagger$ and hence $A^{-1} \in \mathscr{C}$. //

11. Remark: Note that Property 10 could have been derived as a corollary to Property 8 by invoking the relationship between the spectrum and numerical range of an operator.[1] Consider the class of *passive operators* $A \in \mathscr{C}$ for which $A + A^* > 0$. For such operators $\mathrm{Re}(Ax, x) > 0$, $x \neq 0$, and hence $0 \notin w(A)$, implying that $A^{-1} \in \mathscr{C}$. Finally, we note that operators for which $0 \notin w(A)$ have occasionally been termed *definite operators* in the resolution space literature.

12. Property: Let $A \in \mathscr{C}$ and assume that A^n admits a causal inverse for some n. Then $A^{-1} \in \mathscr{C}$.

Proof: If $(A^n)^{-1} = A^{-n} \in \mathscr{C}$, then $A^{-1} = A^{-n}A^{n-1} \in \mathscr{C}$ since the product of causal operators is causal. //

13. Remark: Although Property 12 is quite straightforward it often proves to be convenient in extending the domain of some of the preceding sufficient conditions for causal invertibility. For instance, if A^4 is definite, then $0 \notin w(A^4)$, implying that $A^{-1} \in \mathscr{C}$.

On several occasions in the derivation of the radical factorization theorems we have used the results of Lemma 4.B.7, to the effect that $(I + A)^{-1} \in \mathscr{C}$ if $A \in \mathscr{R}$. Indeed, this follows immediately from Property 7 and the quasinilpotence of $\mathscr{R}$.

14. Property: Let $A \in \mathscr{C}$ and assume that $AK + KA \in \mathscr{R}$ for all $K \in \mathscr{K}_{\mathscr{E}}$. Then $A^{-1} \in \mathscr{C}$ if and only if $[A]_{\mathscr{M}}$ is invertible in $\mathscr{B}$.

Proof: Since $AK - KA \in \mathscr{R}$ for all $K \in \mathscr{K}_{\mathscr{E}}$, $A = [A]_{\mathscr{M}} + [A]_{\mathscr{R}}$ where $[A]_{\mathscr{M}} \in \mathscr{M}$ and $[A]_{\mathscr{R}} \in \mathscr{R}$. If $[A]_{\mathscr{M}}^{-1}$ exists in $\mathscr{B}$, it is actually in $\mathscr{M}$, as shown in Example 2.

$$A = [A]_{\mathscr{M}}(I + [A]_{\mathscr{M}}^{-1}[A]_{\mathscr{R}})$$

is causally invertible since $[A]_{\mathscr{M}}^{-1} \in \mathscr{M}$ and $[A]_{\mathscr{M}}^{-1}[A]_{\mathscr{R}} \in \mathscr{R}$, implying that $(I + [A]_{\mathscr{M}}^{-1}[A]_{\mathscr{R}})^{-1} \in \mathscr{C}$ by Lemma 4.B.7. Conversely, if $A^{-1} \in \mathscr{C}$, then

$$[A]_{\mathscr{M}} = A - [A]_{\mathscr{R}} = A(I + A^{-1}[A]_{\mathscr{R}})$$

is invertible by Lemma 4.B.7 since $A^{-1} \in \mathscr{C}$ and $A^{-1}[A]_{\mathscr{R}} \in \mathscr{R}$. //

15. Example: Consider the operator on $L_2[0, 1]$ with its usual resolution structure defined by the linear differential equation

$$\begin{aligned} \dot{x}(\dagger) &= Ax(\dagger) + Bg(\dagger), \\ & \qquad\qquad\qquad\qquad x(0) = 0 \\ f(\dagger) &= Cx(\dagger) + Dg(\dagger), \end{aligned}$$

where $x(\dagger)$ is an n vector and A, B, C, and D are matrices of appropriate dimension. Since the differential operator is defined over a finite time interval, it decomposes into a memoryless operator defined by

$$f_{\mathscr{M}}(\dagger) = Dg(\dagger)$$

and a strictly causal operator defined by

$$\begin{aligned} \dot{x}(\dagger) &= Ax(\dagger) + Bg(\dagger), \\ & \qquad\qquad\qquad\qquad x(0) = 0 \\ f_{\mathscr{R}}(\dagger) &= Cx(\dagger), \end{aligned}$$

The hypotheses of Property 14 are thus satisfied, and the differential operator admits a causal inverse if and only if D^{-1} exists. In particular, for g and f scalar-valued function in $L_2[0, 1]$, D is a 1×1 matrix, and hence the operator admits a causal inverse if and only if $D \neq 0$. Note that this result is false when the differential operator is defined over an infinite time interval: such an operator breaks up into the sum of a memoryless operator and a strongly causal operator, which may, however, fail to be strictly causal.

16. Remark: Since $\mathscr{R}$ is an ideal in $\mathscr{C}$ one may naturally identify an operator $A \in \mathscr{C}$ and its image $[A] \in \mathscr{C}/\mathscr{R}$. In fact, the causal invertibility of an operator is an invariant of its equivalence class.

17. Property: Suppose that $A \in \mathscr{C}$ and that A^{-1} exists in $\mathscr{B}$. Then $A^{-1} \in \mathscr{C}$ if and only if $[A]$ is invertible in $\mathscr{C}/\mathscr{R}$.

Proof: If $A^{-1} \in \mathscr{C}$, then $[A^{-1}][A] = [A^{-1}] = [I]$. Hence $[A]^{-1} = [A^{-1}]$. Conversely, if $[A]^{-1}$ exists in $\mathscr{C}/\mathscr{R}$, there exists $B \in \mathscr{C}$ such that $[A]^{-1} = [B]$, or equivalently $[AB] = [I]$. Accordingly, there exists $C \in \mathscr{R}$ such that $AB = (I + C)$, and hence

$$A^{-1} = B(I + C)^{-1}$$

by Lemma 4.B.7, as was to be shown. //

18. Remark: If A is strongly strictly causal rather than strictly causal, $(I + A)^{-1}$ may fail to exist. Fortunately, $\mathscr{R}$ is dense in $\mathscr{S}$ in the strong topology, thereby allowing the properties of $(I + A)$ to be deduced from $(I + A_\pi)$ where A_π is a net of strictly causal operators that converges strongly to A.

19. Property: Let A_π be a net of strictly causal operators that is strongly convergent to $A \in \mathscr{C}$. Then $(I + A)^{-1}$ exists and is causal if and only if

$$\limsup_{\pi} \|(I + A_\pi)^{-1}\| < \infty$$

Proof: First note that since $A_\pi \in \mathscr{R}$ $(I + A_\pi)^{-1}$ exists and is causal owing to Lemma 4.B.7. Clearly, if $(I + A)^{-1}$ exists, $(I + A_\pi)^{-1}$ converges strongly to $(I + A)^{-1}$, and hence for any $x \in H$

$$\limsup_{\pi} \|(I + A_\pi)x\| = \|(I + A)^{-1}x\|$$

Therefore the uniform boundedness principle implies that

$$\limsup_{\pi} \|(I + A_\pi)^{-1}\| = \|(I + A)^{-1}\| < \infty$$

Now, suppose that

$$\limsup_{\pi} \|(I + A_\pi)^{-1}\| = M < \infty$$

We shall show that $(I + A)^{-1}$ exists, in which case it then follows that $(I + A)^{-1}$ is causal since it is the strong limit of causal operators. Given $x \in H$ we must show that $\|(I + A)x\|$ and $\|(I + A)^* x\|$ are bounded below. Indeed,

$$\begin{aligned} \|x\| &= \limsup_{\pi} \|x\| = \limsup_{\pi} \|(I + A_\pi)^{-1}(I + A_\pi)x\| \\ &\leq \limsup_{\pi} \|(I + A_\pi)^{-1}\| \, \|(I + A_\pi)x\| \leq M\|(I + A_\pi)x\| \end{aligned}$$

Since A_π converges strongly to A, this implies that $\|x\| \leq M\|(I + A)x\|$ or, equivalently, that

$$\|(I + A)x\| \geq (1/M)\|x\|$$

Of course, a similar argument will verify that

$$\|(I + A)^* x\| \geq (1/M)\|x\|$$

Hence $(I + A)^{-1}$ exists, and being the strong limit of causal operators, it is causal. //

20. Remark: Since $\mathcal{R}$ is strongly dense in $\mathcal{S}$ (Corollary 2.D.8), the test of this proof may be used to determine the characteristics of $(I + A)$, $A \in \mathcal{S}$. Moreover, one may readily verify that $\mathcal{R}$ is strongly dense in $\mathcal{T}$, thereby permitting the test to be applied to $A \in \mathcal{T}$.

21. Remark: Finally, we note that in the proof of Property 19 we did not exploit the special structure of $(I + A_\pi)$ and simply used the fact that $(I + A_\pi) \in \mathcal{C} \cap \mathcal{C}^{-1}$. We thus have actually proved a stronger result. Indeed, if $A_\pi \in \mathcal{C} \cap \mathcal{C}^{-1}$ converges strongly to $A \in \mathcal{C}$, then $A \in \mathcal{C} \cap \mathcal{C}^{-1}$ if and only if

$$\limsup_{\pi} \|A_\pi^{-1}\| < \infty$$

B. TOPOLOGICAL TECHNIQUES

1. Remark: Although some of the sufficient conditions described in Section A characterize both invertibility conditions and causal invertibility, the major part of the causal invertibility problem is the determination of whether the inverse of a causal operator is causal given its existence.

Indeed, the determination of existence conditions for the inverse is a classical problem that need not be considered here. We are interested instead in a causal operator that admits a bounded inverse $A \in \mathscr{C} \cap \mathscr{B}^{-1}$, and we desire to determine whether or not $A \in \mathscr{C} \cap \mathscr{C}^{-1}$. That is, we desire to characterize the subset $\mathscr{C} \cap \mathscr{C}^{-1} \subset \mathscr{C} \cap \mathscr{B}^{-1}$. To this end we desire to investigate the properties of $\mathscr{C} \cap \mathscr{B}^{-1}$ viewed as a topological space in the uniform operator topology.

2. Definition: A *path* in $\mathscr{C} \cap \mathscr{B}^{-1}$ is a uniformly continuous function $f: [0, 1] \rightarrow \mathscr{C} \cap \mathscr{B}^{-1}$.

3. Definition: Two operators $A, B \in \mathscr{C} \cap \mathscr{B}^{-1}$ are said to be *path connected* if there exists a path in $\mathscr{C} \cap \mathscr{B}^{-1}$ such that $f(0) = A$ and $f(1) = B$.

4. Remark: Path connectedness is an *equivalence relation* in $\mathscr{C} \cap \mathscr{B}^{-1}$ the equivalence classes for which are termed the *path connected components* of $\mathscr{C} \cap \mathscr{B}^{-1}$. For reasons that will shortly become apparent, the equivalence class of an operator $A \in \mathscr{C} \cap \mathscr{B}^{-1}$ is denoted $\deg[A]$ and termed the *degree of A*. These equivalence classes form a semigroup with $\deg[A] \deg[B] = \deg[AB]$ in which $\deg[I]$ serves as the unit. Since this semigroup is often (but not always) abelian, we denote the degree of the identity by zero (rather than 1); $\deg[I] = 0$. Although path connectedness would seem to be an abstract concept, it proves to be an extremely powerful tool in our theory since causal invertibility is *degree invariant*. That is, if $\deg[A] = \deg[B]$, then either both A and B are in $\mathscr{C} \cap \mathscr{C}^{-1}$ or neither lie in $\mathscr{C} \cap \mathscr{C}^{-1}$.

5. Property: Causal invertibility is a degree invariant.

Proof: Let $f: [0, 1] \rightarrow \mathscr{C} \cap \mathscr{B}^{-1}$ be a path connecting A and $B \in \mathscr{C} \cap \mathscr{B}^{-1}$. Since $f(\dagger) \in \mathscr{C} \cap \mathscr{B}^{-1}$ for all $\dagger \in [0, 1]$, $f(\dagger)^{-1}$ exists in $\mathscr{B}$ for all $\dagger \in [0, 1]$ and it is continuous since f is continuous. Moreover, since $f(\dagger)^{-1}$ is defined on a compact set $[0, 1]$,

$$\max_{\dagger} \|f(\dagger)^{-1}\| = M < \infty$$

exists. Now, since f is a continuous function defined on a compact set, it is uniformly continuous, and hence there exists $\varepsilon > 0$ such that

$$\|f(\dagger) - f(\ddagger)\| < 1/M$$

whenever $|\dagger - \ddagger| < \varepsilon$. Therefore assume that $f(\dagger)^{-1} \in \mathscr{C}$ and compute $f(\ddagger)^{-1}$, $|\dagger - \ddagger| < \varepsilon$, as

$$f(\ddagger) = f(\dagger) - [f(\dagger) - f(\ddagger)] = f(\dagger)(I - f(\dagger)^{-1}[f(\dagger) - f(\ddagger)])$$

Now

$$\|f(\dagger)^{-1}[f(\dagger) - f(\ddagger)]\| \leq \|f(\dagger)^{-1}\| \, \|f(\dagger) - f(\ddagger)\| < M(1/M) = 1$$

Hence Property A.4 implies that $(I - f(\dagger)^{-1}[f(\dagger) - f(\ddagger)])$ admits a causal inverse. On the other hand $f(\dagger)^{-1} \in \mathscr{C}$ by hypothesis. As such, if $f(\dagger)^{-1} \in \mathscr{C}$ and $|\dagger - \ddagger| < \varepsilon$, then so is $f(\ddagger)^{-1}$. ε is, however, independent of $\dagger$, thus permitting us to piece together a finite number of ε intervals between 0 and 1 and thus to verify that $B = f(1)$ is causally invertible whenever $A = f(0)$ is causally invertible and the converse. The proof is therefore complete. //

6. Corollary: Let $A \in \mathscr{C} \cap \mathscr{B}^{-1}$ and assume that $\deg[A] = 0$. Then $A \in \mathscr{C} \cap \mathscr{C}^{-1}$.

Proof: The proof follows from the proposition together with the observations that $I \in \mathscr{C} \cap \mathscr{C}^{-1}$ and $\deg[I] = 0$. //

7. Remark: Although the requirement that $\deg[A] = 0$ for causal invertibility may at first appear to be rather weak, it is in fact an extremely tight sufficient condition for causal invertibility, and some have conjectured it to be necessary and sufficient. Indeed, all the sufficient conditions formulated in Section A represent classes of operators for which $\deg[A] = 0$. For instance, if $A = I + B$, $B \in \mathscr{R}$, then $f(r) = I + (1 - r)B$ defines a path from A to I. Similarly, if $0 \in \rho_\infty(A)$, one can construct a *spectral mapping* $f(r)$ on A that defines a path between A and I. The one class of operators discussed in Section A known to admit causal inverses that do not obviously have a degree 0 are the memoryless operators. In fact, this is the case, though a more sophisticated spectral argument is required.

8. Property: Let $A \in \mathscr{M} \cap \mathscr{B}^{-1}$ $(= \mathscr{M} \cap \mathscr{M}^{-1})$. Then $\deg[A] = 0$.

Proof: First consider the case in which A is unitary. Then by the spectral theorem for unitary operators there exists a *resolution of the identity* F defined on $[0, 2\pi]$ such that

$$A = \int_0^{2\pi} e^{i\theta} \, dF(\theta)$$

where the integral is the usual weakly convergent spectral integral. Moreover, since $\mathscr{M}$ is a von Neumann algebra and such algebras are closed spectral theoretically, $\mathscr{F} \subset \mathscr{M}$. Therefore

$$f(r) = \int_0^{2\pi} e^{i(1-r)\theta}\, dF(\theta)$$

defines a path from A to I through the memoryless unitary operators, verifying that $\deg[A] = 0$.

Now assume that A is an arbitrary memoryless invertible operator and let $A = UM$ be its polar decomposition, where both U and M lie in $\mathscr{M}$ since $\mathscr{M}$ is a von Neumann algebra (i.e., closed under adjoints and weak limits).[1] Here $\deg[U] = 0$ by the above argument, while

$$f(r) = rI + (1 - r)M$$

defines a path between M and I since $M > 0$. Therefore $\deg[M] = 0$, and hence

$$\deg[A] = \deg[UM] = \deg[U]\deg[M] = \deg[I]\deg[I] = \deg[I] = 0$$

as was to be shown. //

9. Property: Let A be a causal unitary operator. Then $A^{-1} \in \mathscr{C}$ if and only if $\deg[A] = 0$.

Proof: If $\deg[A] = 0$, then $A^{-1} \in \mathscr{C}$ by Corollary 6. On the other hand, if $A^{-1} \in \mathscr{C}$, then since A is unitary $A = (A^{-1})^* \in \mathscr{C}^*$, which, together with the fact that $A \in \mathscr{C}$, implies that $A \in \mathscr{M}$, and hence $\deg[A] = 0$ by Property 8.

10. Example: Consider the *discrete convolution operator* defined on $l_2(-\infty, \infty)$ with its usual resolution structure:

$$a_i = [Gb]_i = \sum_{j=-\infty}^{\infty} G_{i-j} b_j$$

As in Example 1.B.12, this operator is bounded if and only if $\hat{G} \in L_\infty(T)$ where $\hat{G}$ is the *discrete Fourier transform* of the sequence G_j. Moreover, the operator is causal if and only if $\hat{G} \in H_\infty(T)$. Since the inverse operator is represented by the discrete Fourier transform $1/\hat{G}$, the given convolution operator admits a causal inverse if and only if both $\hat{G}$ and $1/\hat{G}$ are in $H_\infty(T)$; i.e., $\hat{G}$ is outer. By invoking Property 9 it is, however, also possible to formulate a topological test for causal invertibility in terms of $\hat{G}$. Indeed, if G is causal, $\hat{G} \in H_\infty(T)$ and hence admits an inner–outer factorization $\hat{G} = \hat{U}\hat{H}$, where $\hat{U}$ is an inner function that is the discrete

Fourier transform of a sequence that defines a unitary convolution operator U, and $\hat{H}$ is an outer function corresponding to a causal causally invertible operator H with $G = UH$. Moreover, since $\hat{H}$ is outer it admits a logarithm $\hat{L}$ corresponding to a convolution operator L, where $\hat{H} = e^{i\hat{L}}$ and $H = e^{iL}$. Accordingly,

$$f(r) = Ue^{i(1-r)L}$$

defines a path between G and U, showing that $\deg[G] = \deg[U]$, and hence Property 5 implies that $G^{-1} \in \mathscr{C}$ if and only if $U^{-1} \in \mathscr{C}$. Since U is unitary Property 9 implies that this is the case if and only if $\deg[G] = \deg[U] = 0$. We have shown that a discrete convolution operator G admits a causal inverse if and only if $\deg[G] = 0$.

Although this criterion for causal invertibility is not readily tested in the most general setting, if one assumes that $\hat{G}(\theta)$ is continuous on T, then it is possible to show that $\deg[G] = \deg_0[\hat{G}]$, where the $\deg_0$ is taken in the usual sense of a function defined on T and may be computed by a classical encirclement argument. Indeed, this is the reason for the adoption of the term "degree" to denote the path connected components of $\mathscr{C} \cap \mathscr{B}^{-1}$.

11. Remark: At the time of this writing it remains an open question whether $\deg[A] = 0$ is a necessary and sufficient condition for causal invertibility. Of course, it is always sufficient, and Example 10 verifies its necessity for the discrete convolution operators. Another quite large class of operators for which the condition is both necessary and sufficient is characterized as follows.

12. Property: Let $A \in \mathscr{C} \cap \mathscr{B}^{-1}$ and assume that AA^* admits a radical factorization. Then $A^{-1} \in \mathscr{C}$ if and only if $\deg[A] = 0$.

Proof: If AA^* admits a radical factorization, then there exist $D \in \mathscr{R}$, $M \in \mathscr{M}$, and $F \in \mathscr{R}^*$ such that

$$AA^* = [I + D]M[I + F]$$

Indeed, since AA^* is positive definite and hermitian, $F = D^*$, and $M = M^* > 0$. As such AA^* admits the spectral factorization

$$AA^* = ([I + D]\sqrt{M})([I + D]\sqrt{M})^*$$

and a little algebra will reveal that

$$A = ([I + D]\sqrt{M})U$$

where $U = ([I + D]\sqrt{M})^{-1}A$ is a causal unitary operator. Now, by Remark 7 $\deg[I + D] = 0$, and by Property 8 $\deg[\sqrt{M}] = 0$. Hence $\deg[A] = \deg[U]$, and $A^{-1} \in \mathscr{C}$ if and only if $U^{-1} \in \mathscr{C}$. Property 9, however, implies that this is the case if and only if $\deg[U] = \deg[A] = 0$. Thus $A^{-1} \in \mathscr{C}$ if and only if $\deg[A] = 0$, as was to be shown. //

PROBLEMS

1. Give an example of causal operator whose inverse is neither causal nor anticausal.

2. Let $\mathscr{A} \subset \mathscr{B}$ be a *-algebra (i.e., if $A \in \mathscr{A}$, then $A^* \in \mathscr{A}$) and show that if A^{-1} exists in $\mathscr{B}$, then $A^{-1} \in \mathscr{A}$.

3. Show that if $A \in \mathscr{C}$ and A has spectral radius less than 1 then $(I + A)^{-1}$ exists and is causal.

4. Show that the numerical range of an operator $A \in \mathscr{B}$ is a convex set whose closure contains the spectrum of A. Use this fact to prove Property A.10 as a corollary to Property A.7.

5. Give an example of a differential operator in the form

$$\dot{x}(\dagger) = Ax(\dagger) + Bg(\dagger), \qquad x(0) = 0$$
$$f(\dagger) = Cx(\dagger),$$

on $L_2[0, \infty)$ with its usual resolution structure that is not strictly causal.

6. Let $\ddagger, \dagger \in \$$ with $\ddagger < \dagger$ and show that $[E^\dagger - E^\ddagger]A[E^\dagger - E^\ddagger] \in \mathscr{R}$ whenever $A \in \mathscr{T}$.

7. Show that $\mathscr{R}$ is strongly dense in $\mathscr{T}$.

8. Show that path connectedness is an equivalence relation in $\mathscr{C} \cap \mathscr{B}^{-1}$.

9. Show that the path connected components of $\mathscr{C} \cap \mathscr{B}^{-1}$ form a well-defined semigroup with identity under the operation $\deg[A]\deg[B] = \deg[AB]$.

10. Let $A \in \mathscr{C}$ and assume that $0 \in \rho_\infty(A)$. Then show that $\deg[A] = 0$.

11. Assume that $A \in \mathscr{C} \cap \mathscr{C}^{-1}$ but $\deg[A] \neq 0$. Then show that there exists no unitary operator $U \in \mathscr{C} \cap \mathscr{B}^{-1}$ such that $\deg[U] = \deg[A]$.

12. Assume that $A \in \mathscr{C}$ and AA^* admits a spectral factorization $AA^* = BB^*$ where $B \in \mathscr{C} \cap \mathscr{C}^{-1}$. Then show that $A = BU$ where U is a causal unitary operator.

13. Show that $x \oplus y$ defines a causal operator if and only if there exists $\dagger \in \$$ such that $x \in H^\dagger$ and $y \in H_\dagger$.

14. Show that every causal compact operator on a continuous Hilbert resolution space is strictly causal.

15. Show that every finite-rank causal operator can be expressed as a sum of rank-1 causal operators.

REFERENCE

1. Yosida, K., "Functional Analysis." Springer-Verlag, Berlin and New York, 1966.

Notes for Part I

1. **Historical Background:** The theory of operators defined on a Hilbert resolution space began in the mid-1960s with the establishment of three separate programs in the United States, the Soviet Union, and Great Britain. The American school was, from the beginning, motivated by the goal of developing an operator theory appropriate to applications in mathematical system theory. This work had its root in the classical paper by Youla *et al.*[34] in which the significance of causality in system theory was first brought to the fore. Motivated by these ideas, the resolution space concept first appeared in the work of Sandberg,[31] Zames,[35] Falb,[10] and Willems[32,33] in the form of an L_2 space into which a family of truncation operators was introduced. This work, in turn, led to the axiomatization of the concept by Porter[9] and Saeks[26] in 1970/1971.

In parallel with their work Gohberg,[16] Krein,[16] Livsic,[20] Brodskii,[3,4] and others working in the Soviet Union developed a new approach to non-self-adjoint spectral theory in which triangular operator representations subsumed the Jordan cononical form. Simultaneously, Ringrose working in Great Britain introduced a new class of operator algebras, the nest algebras.[24] Although motivated by entirely different applications the three approaches all led to the same end point, the theory of operators on a Hilbert resolution space. Indeed, each of the three approaches represents an attempt to extend the theory of triangular matrices to an infinite-dimensional setting, and in each case precisely the same operator algebras resulted.

Not surprisingly, given the tripartite history of the subject the precise origins of many of the concepts and theorems in the theory remain clouded. Indeed, until the late 1970s communication among researchers from the three schools was minimal, when it existed at all. In addition to the obvious problems introduced by translation delays, communications between the three schools was further disrupted by the fact that the British and Soviet researchers were mathematicians, publishing in mathematics journals, while the American school was dominated by engineers, publishing in engineering and applied mathematics journals. It is not surprising that many of the results presented here were discovered independently by each of the three schools, often in widely differing forms. It is for this reason that we have made no attempt to credit results in the main body of the text and have qualified the discussion in these notes.

Strict causality appeared in the earliest works of each of the research programs, in the form of the radical in the nest algebra literature[24] and as the range of an integral of triangular truncation in the Soviet and American programs.[3,16,25] The adoption of the integral approach by the American resolution space community appears to be the first true cross-fertilization of concepts between the three schools. Although strict causality was used by all three research programs from the beginning, the equivalence of the radical and integral formulations has only recently been verified by Erdos and Longstaff.[9] Unlike strict causality, the other hypercausality concepts appear to be unique to the American school. Strong strict causality was introduced by Saeks[10] in 1970, and strong causality has its origins in the work of Zames[35] and Willems.[32]

The uniqueness and characterization theorems for additive decomposition were apparently discovered independently by Brodskii *et al.*[4] and by Saeks.[25] The causal additive decomposition theorem is due to Larson, who used a construction derived from the work of Johnson and Parrott.[17] Finally, the additive decomposition theory for compact operators was developed by Brodskii *et al.*[16] The results presented here thus represent the work of all three research programs integrated into a common package.

Indeed, this is also the case with the factorization theory. The discrete factorization theorem was developed in a nest algebra context by Arveson,[2] whereas the radical factorization theorem and the Hilbert–Schmidt factorization theorem are products of the Soviet school.[16] In contrast, the major work on causal invertibility was done by the American school, motivated by the feedback system stability problem. An exception was the characterization theorem for the inverse of $I + A$ with A strictly causal, which was obtained independently by Brodskii *et al.*,[4] and DeSantis.[6] The topological approach to the causal invertibility problem is rooted in the work of Saeks.

2. Categorical Concepts: If one fixes the time set $, it is possible to define a natural *category of resolution spaces* in which the objects are resolution spaces $(H, \mathscr{E})$ and $\mathrm{Mor}[(H, \mathscr{E}); (\tilde{H}, \tilde{\mathscr{E}})]$ is the space of linear bounded causal operators mapping $(H, \mathscr{E})$ to $(\tilde{H}, \tilde{\mathscr{E}})$. Here, a *causal map* A is defined by the equality

$$\tilde{E}^{\dagger} A = \tilde{E}^{\dagger} A E^{\dagger}, \qquad \dagger \in \$$$

The definition of causality for a mapping between distinct resolution spaces is the natural generalization of that used in the text for maps from a space to itself, and most of the theory derived here carries over to the generalization.

Within this category of resolution spaces one can define all of the usual categorical concepts, such as subobject, quotient object, domination, and equivalence. In particular, for $(H, \mathscr{E})$ and $(\tilde{H}, \tilde{\mathscr{E}})$ to be equivalent one must have a mapping in $\mathrm{Mor}[(H, \mathscr{E}); (\tilde{H}, \tilde{\mathscr{E}})]$ that is a Hilbert space equivalence (i.e., a unitary operator whose inverse is in $\mathrm{Mor}[(\tilde{H}, \tilde{\mathscr{E}}); (H, \mathscr{E})]$). Now, since the inverse of a causal unitary operator is its adjoint, which is anticausal, the inverse will lie in $\mathrm{Mor}[(\tilde{H}, \tilde{\mathscr{E}}); (H, \mathscr{E})]$ if and only if the given unitary operator is memoryless.[29] A resolution space equivalence is therefore a memoryless unitary operator, and $(H, \mathscr{E})$ and $(\tilde{H}, \tilde{\mathscr{E}})$ are equivalent if and only if there exists a unitary operator $U: H \to H$ such that

$$\tilde{E}^{\dagger} = U E^{\dagger} U^{-1} \qquad \dagger \in \$$$

Alternatively, we may identify hermitian operators

$$S = \int_{\$} \dagger \, dE(\dagger) \qquad \text{and} \qquad \tilde{S} = \int_{\$} \dagger \, d\tilde{E}(\dagger)$$

with $\mathscr{E}$ and $\tilde{\mathscr{E}}$ by means of the special theorem, in which case $(H, \mathscr{E})$ and $(\tilde{H}, \tilde{\mathscr{E}})$ are equivalent if and only if S and $\tilde{S}$ are unitarily equivalent. In this way the equivalence theory for Hilbert resolution space is identical to the unitary equivalence problem for (possibly unbounded) hermitian operators and may therefore be resolved by the Hahn–Helinger theorem or its equivalents.[8] In particular, unlike classical Hilbert spaces, two resolution spaces of the same dimension may fail to be equivalent. Indeed, $L_2(-\infty, \infty)$ and $l_2(-\infty, \infty)$ with their usual resolution structures provide an example. To verify this one simply observes that the spectrum of S_{L_2} is the entire real line, whereas the spectrum of S_{l_2} is the integers, so they cannot be unitarily equivalent.

Our dual concepts, such as causality and anticausality, also follow naturally from the categorical viewpoint. Here one constructs a functor from the category of Hilbert resolution spaces over $\$$ to the category of Hilbert resolution spaces over $\$^*$ by identifying $(H, \mathscr{E})$ with $(H, \mathscr{E}^*)$ and $A \in \text{Mor}[(H, \mathscr{E}); (\tilde{H}, \tilde{\mathscr{E}})]$ with $A^* \in \text{Mor}[(\tilde{H}, \tilde{\mathscr{E}}^*); (H, \mathscr{E}^*)]$.

3. Uniform Resolution Space: Although causality is the most important time-related concept encountered in system theory, occasionally it is also desirable to work with time-invariant operators. On $l_2(-\infty, \infty)$ an operator A is said to be time-invariant if it commutes with the bilateral shift, whereas on $L_2(-\infty, \infty)$ an operator is said to be time invariant if it commutes with $\sigma^{\ddagger}$, $\ddagger \in R$, where $\sigma^{\ddagger}$ is the delay operator[25] defined by

$$[\sigma^{\ddagger} f](\dagger) = f(\dagger - \ddagger), \qquad \dagger \in R$$

Although time invariance is well defined in these spaces, it is not well defined in an abstract Hilbert resolution space. However, we can define a restricted class of Hilbert resolution spaces, termed *uniform resolution spaces*, in which time invariance is well defined. To this end we let $\$$ be an ordered locally compact abelian group and define a uniform resolution space to be a triple $(H, \mathscr{E}, \mathscr{U})$ where H is a Hilbert space, $\mathscr{E}$ is a resolution of the identity in H defined on $\$$, and $\mathscr{U}$ is a unitary representation of $\$$ in $\mathscr{B}$ such that

$$U^{\ddagger} E^{\dagger} = E^{\dagger + \ddagger} U^{\ddagger}, \qquad \dagger, \ddagger \in \$$$

Of course, all of the usual resolution space concepts are well defined on $(H, \mathscr{E}, \mathscr{U})$, and we may say that an operator $A \in \mathscr{B}$ is time invariant on $(H, \mathscr{E}, \mathscr{U})$ if

$$U^{\ddagger} A = A U^{\ddagger}, \qquad \ddagger \in \$$$

A little algebra will reveal that the time-invariant operators on $(H, \mathscr{E}, \mathscr{U})$ form a von Neumann algebra in $\mathscr{B}$ and are characterized by properties similar to those of the memoryless operators.[26,29]

By invoking Stone's theorem one can formulate an interesting "duality" structure for the uniform resolution spaces.[4] Recall that every locally compact abelian group \$ admits a character group $\hat{\$}$ where $\hat{\$}$ is the set of continuous group homomorphisms of \$ into T (the multiplicative group of complex number with magnitude 1). In particular, if $\gamma \in \hat{\$}$, then the action of γ on $\dagger \in \$$ is denoted $(\gamma, \dagger) \in T$. Now, given a uniform resolution space $(H, \mathscr{E}, \mathscr{U})$, defined on \$ we may identify a *character space* $(H, \hat{\mathscr{E}}, \hat{\mathscr{U}})$ defined on $\hat{\$}$ where

$$\hat{U}^{\gamma} = \int_{\$} (\gamma, \dagger)\, dE(\dagger) \qquad \text{and} \qquad U^{\dagger} = \int_{\hat{\$}} (\gamma, \dagger)\, d\hat{E}(\gamma)$$

define $\hat{\mathscr{U}}$ and $\hat{\mathscr{E}}$ by means of Stone's theorem. Here $\hat{\mathscr{E}}$ is a resolution of the identity in H defined on $\hat{\$}$ and $\hat{\mathscr{U}}$ is a unitary representation of $\hat{\$}$ in $\mathscr{B}$. The character space exhibits many of the properties of a uniform resolution space and, indeed, fails to be a uniform resolution space only by virtue of the fact that $\hat{\$}$ is not ordered. Hence resolution space properties that do not make use of the ordering on \$ are well defined on the character space. Indeed, A is memoryless on $(H, \mathscr{E}, \mathscr{U})$ if and only if it commutes with $\hat{\mathscr{U}}$ and is time invariant on $(H, \mathscr{E}, \mathscr{U})$ if and only if it commutes with $\hat{\mathscr{E}}$. When translated to the character space setting, the defining equality for the uniform resolution space takes the form[4]

$$\hat{U}^{\gamma}\hat{E}^{\delta} = \hat{E}^{\delta-\gamma}\hat{U}^{\gamma}, \qquad \gamma, \delta \in \hat{\$}$$

while the given space and the character space are related by

$$U^{\dagger}\hat{U}^{\gamma} = (\gamma, \dagger)\hat{U}^{\gamma}U^{\dagger}, \qquad \dagger \in \$, \quad \gamma \in \hat{\$}$$

With the aid of the above equalities and the Mackey imprimitivity theorem one can construct two alternative representations for any given uniform resolution space, in $L_2(\$, K)$ and $L_2(\hat{\$}, \hat{K})$.[29] Here $L_2(\$, K)$ is the space of square integrable functions defined on \$ relative to the Haar measure with values in a Hilbert space K. $L_2(\hat{\$}, \hat{K})$ is defined similarly. Indeed, it follows from the imprimitivity theorem[5] that there exists a unitary operator $M : H \to L_2(\$, K)$ such that

$$E^{\dagger}_{L_2} = ME^{\dagger}M^{-1}, \qquad \dagger \in \$$$

and

$$\sigma^{\ddagger} = MU^{\ddagger}M^{-1}, \qquad \ddagger \in \$$$

where $\mathscr{E}_{L_2}$ denotes the usual resolution structure for $L_2(\$, K)$ and $\sigma^{\ddagger}$ is the delay operator on $L_2(\$, K)$. Similarly, by interchanging the role of $\mathscr{E}$ and $\mathscr{U}$ with $\hat{\mathscr{E}}$ and $\hat{\mathscr{U}}$ the imprimitivity theorem may be used to verify the existence of a unitary operator $\hat{M} : H \to L_2(\hat{\$}, \hat{K})$ such that

$$\hat{E}^{\gamma}_{L_2} = \hat{M}\hat{E}^{\gamma}\hat{M}^{-1}, \qquad \gamma \in \hat{\$}$$

and

$$\hat{\sigma}^{\delta} = \hat{M}\hat{U}^{\delta}\hat{M}^{-1}, \qquad \delta \in \hat{\$}$$

Moreover, the following diagram commutes:

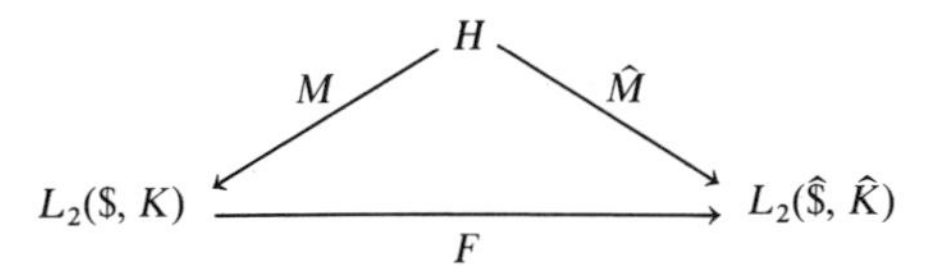

where F is the classical Fourier transform operator on $L_2(G, K)$. In classical system theoretic notation M may be viewed as a mapping from an abstract space into a time-domain representation while $\hat{M}$ maps the abstract space to a frequency-domain representation. Of course, if one starts with $L_2(R, K)$, as is often the case in classical system theory, then $M = I$ and $\hat{M} = F$, in which case the above diagram reduces to the classical time-frequency domain setup.[34]

If one now applies a classical theorem of Foures and Segal, it is found that a time-invariant operator on $(H, \mathscr{E}, \mathscr{U})$ is mapped into a multiplication operator on $L_2(\hat{\$}, \hat{K})$ and into a convolution operator on $L_2(\$, K)$. Similarly, a memoryless operator on $(H, \mathscr{E}, \mathscr{U})$ is mapped into a multiplication operator on $L_2(\$, K)$ and a convolution operator on $L_2(\hat{\$}, \hat{K})$, illustrating the essential "duality" between memoryless and time-invariant operators.[5]

Although it is most natural to let \$ be a group when studying time invariance, it is occasionally desirable to formulate time invariance for \$ taken to be the positive semigroup of an ordered locally compact abelian group, say as a model of $l_2[0, \infty)$.[30] In this case we define a *semiuniform resolution space* to be a triple $(H, \mathscr{E}, \mathscr{V})$ where H is a Hilbert space, $\mathscr{E}$ is a resolution of the identity in H defined on \$, and $\mathscr{V}$ is an isometric representation of \$ in $\mathscr{B}$ satisfying

$$V^{\ddagger}E^{\dagger} = E^{\dagger+\ddagger}V^{\ddagger}, \qquad \dagger, \ddagger \in \$$$

As before, a semiuniform resolution space is a resolution space and all of the usual causality related concepts are well defined thereon. Surprisingly, however, the definition of time invariance takes the form

$$V^{\dagger}A = V^{\dagger}V^{\dagger *}AV^{\dagger}, \qquad \dagger \in \$$$

in a semiuniform resolution space. Of course, in a uniform resolution space $U^{\dagger}$ is unitary and $U^{\dagger}U^{\dagger *} = I$, so that time invariance reduces to its previous definition.

4. Banach Resolution Space: Although one can define a projection on a Banach space by the equality $E^2 = E$, there is no analog of an orthogonal projection on a Banach space, and so the correct generalization of a resolution of the identity to Banach space is not obvious. In some cases it suffices to require that $\|E\| = 1$, though this does not always prove to be sufficient. Rather, we say that a family of projections $\mathscr{E} = \{E^{\dagger} : \dagger \in \$\}$ on a Banach space B is a resolution of the identity if

(i) $E^{\ddagger}(B) \subset E^{\dagger}(B)$ whenever $\ddagger < \dagger$, $\dagger, \ddagger \in \$$;
(ii) $E^{\dagger_0} = 0$, $E^{\dagger_\infty} = I$;
(iii) $\mathscr{E}$ is closed under strong limits; and
(iv) $\mathscr{E}$ is of *bounded variation*, in the sense that if $\mathscr{P} = \{E^i : i = 0, 1, \ldots, n\}$ is a partition of $\mathscr{E}$ and A^i, $i = 1, 2, \ldots, n$, are bounded linear operators on B, then

$$\left\| \sum_{i=1}^{n} \Delta^i A^i \Delta^i \right\| = \max_i \|\Delta^i A^i \Delta^i\|$$

The bounded variation condition implies that $\|E^{\dagger}\| = 1$ (or 0), though it is in fact a stronger condition.[15] Although it represents a significant restriction on the class of Banach resolution spaces, with the aid of this assumption most of the basic Hilbert resolution space theores carries through to the Banach space case. In particular, strict causality is well defined and is characterized by a theory quite similar to that derived in the Hilbert space case. Moreover, one can derive viable decomposition, factorization, and causal invertibility theorems. Of course, the bounded variation assumption holds in a Hilbert resolution space by Lemma 2.A.4. Indeed, even in Hilbert space the bounded variation assumption can be shown to characterize the classical resolutions of the identity up to a similarity transformation, and thus it is a natural assumption in our Banach resolution space setting.[15] Finally, we note that the classical L_p and l_p, $1 \leq p \leq \infty$, spaces with their usual resolution structure satisfy the bounded variation assumption.

5. Multiplicative Integrals: Rather than working with the additive operator integrals defined in Chapter 1, we could have formulated much of our theory in terms of *multiplicative integrals.*[3,21] Given an operator-valued function $F: \$ \to \mathscr{B}$ we let

$$(M) \int^{\leftarrow} (I + dE(\dagger)\, F(\dagger)) = \lim_{\mathscr{P}} [I + \Delta^n F(n)][I + \Delta^{n-1} F(n-1)] \cdots [I + \Delta^1 F(1)]$$

where the limit is taken in the uniform operator topology over the net of partitions $\mathscr{P}$ of $\mathscr{E}$. Since the factors in the partial products that define the integrals are noncommutative, one may also define an analogous integral in which the order of the factors is reversed:

$$(M) \int^{\rightarrow} (I + dE(\dagger)\, F(\dagger)) = \lim_{\mathscr{P}} [I + \Delta^1 F(1)][I + \Delta^2 F(2)] \cdots [I + \Delta^n F(n)]$$

Of course, as with the additive integrals, one may define an entire family of multiplicative integrals including strongly convergent integrals, (m) integrals, integrals with the measure operating on the right of the function, and integrals with the measure operator on both the right and left side of the function.

Interestingly, from the point of view of most of our applications the multiplicative integrals are in one-to-one correspondence with the various integrals of triangular truncation and thus may be bypassed in our theory. They may, however, simplify the theory.[4,7] For instance, if A is strictly causal, then

$$(I + A)^{-1} = (m) \int^{\leftarrow} (I - dE(\dagger)\, AE^{\dagger})$$

The precise relationship between the integrals of triangular truncation and the various multiplicative integrals is given by[7]

$$(m)\int^{\rightarrow}(I + dE(\dagger)\,AE^{\dagger}) = I + (m)\int dE(\dagger)\,AE^{\dagger}$$

$$(M)\int^{\rightarrow}(I + dE(\dagger)\,AE^{\dagger}) = I + (M)\int dE(\dagger)\,AE^{\dagger}$$

$$(M)\int^{\leftarrow}(I + dE(\dagger)\,AE_{\dagger}) = I + (M)\int dE(\dagger)\,AE_{\dagger}$$

$$(m)\int^{\leftarrow}(I + dE(\dagger)\,AE_{\dagger}) = I + (m)\int dE(\dagger)\,AE_{\dagger}$$

$$(m)\int^{\leftarrow}(I + dE(\dagger)\,AE^{\dagger}) = [I - (m)\int dE(\dagger)\,AE^{\dagger}]^{-1}$$

$$(M)\int^{\rightarrow}(I + dE(\dagger)\,AE_{\dagger}) = [I - (M)\int dE(\dagger)\,AE_{\dagger}]^{-1}$$

$$\begin{aligned}(M)\int^{\rightarrow}(I + dE(\dagger)\,A\,dE(\dagger)) &= (m)\int^{\rightarrow}(I + dE(\dagger)\,A\,dE(\dagger))\\ &= (M)\int^{\leftarrow}(I + dE(\dagger)\,A\,dE(\dagger))\\ &= (m)\int^{\rightarrow}(I + dE(\dagger)\,A\,dE(\dagger))\\ &= I + (M)\int dE(\dagger)\,A\,dE(\dagger)\\ &= I + (m)\int dE(\dagger)\,A\,dE(\dagger)\end{aligned}$$

6. **Factorization:** Although a regular factorization may fail to exist, if one weakens the concept by permitting an operator to be factored through an arbitrary factor space, then it can be shown that every positive definite hermitian operator admits a *miniphase spectral factorization.*[27] Indeed, if $A: H \rightarrow H$ is positive definite and hermitian, one can construct a new Hilbert resolution space $(H', \mathscr{E}')$ and an operator $K: H' \rightarrow H$ that is causal and admits a causal inverse such that $A = KK^*$. Moreover,

$(H', \mathscr{E}')$ and K are unique in the sense that if $\tilde{K}: \tilde{H}' \to H$ defines an alternative miniphase spectral factorization of A, then there exists a Hilbert resolution space equivalence (i.e., a memoryless unitary operator) $U: H' \to \tilde{H}'$ such that the diagram

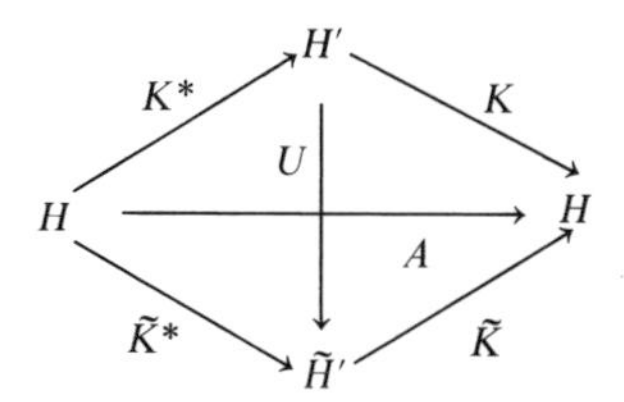

is commutative.

Interestingly, the equivalence class of Hilbert resolution spaces defined by the miniphase spectral factorization admits a natural representation in the form of a *reproducing kernel resolution space*[30] $(H_A, \mathscr{E}_A)$. The Hilbert space H_A for the reproducing kernel resolution space is a renormed version of H with an alternative resolution structure $\mathscr{E}_A$, while the resulting cononical representation for the miniphase spectral factorization takes the form of the identity operator mapping H_A (with its new norm and resolution structure) to H (with its old norm and resolution structure). Moreover, the new norm is chosen to ensure that the adjoint of the identity relative to these topologies is A. We thus have the required factorization $KK^* = IA = A$.

From a practical point of view this factorization does not prove to be especially useful since one has no control over the factor space. It is, however, unique, and hence if a factorization of A through H exists, it coincides with the miniphase spectral factorization up to a resolution space equivalence and exists if and only if $(H_A, \mathscr{E}_A)$ is equivalent to $(H, \mathscr{E})$. Alternatively, such a factorization exists if and only if

$$S = \int_{\$} \dagger \, dE(\dagger)$$

is unitarily equivalent to

$$S_A = \int_{\$} \dagger \, dE_A(\dagger)$$

Conceptually, the general spectral factorization problem thus reduces to a unitary equivalence problem for (unbounded) hermitian operators.

If this unitary equivalence criterion for the existence of a spectral factorization is to be implemented, one requires a characterization of $\mathscr{E}_A$ (or equivalently of S_A). Indeed, it can be shown[13] that

$$E_A^{\dagger} = A^{1/2}E^{\dagger}[E_{\dagger} + E^{\dagger}AE^{\dagger}]^{-1}E^{\dagger}A^{1/2}$$

where the existence of the inverse operator $[E_{\dagger} + E^{\dagger}AE^{\dagger}]^{-1}$ for all $\dagger \in \$$ is just the necessary condition for the existence of a factorization (Property 4.A.11). As such, a necessary and sufficient condition for the existence of a spectral factorization of $A = A^* > 0$ is that $[E_{\dagger} + E^{\dagger}AE^{\dagger}]^{-1}$ exist for all $\dagger \in \$$ and that $A^{1/2}E^{\dagger}[E_{\dagger} + E^{\dagger}AE^{\dagger}]^{-1}E^{\dagger}A^{1/2}$ be unitarily equivalent to $E^{\dagger}$.

Although the question was open for many years, Larsen[19] has recently shown that there exist positive definite hermitian operators that do not admit a spectral factorization. On the other hand, the criterion reduces the spectral factorization problem to a "notorious" problem of classical operator theory. It is doubtful that a simple characterization of the class of positive definite hermitian operators that admit a spectral factorization will be possible. Certainly no such characterization exists at the time of this writing.

7. Nonlinear Operators: Although the Hilbert resolution space concept was motivated by linear problems, a surprisingly powerful theory of nonlinear operators on a Hilbert resolution space has developed. One can define either of two alternative norms

$$\|A\|_G = \sup_{x \neq 0} \|Ax\|/\|x\| \qquad \text{or} \qquad \|A\|_L = \sup_{x \neq y} \|Ax - Ay\|/\|x - y\|$$

and then work with the algebra of *finite gain operators* satisfying $\|A\|_G < \infty$ or the algebra of *Lipschitz continuous operators* satisfying $\|A\|_L < \infty$. Both of these nonlinear operator algebras satisfy all of the axioms for a Banach algebra with the exception of right distributivity ($A[B + C] \neq AB + AC$) and are thus termed *left-distributive Banach algebras*[33] or *nearrings*.

The key to the development of a causality theory for these nonlinear operators is the observation that the equality

$$E^{\dagger}A = E^{\dagger}AE^{\dagger}, \qquad \ddagger \in \$$$

is the correct definition for causality in the nonlinear case. Note that in the nonlinear case this definition is no longer equivalent to the invariant subspace criterion that $A(H_{\dagger}) \subset H_{\dagger}$, $\dagger \in \$$, and hence the causal nonlinear operators do not define a nest algebra. Surprisingly, however, much of our causality theory does carry through to the nonlinear case. In particular, one can formulate integrals of triangular truncation, though one must deal only with integrals in which the measure operates on the left of the function.[6] Furthermore, when the appropriate integrals exist they can be used to define a unique decomposition of a nonlinear operator into the sum of four operators: a strictly causal operator, a strictly anticausal operator, a memoryless operator, and a cross-causal operator.[6] The latter has no linear analog and corresponds to an operator whose output is simultaneously dependent on both the past and future input. For instance, the operator on $L_2(-\infty, \infty)$ with its usual resolution structure defined by

$$[Af](\dagger) = f(\dagger + 1)f(\dagger - 1)$$

is cross-causal.

The most powerful class of results for nonlinear operators defined on a Hilbert resolution space deal with the causal invertability problem. Indeed, this problem motivated many of the early researchers in the field who considered nonlinear operators from the inception.[31,35] The key result in this area is the so-called *small-gain theorem.* In the Lipschitz norm this follows from the contraction mapping theorem and states that $[I + A]^{-1}$ exists and is causal if $\|A\|_L < 1$ and A is causal. When $\|A\|_G < 1$ and A is causal, the theorem is more subtle and states that $[I - A]^{-1}$ is causal when it exists in an appropriate (weak) sense.[32] This theorem, in turn, leads to a *topological criterion for causal invertibility* to the effect that A^{-1} is causal if $\deg[A] = \deg[I]$. Here the degree of a nonlinear operator is defined in a manner similar to that used in the linear case.[28] As with the linear case a variety of sufficient conditions for causal invertibility follow from the degree condition. For instance, if $0 \in \rho_\infty(A)$, then A^{-1} exists and is causal. Moreover, one can formulate a causal invertibility criterion similar to the circle criterion of classical stability theory.[33]

8. Weakly Additive Operators: An interesting class of nonlinear operators defined on a Hilbert resolution space that has no analog in a classical Hilbert space setting consists of the *weakly additive operators*, which satisfy

$$A = AE^\dagger + AE_\dagger, \qquad \dagger \in \$$$

Since these operators are right distributive with respect to $\mathscr{E}$, they exhibit a number of the properties associated with linear operators on $(H, \mathscr{E})$ that do not hold for general nonlinear operators. In particular, $A(H_{\dagger}) \subset H_\dagger$ for causal weakly additive operators. The most common examples of weakly additive operators are the operators of the form $A = LM$ where L is linear and M is memoryless but (possibly) nonlinear. Since these operators are commonly encountered in system theoretic application, the weakly additive operators play a significant role in our nonlinear theory. For instance, if A is strictly causal and weakly additive, then $[I + A]^{-1}$ exists and is causal.

9. Polynomic Operators: Another important class of nonlinear operators are the *polynomic operators*, defined as follows.[22] For a given Hilbert resolution space $(H, \mathscr{E})$ we construct a *Hilbert scale*

$$S = \sum_{i=0}^{n} \otimes H^i$$

and identify each element $x \in H$ with

$$\tau(x) = \sum_{i=0}^{n} \otimes^i x \in S$$

Now, with some care one can construct a Hilbert resolution space K such that the range of τ is contained in $K \subset S$.[23] Then for any linear operator $A: K \to H$ we construct a polynomic operator $P: H \to H$ by

$$P(x) = A\tau(x)$$

Since the class of polynomic operators are constructed from a fixed memoryless operator and a linear operator, all the usual resolution space concepts are inherited by P from A.[22] Moreover, many problems in polynomic system theory can be transformed into linear problems by "lifting" them from H to K.[23]

10. Compact Perturbations of the Causals: A major new trend in resolution space theory at the time of this writing is the study of *compact perturbations of the causal operators*. This theory is built around the space $\mathscr{C} + \mathscr{K}$, which is in fact a Banach algebra.[1,11] Although two resolution spaces of the same dimension may not be equivalent, and thus the corresponding Banach algebras of causal operators may also fail to be isomorphic, the surprising result of this theory is that the algebras $\mathscr{C} + \mathscr{K}$ of causal operators perturbed by compacts are isomorphic.[1] This, in turn, yields a number of significant results. In particular, Larsen uses the results to "construct" a positive definite hermitian operator that does not admit a spectral factorization.[19] Additionally, Feintuch and Lambert[14] have used the properties of the Banach algebra $\mathscr{C} + \mathscr{K}$ to show that for a continuous resolution space the inverse of a causal operator is in $\mathscr{C} + \mathscr{K}$ if and only if it is in $\mathscr{C}$. That is, if A^{-1} fails to be causal, it cannot be a compact perturbation of a causal operator; if it is not causal, it is not "close to" a causal.

Although the study of the space $\mathscr{C} + \mathscr{K}$ is only just the beginning, the fact that $\mathscr{C} + \mathscr{K}$ is better behaved than $\mathscr{C}$ suggests that many results that fail when formulated in terms of causal operators may hold for compact perturbations of causal operators.

For instance, it is conjectured that every positive definite hermitian operator A admits a factorization in the form $A = BB^*$ where $B \in (\mathscr{C} + \mathscr{K}) \cap (\mathscr{C} + \mathscr{K})^{-1}$. Although the relevance of such results to our system theoretic application is open to question, it appears that a viable theory can be formulated around the properties of the Banach algebra $\mathscr{C} + \mathscr{K}$.

REFERENCES

1. Anderson, N., Ph.D. Dissertation, Univ. of California at Berkeley, Berkeley, California (1979).
2. Arveson, W. A., Interpolation problems in nest algebras, *J. Funct. Anal.* **20**, 208–233 (1975).
3. Brodskii, M. S., "Triangular and Jordan Representations of Linear Operators." American Mathematical Society, Providence, Rhode Island, 1971.
4. Brodskii, M. S., Gohberg, I. C., and Krein, M. G., General theorems on triangular representations of linear operators and multiplications of their characteristic functions, *J. Funct. Anal.* **3**, 1–27 (1969).
5. DeCarlo, R. A., Saeks, R., and Strauss, M. J., The Fourier transform of a resolution space and a theorem of Masani, *Proc. Internat. Symp. Operator Theory of Networks and Systems, Montreal, August* pp. 69–74 (1975).
6. DeSantis, R. M., Ph.D. Dissertation, Univ. of Michigan, Ann Arbor, Michigan (1971).
7. DeSantis, R. M., and Feintuch, A., Causality theory and multiplicative transformators, *J. Math. Anal. Appl.* **75**, 411–416 (1980).
8. Dunford, N., and Schwartz, J. T., "Linear Operators." Wiley (Interscience), New York, 1958.
9. Erdos, J. A., and Longstaff, N. A., The convergence of triangular integrals of operators on Hilbert space, *Indiana Math. J.* **22**, 929–938 (1973).
10. Falb, P. L., "On a Theorem of Bochner," Publication Mathematiques No. 37, Inst. des Hautes Etudes Scientifiques, 1969.
11. Fall, T., Arveson, W., and Muhly, P., Perturbations of nest algebras, *J. Operator Theory* **1**, 137–150 (1979).
12. Feintuch, A., Strictly and strongly strictly causal linear operator, *SIAM J. Math. Anal.* **10**, 603–613 (1979).
13. Feintuch, A., unpublished notes, Ben Gurion Univ., Beer Sheva (1980).

14. Feintuch, A., and Lambert, A. Invertibility in Nest Algebras, unpublished notes, Ben Gurion Univ., Beer Sheva (1980).
15. Gohberg, I. C., and Barkar, M. A., Factorization of operators in a Banach space, *Mat. Issued*, **1**, 98–129 (1966).
16. Gohberg, I. C., and Krein, M. G., "Theory and Application of Volterra Operators in Hilbert Space." American Mathematical Society, Providence, Rhode Island, 1970.
17. Johnson, B., and Parrott, S., "Operators commuting with a von Neumann algebra modulo the set of compact operators, *J. Funct. Anal.* **11**, 39–61 (1971).
18. Larsen, D., Ph.D. Dissertation, Univ. of California at Berkeley, Berkeley, California (1976).
19. Larsen, D., unpublished notes, Univ. of Nebraska, Lincoln, Nebraska (1980).
20. Livsic, M. S., "Operators, Oscillations, Waves." American Mathematical Society, Providence, Rhode Island, 1970.
21. Porter, W. A., Some circuit theory concepts revisited, *Internat. J. Control* **12**, 433–448 (1970).
22. Porter, W. A., An overview of polynomic system theory, *Proc. IEEE* **64**, 18–23 (1976).
23. Porter, W. A., On factoring the polyvariance operator, *Math. Systems Theory* **14**, 67–82 (1981).
24. Ringrose, J. R., On some algebras of operators, *Proc. London Math. Soc.* **3**, 61–83 (1965).
25. Saeks, R., Causality in Hilbert space, *SIAM Rev.* **12**, 357–383 (1970).
26. Saeks, R., Fourier analysis in Hilbert space, *SIAM Rev.* **15**, 605–638 (1973).
27. Saeks, R., Reproducing kernel resolution space and its applications, *J. Franklin Inst.* **302**, 331–355 (1976).
28. Saeks, R., Stability and homotopy, *in* "Alternatives for Linear Multivariable Control," pp. 247–252. NEC, Chicago, 1978.
29. Saeks, R., Causal factorization, shift operators, and the spectral multiplicity function, *in* "Vector and Operator Valued Measures and Applications" (D. H. Tucker and H. B. Maynard, eds.), pp. 319–335. Academic Press, New York, 1973.
30. Saeks, R., and Leake, R. J., On semi-uniform resolution space, *Proc. Midwest Symp. Circuit Theory, 14th, Univ. of Denver* Paper IV-5 (May 1971).
31. Sandberg, I. W., On the L_2 boundedness of solutions of nonlinear functional equations, *Bell Systems Tech. J.* **43**, 1601–1608 (1965).
32. Willems, J. C., Stability, instability, invertibility, and causality, *SIAM J. Control* **7**, 645–671 (1969).
33. Willems, J. C., "Analysis of Feedback Systems." MIT Press, Cambridge, Massachusetts, 1971.
34. Youla, D. C., Carlin, H. J., and Castriota, L. J., Bounded real scattering matrices and the foundations of linear passive network theory, *IRE Trans. Circuit Theory* **CT-6**, 102–124 (1959).
35. Zames, G., Functional analysis applied to nonlinear feedback systems, *IEEE Trans. Circuit Theory* **CT-10**, 392–402 (1963).

PART II
State Space Theory

In addition to causality, there is a second time-related operator concept that is well defined in a Hilbert resolution space setting but has no analog in a classical Hilbert space setting—the state concept. Intuitively, the state of a system at time † corresponds to the information about the input prior to time † that effects the output after time †. Historically, the concept was introduced in an automata theoretic setting as an abstraction of computer memory. In the computer the output after † is determined entirely by the input after time † and the contents of memory at time †. Hence the contents of memory at time † represent the only information about the input prior to time † that effects the output after time †. In Chapter 6 our abstract state model for operators defined on a Hilbert resolution space is formulated. The relationship between the state model and the given operator is developed in Chapter 7 in the form of a state trajectory factorization.

6

State Models

A. STATE DECOMPOSITION

1. Definition: A *state decomposition* for $A \in \mathscr{B}$ is a family of triples $\{(S_\dagger, \lambda_\dagger, \theta_\dagger) : \dagger \in \$\}$ where for each $\dagger \in \$$

(i) $S_\dagger$ is a Hilbert space,
(ii) $\lambda_\dagger : H \to S_\dagger$ is a bounded linear operator satisfying $\lambda_\dagger = \lambda_\dagger E^\dagger$,
(iii) $\theta_\dagger : S_\dagger \to H$ is a bounded linear operator satisfying $\theta_\dagger = E_\dagger \theta_\dagger$, and
(iv) $E_\dagger A E^\dagger = \theta_\dagger \lambda_\dagger$.

2. Remark: Intuitively, $S_\dagger$ is the *state space* at time $\dagger \in \$$, $\lambda_\dagger$ is the *input-to-state map*, and $\theta_\dagger$ is the *state-to-free-response map*. Furthermore,

$$E_\dagger A = E_\dagger A[E^\dagger + E_\dagger] = E_\dagger A E^\dagger + E_\dagger A E_\dagger = \theta_\dagger \lambda_\dagger + E_\dagger A E_\dagger$$

Thus if an input $u \in H$ is applied to an operator A, the response of A to u

after time † is given by

$$E_\dagger Au = \theta_\dagger x(\dagger) + E_\dagger AE_\dagger u$$

where $x(\dagger) = \lambda_\dagger u \in S_\dagger$ is the *state* at time †. Accordingly, the output after time † is entirely determined by the input after time † and the state at time †, verifying that the abstract state decomposition of Definition 1 is compatible with the intuition of the introduction.[1,2]

Finally, we note that condition (iv) of Definition 1 is equivalent to the requirement that the diagram

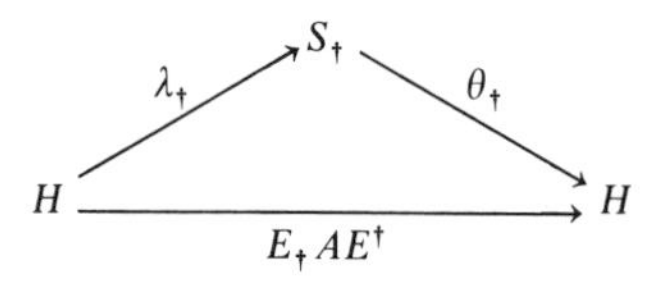

be commutative.

3. Example: Every $A \in \mathscr{B}$ admits the *trivial state decomposition* $\{(H, E^\dagger, E_\dagger A) : \dagger \in \$\}$. In general, this does not prove to be an especially useful state decomposition, since H is "too large" a state space.

4. Example: Consider the differential operator

$$\dot{x}(\dagger) = Ax(\dagger) + Bf(\dagger), \qquad x(0) = 0$$

$$g(\dagger) = Cx(\dagger),$$

defined on $L_2[0, 1]$ with its usual resolution structure. Here $f, g \in L_2[0, 1]$, $x(\dagger) \in R^n$, A is an $n \times n$ matrix, B is $n \times 1$, and C is $1 \times n$. It then follows from classical differential equation theory that

$$g(\dagger) = \int_0^\dagger Ce^{A(\dagger - \ddagger)}Bf(\ddagger)\, d\ddagger = [Af](\dagger)$$

defines a bounded linear operator $B \in \mathscr{B}$. To construct a state decomposition for this operator we let $S_\dagger = R^n$, $\dagger \in [0, 1]$, define $\lambda_\dagger : L_2[0, 1] \to R^n$ by

$$x(\dagger) = \lambda_\dagger f = \int_0^\dagger e^{A(\dagger - \ddagger)}Bf(\ddagger)\, d\ddagger$$

and define $\theta_\dagger : R^n \to L_2[0, 1]$ by

$$[\theta_\dagger x(\dagger)](\ddagger) = \begin{cases} Ce^{A(\ddagger - \dagger)}x(\dagger), & \ddagger \geq \dagger \\ 0, & \ddagger < \dagger \end{cases}$$

Of course, similar constructions can be given for the case of a difference operator or a differential operator with $x(\dagger)$ in a Hilbert space (whenever A defines an appropriate semigroup of operators on that Hilbert space).

5. Example: We now consider a nontrivial example in which the state space is time varying. Let H be the sequence space $l_2[0, \infty)$ with its usual resolution structure and let $\{e_i\}_0^\infty$ be its standard basis. Now rather than specifying $A \in \mathscr{B}$ we specify $E_n AE^n$, $n \in Z^+$, by

$$E_n AE^n a = \sum_{i=0}^{n} (a, e_i)\hat{e}_{n+1+i}$$

where $\hat{e}_i = e_i/e^i$, $i = 0, 1, \ldots$.

As we shall see, this uniquely specifies a strongly strictly causal $A \in \mathscr{B}$. We now construct a state decomposition for A. For $n = 0, 1, 2, \ldots$ we let $S_n = R^{n+1}$ and define $\lambda_n: l_2[0, \infty) \to R^{n+1}$ by

$$\lambda_n a = \operatorname{col}[(a, e_0), (a, e_1), \ldots, (a, e_n)]$$

and $\theta_n: R^{n+1} \to l_2[0, \infty)$ by

$$\theta_n(\operatorname{col}[x_0, x_1, \ldots, x_n]) = \sum_{i=0}^{n} x_i \hat{e}_{n+1+i}$$

A little algebra will now verify that this is, indeed, a state decomposition for A. Moreover, it can be shown that A does not admit a state decomposition with a single time-invariant finite-dimensional state space.[3] Thus the example verifies the necessity of allowing a variable state space, at least when one desires that the state space be "small" in some sense. Of course, one could always construct a trivial state decomposition for this operator with the time-invariant but infinite-dimensional state space $l_2[0, \infty)$, as in Example 3.

6. Remark: Although A is arbitrary in Definition 1, $\lambda_\dagger \theta_\dagger = E_\dagger AE^\dagger$ and therefore the state decomposition is dependent in some sense on the causal part of A. This is made precise in the following.

7. Property: Let $A \in \mathscr{B}$ and assume that A admits a decomposition $A = [A]_{\mathscr{C}^*} + [A]_{\mathscr{S}}$ where $[A]_{\mathscr{C}^*} \in \mathscr{C}^*$ and $[A]_{\mathscr{S}} \in \mathscr{S}$. Then a state decomposition $\{(S_\dagger, \lambda_\dagger \theta_\dagger) : \dagger \in \$\}$ for A is determined entirely by $[A]_{\mathscr{S}}$ and $[A]_{\mathscr{S}}$ is uniquely determined by the state decomposition.

Proof: Since

$$\begin{aligned}\lambda_\dagger \theta_\dagger &= E_\dagger AE^\dagger = E_\dagger([A]_{\mathscr{C}^*} + [A]_{\mathscr{S}})E^\dagger = E_\dagger[A]_{\mathscr{C}^*}E^\dagger + E_\dagger[A]_{\mathscr{S}}E^\dagger \\ &= E_\dagger[A]_{\mathscr{C}^*}E_\dagger E^\dagger + E_\dagger[A]_{\mathscr{S}}E^\dagger = E_\dagger[A]_{\mathscr{S}}E^\dagger\end{aligned}$$

the state decomposition is entirely determined by $[A]_{\mathscr{S}}$. To verify the converse consider the integral

$$\begin{aligned} \mathrm{s}\,(m) \int dE(\dagger)\,\lambda_{\dagger}\theta_{\dagger} &= \mathrm{s}\lim_{\mathscr{P}} \left[\sum_{i=1}^{n} \Delta^i \lambda_{i-1}\theta_{i-1} \right] \\ &= \mathrm{s}\lim_{\mathscr{P}} \left[\sum_{i=1}^{n} \Delta^i E_{i-1} A E^{i-1} \right] \\ &= \mathrm{s}\lim_{\mathscr{P}} \left[\sum_{i=1}^{n} \Delta^i A E^{i-1} \right] = \mathrm{s}\,(m) \int dE(\dagger)\, A E^{\dagger} = [A]_{\mathscr{S}} \end{aligned}$$

which exists since $[A]_{\mathscr{S}}$ is assumed to be well defined and given by $\mathrm{s}\,(m) \int dE(\dagger)\, AE^{\dagger}$ by Property 3.A.4. Hence the state decomposition determines $[A]_{\mathscr{S}}$ and the proof is complete. //

8. Remark: As is our usual practice, we can define a concept dual to the state decomposition by working with the dual Hilbert resolution space $(H, \mathscr{E}^*)$ defined on $\$^*$.

9. Definition: A costate decomposition for $A \in \mathscr{B}$ is a state decomposition for A defined with respect to the dual Hilbert resolution space $(H, \mathscr{E}^*)$.

10. Property: $(\tilde{S}_{\dagger}, \tilde{\lambda}_{\dagger}, \tilde{\theta}_{\dagger})$ is a costate decomposition for $A \in \mathscr{B}$ if and only if $(\tilde{S}_{\dagger}, \tilde{\theta}_{\dagger}^*, \tilde{\lambda}_{\dagger}^*)$ is a state decomposition for A^*.

Proof: If $(\tilde{S}_{\dagger}, \tilde{\lambda}_{\dagger}, \tilde{\theta}_{\dagger})$ is a costate decomposition for A, it satisfies the equalities

$$\tilde{\lambda}_{\dagger} = \tilde{\lambda}_{\dagger} E_{\dagger}, \qquad \tilde{\theta}_{\dagger} = E^{\dagger}\tilde{\theta}_{\dagger}, \qquad \text{and} \qquad \tilde{\theta}_{\dagger}\tilde{\lambda}_{\dagger} = E^{\dagger} A E_{\dagger}$$

where $\tilde{\lambda}_{\dagger}: H \to \tilde{S}_{\dagger}$ and $\tilde{\theta}_{\dagger}: \tilde{S}_{\dagger} \to H$. Hence $\tilde{\lambda}_{\dagger}^*: \tilde{S}_{\dagger} \to H$ and $\tilde{\theta}_{\dagger}^*: H \to \tilde{S}_{\dagger}$ where

$$\tilde{\lambda}_{\dagger}^* = E_{\dagger}\tilde{\lambda}_{\dagger}^*, \qquad \tilde{\theta}_{\dagger}^* = \tilde{\theta}_{\dagger}^* E^{\dagger}, \qquad \text{and} \qquad \tilde{\lambda}_{\dagger}^*\tilde{\theta}_{\dagger}^* = E_{\dagger} A^* E^{\dagger}$$

which are precisely the axioms for a state decomposition of A^*. Of course, a similar argument will transform a state decomposition for A^* into a costate decomposition for A, thereby completing our proof. //

11. Remark: Consistent with Property 10 our entire state decomposition theory can be translated into a costate decomposition theory by the simple artifice of taking adjoints. For instance, if A admits a decomposition in the form

$$A = [A]_{\mathscr{C}} + [A]_{\mathscr{S}^*}$$

then its costate decomposition is entirely determined by $[A]_{\mathscr{S}*}$ and, conversely, $[A]_{\mathscr{S}*}$ is uniquely determined by a costate decomposition in a manner dual to that developed in Property 7. We shall not formulate the details of the costate theory, leaving appropriate translations required to the reader.

B. CONTROLLABILITY AND OBSERVABILITY

1. Definition: A state decomposition $\{(S_\dagger, \lambda_\dagger, \theta_\dagger) : \ddagger \in \$\}$ is *completely controllable* if the image of $\lambda_\dagger$ is dense in $S_\dagger$ for all $\dagger \in \$$, it is *completely observable* if for each $\dagger \in \$$ there exists $\varepsilon > 0$ such that $\|\theta_\dagger x\| \geq \varepsilon\|x\|$ for all $x \in S_\dagger$, and it is *minimal* if it is completely controllable and completely observable.

2. Remark: It should be noted that there are numerous variations on the concepts of controllability and observability. One can require, for example, that $\lambda_\dagger$ be onto, that ε be independent of $\dagger$, and that $\theta_\dagger$ be uniformly bounded. Although these concepts play a nontrivial role in state space theory, for the present purposes Definition 1 of complete controllability and complete observability will suffice.

3. Remark: The concepts of controllability and observability were originally developed in a control theoretic context, where they have deep physical interpretations. Indeed, if an operator is characterized by a completely controllable state decomposition, then given any $\dagger \in \$$, $x \in S_\dagger$, and $\varepsilon > 0$, there exists a $u \in H^\dagger$ such that

$$\|x - \lambda_\dagger u\| < \varepsilon$$

That is, it is possible to find an input that will drive the state of the system arbitrarily close to any desired state. Similarly, complete observability implies that the state of the system at time $\dagger \in \$$ can be uniquely determined by observation of the system input and output after time $\dagger$. Indeed, the required observer is given by the bounded linear operator

$$x(\dagger) = \theta_\dagger^{-L}[E_\dagger y - E_\dagger A E_\dagger u]$$

where $y = Au$ and $\theta_\dagger^{-L} = [\theta_\dagger^* \theta_\dagger]^{-1}\theta_\dagger^*$ exists owing to complete observability (see Property 5).

4. Example: For the system of Example A.5 the input

$$a = \sum_{i=0}^{n} x_i e_i$$

is mapped by λ_n to the state vector $x(n) = \text{col}[x_0, x_1, \ldots, x_n]$, verifying that λ_n is onto for all $n \in Z^+$ and that the state decomposition of Example A.5 is completely controllable. On the other hand for $x = \text{col}[x_0, x_1, \ldots, x_n] \in R^{n+1}$

$$\theta_n x = \sum_{i=0}^{n} x_i \hat{e}_{n+1+i}$$

Hence

$$\|\theta_n x\|^2 = \left\| \sum_{i=0}^{n} x_i \hat{e}_{n+1+i} \right\|^2 = \sum_{i=0}^{n} \|x_i \hat{e}_{n+1+i}\|^2 = \sum_{i=0}^{n} |x_i|^2/e^{n+1+i}$$
$$\geq \|x\|^2/e^{2n+1}$$

verifying that the state decomposition is completely observable. Thus the state decomposition of Example A.5 is minimal.

5. Property: A state decomposition $\{(S_\dagger, \lambda_\dagger, \theta_\dagger) : \ddagger \in \$\}$ is completely controllable if and only if $\lambda_\dagger \lambda_\dagger^* > 0$ for all $\dagger \in \$$. It is completely observable if and only if $\theta_\dagger^* \theta_\dagger > 0$ for all $\dagger \in \$$.

Proof: If $\lambda_\dagger$ has dense range in $S_\dagger$, then $\lambda_\dagger^*$ is one-to-one, and hence for $x \neq 0$

$$0 < \|\lambda_\dagger x\|^2 = (\lambda_\dagger^* x, \lambda_\dagger^* x) = (\lambda_\dagger \lambda_\dagger^* x, x)$$

verifying that $\lambda_\dagger \lambda_\dagger^* > 0$ and conversely, while a similar argument applies to $\theta_\dagger^* \theta_\dagger$. //

6. Remark: Note that both $\lambda_\dagger \lambda_\dagger^*$ and $\theta_\dagger^* \theta_\dagger$ map $S_\dagger$ to itself, and since $S_\dagger$ is often of lower dimension than H, these tests for complete controllability and complete observability may be simpler than a direct test on $\lambda_\dagger$ or $\theta_\dagger$. In particular, this is the case for the differential operator of Example A.4, where $S_\dagger = R^n$, $\dagger \in \$$, and $H = L_2[0, 1]$.

7. Example: For the differential operator of Example A.4, $S_\dagger = R^n$, $\dagger \in [0, 1]$, and $H = L_2[0, 1]$ with its usual resolution structure, while

$$[\theta_\dagger x(\dagger)](\ddagger) = \begin{cases} Ce^{A(\ddagger - \dagger)} x(\dagger), & \ddagger \geq \dagger \\ 0, & \ddagger < \dagger \end{cases}$$

Thus, if $x \in R^n$ and $g \in L_2[0, 1]$,

$$(g, \theta_\dagger x)_{L_2} = \int_\dagger^1 g(\ddagger) C e^{A(\ddagger - \dagger)} x \, d\ddagger = \left[\int_\dagger^1 g(\ddagger) C e^{A(\ddagger - \dagger)} \, d\ddagger \right] x$$
$$= \left[\int_\dagger^1 e^{A^t(\ddagger - \dagger)} C^t g(\ddagger) \, d\ddagger \right]^t x = (\theta_\dagger^* g, x)_{R^n}$$

where t denotes matrix transposition. Then $\theta_\dagger^*: L_2[0, 1] \to R^n$ is defined by

$$\theta_\dagger^* g = \int_\dagger^1 e^{A^t(\ddagger - \dagger)} C^t g(\ddagger)\, d\ddagger$$

Letting $g = \theta_\dagger x = Ce^{A(\ddagger - \dagger)}x$, $\ddagger \geq \dagger$, we then find that $\theta_\dagger^* \theta_\dagger$ is characterized by the matrix representation

$$\theta_\dagger^* \theta_\dagger = \int_\dagger^1 e^{A^t(\ddagger - \dagger)} C^t C e^{A(\ddagger - \dagger)}\, d\ddagger$$

Therefore the state decomposition is completely observable if and only if this matrix is positive definite for all $\dagger \in [0, 1]$. Letting $z \in R^n$, we then obtain

$$(z, \theta_\dagger^* \theta_\dagger z) = \int_\dagger^1 [Ce^{A(\ddagger - \dagger)}z]^t [Ce^{A(\ddagger - \dagger)}z]\, d\ddagger$$

which is zero if and only if $Ce^{A(\ddagger - \dagger)}z = 0$ for all $\ddagger \in [\dagger, 1]$. The given differential operator fails to be completely observable if and only if $Ce^{A\dagger}z = 0$ for some $z \in R^n$ and all $\dagger \in [0, 1]$. With $Ce^{A\dagger}z$ expanded in a Taylor series, this implies that the differential operator is completely observable if and only if each of the coefficients $CA^i z$, $i = 0, 1, \ldots$, in the expansion

$$Ce^{A\dagger}z = \sum_{i=0}^{n} [CA^i z]\dagger^i / L^i$$

is zero for some $z \in R^n$. The Caley–Hamilton theorem, however, implies that, for any matrix A, A^i, $i \geq n$, is a linear combination of A^i, $0 \leq i < n$. Thus $CA^i z = 0$, $i = 0, 1, \ldots$, if and only if $CA^i z = 0$, $i = 0, 1, \ldots, n - 1$. The given differential operator is therefore completely observable if and only if there exists no $z \in R^n$ such that $CA^i z = 0$, $i = 0, 1, \ldots, n - 1$. Equivalently, the differential operator is completely observable if and only if the *observability matrix*

$$Q = \begin{bmatrix} C \\ \hdashline CA \\ \hdashline CA^2 \\ \hdashline CA^3 \\ \hdashline \vdots \\ \hdashline CA^{n-1} \end{bmatrix}$$

has rank n. This is, of course, the classical observability matrix for a differential operator, verifying that our general theory coincides with the classical theory.

By a parallel argument one may verify that $\lambda_\dagger^*: R^n \to L_2[0, 1]$ is given by

$$\lambda^* x(\dagger) = \begin{cases} B^t e^{A^t(\dagger - \ddagger)}, & \ddagger \le \dagger \\ 0, & \ddagger > \dagger \end{cases}$$

and $\lambda_\dagger \lambda_\dagger^*$ has the matrix representation

$$\lambda_\dagger \lambda_\dagger^* = \int_0^\dagger e^{A(\dagger - \ddagger)} B B^t e^{A^t(\dagger - \ddagger)} \, d\ddagger$$

Therefore, the state decomposition is completely controllable if and only if this matrix is positive definite for all $\dagger \in [0, 1]$. As before, the application of some linear algebra and the Cayley–Hamilton theorem will reveal that this is equivalent to the requirement that the controllability matrix

$$R = [B \mid BA \mid BA^2 \mid BA^3 \mid \cdots \mid BA^{n-1}]$$

have rank n, again replicating the classical theory.

8. Property: Every bounded linear operator on a Hilbert resolution space admits a minimal state decomposition.

Proof: For fixed $\dagger \in \$$ define a relation on $H^\dagger = E^\dagger(H)$ by $x \simeq y$ if and only if $E_\dagger A E^\dagger x = E_\dagger A E^\dagger y$. The equivalence classes $[x]^\dagger$defined by this relation form a vector space with

$$[x + y]^\dagger = [x]^\dagger + [y]^\dagger \qquad \text{and} \qquad c[x]^\dagger = [cx]^\dagger$$

for any scalar c. Moreover, we may define an inner product on this vector space:

$$([x]^\dagger, [y]^\dagger) = (E_\dagger A E^\dagger x, E_\dagger A E^\dagger y)$$

Completing this vector space in the topology defined by the inner product now defines the desired state space $S_\dagger$. To construct the desired state decomposition we let $\lambda_\dagger x = [E^\dagger x]^\dagger$ while $\theta_\dagger [x]^\dagger = E_\dagger A E^\dagger x$ for all $[x]^\dagger$. We now show that $\theta_\dagger$ is bounded, in which case it may be extended to all of $S_\dagger$ by continuity. Indeed,

$$\|\theta_\dagger [x]^\dagger\|^2 = \|E_\dagger A E^\dagger x\|^2 = (E_\dagger A E^\dagger x, E_\dagger A E^\dagger x) = \|[x]^\dagger\|_{S_\dagger}^2$$

As such, $\theta_\dagger$ is an isometry, and hence the state decomposition is completely observable. For $\lambda_\dagger$ we then have

$$\|\lambda_\dagger y\| = \|\theta_\dagger \lambda_\dagger y\| = \|E_\dagger A E^\dagger y\| \le \|A\| \, \|y\|$$

showing that it is bounded, while it has dense range by construction. As such, $\{(S_\dagger, \lambda_\dagger, \theta_\dagger) : \ddagger \in \$\}$ is a minimal state decomposition for an arbitrary $A \in \mathscr{B}$. //

9. Remark: The equivalence classes used to construct the minimal state decomposition in Property 8 are sometimes termed *Nerode equivalence classes* and represent a Hilbert resolution space formulation of a construction originally used by Nerode in an automata theoretic setting.[1] Note also that the state decomposition is observable in a much stronger sense than required since $\theta_\dagger$ is isometric for all $\dagger \in \$$.

10. Definition: Let $A \in \mathscr{B}$ admit two state decompositions $\{(S_\dagger, \lambda_\dagger, \theta_\dagger): \dagger \in \$\}$ and $\{(S'_\dagger, \lambda'_\dagger, \theta'_\dagger) : \dagger \in \$\}$. The two decompositions are said to be *equivalent* if there exists a family of bounded invertible operators $\pi_\dagger : S_\dagger \to S'_\dagger$ such that $\lambda'_\dagger = \pi_\dagger \lambda_\dagger$ and $\theta'_\dagger = \theta_\dagger \pi_\dagger^{-1}$ for all $\dagger \in \$$.

11. Remark: The equalities of Definition 10 correspond to the requirement that the diagram

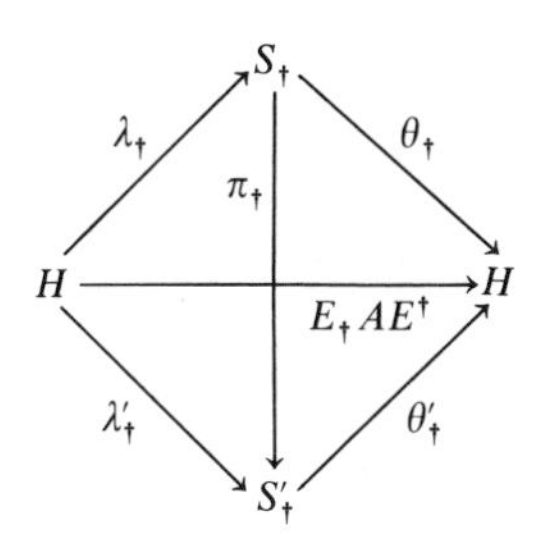

be commutative.

12. Property: Suppose that $A \in \mathscr{B}$ admits two minimal state decompositions $\{(S_\dagger, \lambda_\dagger, \theta_\dagger) : \dagger \in \$\}$ and $\{(S'_\dagger, \lambda'_\dagger, \theta'_\dagger) : \dagger \in \$\}$. Then they are equivalent.

Proof: Fix $\dagger \in \$$. If x is in the range of $\lambda_\dagger$, then $x = \lambda_\dagger u$ for some $u \in H$ and we may define $\pi_\dagger x$ by

$$\pi_\dagger x = \lambda'_\dagger u$$

Now, $\pi_\dagger x$ is well defined, for if $x = \lambda_\dagger u = \lambda_\dagger v$, $u, v \in H$, then

$$\theta'_\dagger \lambda'_\dagger u = E_\dagger A E^\dagger u = \theta_\dagger \lambda_\dagger u = \theta_\dagger \lambda_\dagger v = E_\dagger A E^\dagger v = \theta'_\dagger \lambda'_\dagger v$$

which implies that $\lambda'_\dagger u = \lambda'_\dagger v$ since $\theta'_\dagger$ is one-to-one. Moreover, $\pi_\dagger$ is clearly linear and satisfies $\pi_\dagger \lambda_\dagger = \lambda'_\dagger$ and $\theta'_\dagger \pi_\dagger = \theta_\dagger$. Up to this time, however, $\pi_\dagger$ has been defined only on the range of $\lambda_\dagger$, which is dense in $S_\dagger$ but may fail to be onto. We therefore show that $\pi_\dagger$ is bounded, allowing us to extend its domain to all of $S_\dagger$. Indeed, let $x = \lambda_\dagger u$ for some $u \in H$ and let $\varepsilon > 0$ be chosen so that $\|\theta_\dagger y\| \geq \varepsilon\|y\|$, $y \in S'_\dagger$. Then

$$\|\pi_\dagger x\| \leq (1/\varepsilon)\|\theta'_\dagger \pi_\dagger x\| = (1/\varepsilon)\|\theta_\dagger x\| \leq (\|\theta_\dagger\|/\varepsilon)\,\|x\|$$

showing that $\pi_\dagger$ is bounded.

Finally, if we choose $\varepsilon > 0$ so that $\|\theta_\dagger y\| \geq \varepsilon\|y\|$, $y \in S_\dagger$, a similar argument yields

$$\|\pi_\dagger\| \geq (\varepsilon/\|\theta'_\dagger\|)\,\|x\|$$

showing that $\pi_\dagger$ is bounded below. Now, since $\pi_\dagger \lambda_\dagger = \lambda'_\dagger$ and $\lambda'_\dagger$ has dense range, so does $\pi_\dagger$. Thus $\pi_\dagger$ is bounded both above and below and has dense range. Thus, its admits a bounded inverse as required. //

13. Example: The proposition implies that any two minimal state decompositions for the same operator are equivalent in the sense that there exists a family of invertible mappings $\pi_\dagger$ such that $x'(\dagger) = \pi_\dagger x(\dagger)$, where $x(\dagger)$ and $x'(\dagger)$ are the state trajectories resulting from the application of a given input in the two decompositions. Indeed, if one is given two differential operators, such as employed in Example A.4, that define the same operator on $L_2[0, 1]$, a little linear algebra and application of the Cayley–Hamilton theorem reveals that $\pi_\dagger$ is defined by a matrix T satisfying

$$C' = CT, \qquad B' = TB, \qquad \text{and} \qquad A' = TAT^{-1}$$

Note in this example that the time invariance of the differential operator and the associated state space allows one to construct an equivalence between the two systems that also is time invariant.[3]

14. Remark: For the differential operator of Example A.4 the semigroup of matrices $e^{A(\dagger - \ddagger)}$ characterizes the relationship between the state at time $\ddagger$ and the state at time $\dagger$ by

$$x(\dagger) = e^{A(\dagger - \ddagger)} x(\ddagger)$$

given that the input is zero in the time interval $[\ddagger, \dagger]$. In our general state model the same role is played by the *transition operators* $\Phi(\dagger, \ddagger)$, formulated as follows.

15. Property: Let $\{(S_\dagger, \lambda_\dagger, \theta_\dagger) : \dagger \in \$\}$ be a minimal state decomposition for $A \in B$. Then for all $\dagger, \ddagger \in \$$, $\dagger > \ddagger$, there exists a bounded linear operator $\Phi(\dagger, \ddagger): S_{\ddagger} \to S_{\dagger}$ such that

(i) $\theta_{\dagger} \lambda_{\dagger} E^{\ddagger} = \theta_{\dagger} \Phi(\dagger, \ddagger) \lambda_{\ddagger} = E_{\dagger} \theta_{\ddagger} \lambda_{\ddagger}$,
(ii) $\Phi(\dagger, \dagger) = I$, and
(iii) $\Phi(\dagger, \ddagger)\Phi(\ddagger, \times) = \Phi(\dagger, \times)$ whenever $\times < \ddagger < \dagger$.

Proof: If x is in the range of $\lambda_{\ddagger}$, i.e., $x = \lambda_{\ddagger} y$ for some $y \in H$, then we define $\Phi(\dagger, \ddagger)x$ by

$$\Phi(\dagger, \ddagger)x = \lambda_{\dagger} E^{\ddagger} y$$

This is well defined for $y, z \in H$ and satisfies $x = \lambda_{\ddagger} y = \lambda_{\ddagger} z$. Thus

$$\begin{aligned} \theta_{\dagger} \lambda_{\dagger} E^{\ddagger} y &= E_{\dagger} A E^{\dagger} E^{\ddagger} y = E_{\dagger} A E^{\ddagger} y = E_{\dagger} E_{\ddagger} A E^{\ddagger} y = E_{\dagger} \theta_{\ddagger} \lambda_{\ddagger} y = E_{\dagger} \theta_{\ddagger} \lambda_{\ddagger} z \\ &= E_{\dagger} E_{\ddagger} A E^{\ddagger} z = E_{\dagger} A E^{\ddagger} z = E_{\dagger} A E^{\dagger} E^{\ddagger} z = \theta_{\dagger} \lambda_{\dagger} E^{\ddagger} z \end{aligned}$$

and since $\theta_{\dagger}$ is one-to-one this implies that $\lambda_{\dagger} E^{\ddagger} y = \lambda_{\dagger} E^{\ddagger} z$. $\Phi(\dagger, \ddagger)$ is clearly linear and

$$\theta_{\dagger} \Phi(\dagger, \ddagger) \lambda_{\ddagger} y = \theta_{\dagger} \lambda_{\dagger} E^{\ddagger} y = E_{\dagger} A E^{\dagger} E^{\ddagger} y = E_{\dagger} E_{\ddagger} A E^{\ddagger} y = E_{\dagger} \theta_{\ddagger} \lambda_{\ddagger} y$$

verifying (i). Now, since the state decomposition is completely observable, there exists $\varepsilon > 0$ such that $\|\theta_{\dagger} v\| \geq \varepsilon \|v\|$; hence for $v = \Phi(\dagger, \ddagger)x$, x in the range of $\lambda_{\ddagger}$,

$$\varepsilon \|\Phi(\dagger, \ddagger)x\| \leq \|\theta_{\dagger} \Phi(\dagger, \ddagger)x\| = \|E_{\dagger} \theta_{\ddagger} x\| \leq \|\theta_{\ddagger}\| \, \|x\|$$

verifying that $\Phi(\dagger, \ddagger)$ is bounded with norm less than or equal to $\|\theta_{\ddagger}\|/\varepsilon$. Since $\Phi(\dagger, \ddagger)$ is bounded and the complete controllability for the state decomposition implies that the range of $\lambda_{\ddagger}$ is dense in $S_{\ddagger}$, the domain of $\Phi(\dagger, \ddagger)$ may thus be extended to all of $S_{\ddagger}$. To verify (ii) we note that for $\dagger = \ddagger$ (i) implies that

$$\theta_{\dagger} \Phi(\dagger, \dagger) \lambda_{\dagger} = \theta_{\dagger} \lambda_{\dagger}$$

which together with the fact that $\theta_{\dagger}$ is one-to-one and $\lambda_{\dagger}$ has dense range implies that $\Phi(\dagger, \dagger) = I$. For (iii), observe from (i) that

$$\begin{aligned} \theta_{\dagger} \Phi(\dagger, \times) \lambda_{\times} &= \theta_{\dagger} \lambda_{\dagger} E^{\times} = \theta_{\dagger} \lambda_{\dagger} E^{\ddagger} E^{\times} = \theta_{\dagger} \Phi(\dagger, \ddagger) \lambda_{\ddagger} E^{\times} \\ &= \theta_{\dagger} \Phi(\dagger, \ddagger) \Phi(\ddagger, \times) \lambda_{\times} \end{aligned}$$

As before, the fact that $\theta_{\dagger}$ is one-to-one and $\lambda_{\times}$ has dense range now implies that $\Phi(\dagger, \times) = \Phi(\dagger, \ddagger)\Phi(\ddagger. \times)$, as required. //

16. Remark: Note that condition (i) of Property 15 is equivalent to the requirement that the diagram

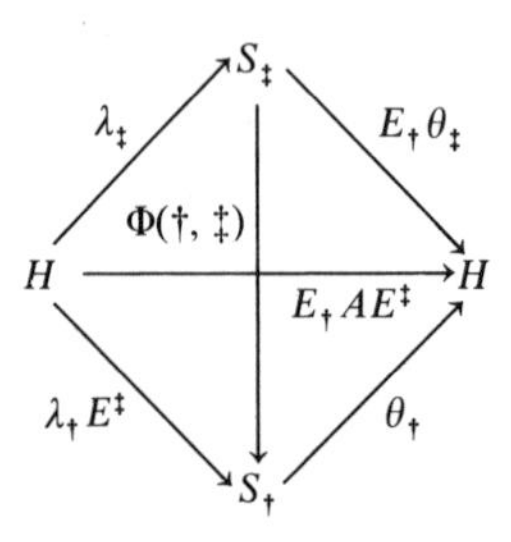

be commutative, condition (ii) can be obtained from the corresponding diagram for $\Phi(\dagger, \dagger)$, and condition (iii) can be obtained by piecing together the diagrams for $\Phi(\dagger, \ddagger)$ and $\Phi(\ddagger, \text{✗})$ to obtain the diagram for $\Phi(\dagger, \text{✗})$.

17. Example: For the differential operator of Example A.4, $E_\dagger\theta_\ddagger x(\ddagger)$ is given by

$$[E_\dagger\theta_\ddagger x(\dagger)](\text{✗}) = \begin{cases} Ce^{A(\text{✗} - \ddagger)}x(\ddagger), & \text{✗} \geq \dagger \\ 0, & \text{✗} < \dagger \end{cases}$$

while $\theta_\dagger \Phi(\dagger, \ddagger)x(\ddagger)$ is given by

$$[\theta_\dagger \Phi(\dagger, \ddagger)x(\ddagger)](\text{✗}) = \begin{cases} Ce^{A(\text{✗} - \dagger)}\Phi(\dagger, \ddagger)x(\ddagger), & \text{✗} \geq \dagger \\ 0, & \text{✗} < \dagger \end{cases}$$

Therefore

$$Ce^{A(\text{✗} - \dagger)}e^{A(\dagger - \ddagger)}x(\ddagger) = Ce^{A(\text{✗} - \ddagger)}x(\ddagger) = Ce^{A(\text{✗} - \dagger)}\Phi(\dagger, \ddagger)x(\dagger), \qquad \text{✗} \geq \dagger$$

for all $x(\ddagger) \in S_\ddagger$. Hence with the aid of the observability assumption we obtain

$$e^{A(\dagger - \ddagger)}x(\ddagger) = \Phi(\dagger, \ddagger)x(\ddagger)$$

for all $x(\ddagger) \in S_\ddagger$, verifying our initial intuition to the effect that $\Phi(\dagger, \ddagger) = e^{A(\dagger - \ddagger)}$ for a differential operator.

18. Example: Unlike in the above situation in which $e^{A(\dagger - \ddagger)}$ is invertible for all $\dagger$ and $\ddagger$ and hence extendable to a group of operators, in general $\Phi(\dagger, \ddagger)$ may fail to be invertible. Indeed, for Example A.5, $\Phi(n, m): R^{m+1} \to R^{n+1}$, $n > m$, and thus cannot be invertible. Indeed, a little algebra with the defining equalities for $\Phi(m, n)$, $m \geq n$, reveals that $\Phi(n, m)$ is represented by the $(n + 1) \times (m + 1)$ matrix

$$\Phi(n, m) = \begin{bmatrix} 1 & 0 & 0 & 0 & \cdots & 0 & 0 \\ 0 & 1 & 0 & 0 & \cdots & 0 & 0 \\ 0 & 0 & 1 & 0 & \cdots & 0 & 0 \\ 0 & 0 & 0 & 1 & \cdots & 0 & 0 \\ \vdots & \vdots & \vdots & \vdots & & \vdots & \vdots \\ 0 & 0 & 0 & 0 & \cdots & 1 & 0 \\ 0 & 0 & 0 & 0 & \cdots & 0 & 1 \\ 0 & 0 & 0 & 0 & \cdots & 0 & 0 \\ \vdots & \vdots & \vdots & \vdots & \cdots & \vdots & \vdots \\ 0 & 0 & 0 & 0 & \cdots & 0 & 0 \end{bmatrix}$$

C. CHARACTERIZATION THEORY

1. Remark: For Property B.8 it was shown that every causal linear operator A admits a minimal state decomposition $\{(S_\dagger, \lambda_\dagger \theta_\dagger) : \dagger \in \$\}$ such that

(i) $\lambda_\dagger = \lambda_\dagger E^\dagger$ has dense range for all $\dagger \in \$$;

(ii) $\theta_\dagger = E_\dagger \theta_\dagger$, and for each $\dagger \in \$$ there exists $\varepsilon > 0$ such that $\|\theta_\dagger x\| \geq \varepsilon \|x\|$, $x \in S_\dagger$;

(iii) a set of transition operators satisfying conditions (i)–(iii) of Property B.15 exists for all $\dagger, \ddagger \in \$$; and

(iv) there exists $M > 0$ such that for any partition $\mathscr{P} = \{E^i : i = 0, 1, \ldots, n\}$

$$\left\| \sum_{i=1}^{n} \Delta^i \theta_{i-1} \lambda_{i-1} \right\| < M$$

(i) and (ii) follow from Property B.8, (iii) follows from B.15, and (iv) may be verified by

$$\begin{aligned} \left\| \sum_{i=1}^{n} \Delta^i \theta_{i-1} \lambda_{i-1} \right\| &= \left\| \sum_{i=1}^{n} \Delta^i E_{i-1} A E^{i-1} \right\| = \left\| \sum_{i=1}^{n} \Delta^i A E^{i-1} \right\| \\ &= \left\| \sum_{i=1}^{n} \Delta^i A E^i - \sum_{i=1}^{n} \Delta^i A \Delta^i \right\| = \left\| A - \sum_{i=1}^{n} \Delta^i A \Delta^i \right\| \\ &\leq \|A\| + \left\| \sum_{i=1}^{n} \Delta^i A \Delta^i \right\| \leq 2\|A\| = M \end{aligned}$$

where we have invoked the fact that $\sum_{i=1}^{n} \Delta^i A E^i = A$ and Lemma 2.A.4. The purpose of the present section is to verify that any family of Hilbert spaces and operators $\{(S_\dagger, \lambda_\dagger \theta_\dagger) : \dagger \in \$\}$ satisfying (i)–(iv) is the state decomposition for some causal operator. In particular, this will justify Example A.5, in which we specified a state decomposition but not the operator to which it applies.

2. Property: Given $\{(S_\dagger, \lambda_\dagger, \theta_\dagger) : \dagger \in \$\}$ satisfying (i)–(iv) of Remark 1, there exists a causal $A \in \mathscr{C}$ such that $\{(S_\dagger, \lambda_\dagger, \theta_\dagger) : \dagger \in \$\}$ is a state decomposition for A. Moreover, if $\{(S_\dagger, \lambda_\dagger, \theta_\dagger) : \dagger \in \$\}$ is also a state decomposition for $A' \in \mathscr{C}$, then $A - A' \in \mathscr{M}$.

Proof: Given a partition $\mathscr{P} = \{E^i : i = 0, 1, \ldots, n\}$, define an operator by

$$A_{\mathscr{P}} = \sum_{i=1}^{n} \Delta^i \theta_{i-1} \lambda_{i-1}$$

Now $\|A_{\mathscr{P}}\| \leq M$ by condition (iv); hence the net $A_{\mathscr{P}}$ admits a weakly convergent subnet $A_{\mathscr{P}_2}$ with limit point A, $\|A\| \leq M$. Since $A_{\mathscr{P}_2}$ converges to A the subnet $A_{\mathscr{P}_2}$ of $A_{\mathscr{P}_{2\dagger}}$, defined by those partitions that contain $E^\dagger$, also converges to A. We therefore have

$$\begin{aligned} E^\dagger A &= E^\dagger[\text{w}\lim_{\mathscr{P}_2} A_{\mathscr{P}_2}] = \text{w}\lim_{\mathscr{P}_2}[E^\dagger A_{\mathscr{P}_2}] = \text{w}\lim_{\mathscr{P}_2}[E^\dagger A_{\mathscr{P}_2} E^\dagger] \\ &= E^\dagger[\text{w}\lim_{\mathscr{P}_2} A E^\dagger] = E^\dagger A E^\dagger \end{aligned}$$

showing that $A \in \mathscr{C}$. Note, in the above derivation we have used the fact that

$$\begin{aligned} E^\dagger A_{\mathscr{P}_2} &= E^\dagger\left[\sum_{i=1}^{n} \Delta^i \theta_{i-1} \lambda_{i-1}\right] = E^\dagger\left[\sum_{i=1}^{n} \Delta^i A E^{i-1}\right] = E^\dagger\left[\sum_{i=1}^{n} \Delta^i A E^{i-1}\right] E^\dagger \\ &= E^\dagger A_{\mathscr{P}_2} E^\dagger \end{aligned}$$

for any partition $\mathscr{P}_2$ that contains $E^\dagger$.

It remains to show that $\{(S_\dagger, \lambda_\dagger, \theta_\dagger) : \dagger \in \$\}$ is a state decomposition for A, i.e., that $E_\dagger A E^\dagger = \theta_\dagger \lambda_\dagger$. By construction it follows that for any $x, y \in H$ and $\varepsilon > 0$ there exists a partition $\mathscr{P}_2$ in the subset such that, for any partition $\mathscr{P}_2 = \{E^i : i = 0, 1, \ldots, n\}$ in the subnet that refines $\mathscr{P}_2$,

$$\left| \left(\left[A - \sum_{i=1}^{n} \Delta^i \theta_{i-1} \lambda_{i-1} \right] x, y \right) \right| < \varepsilon$$

Thus if we let $y = E_\dagger y'$ and $x = E^\dagger x'$, $y', x' \in H$, we have

$$\varepsilon > \left|\left(\left[A - \sum_{i=1}^{n} \Delta^i \theta_{i-1}\lambda_{i-1}\right]E^\dagger x', E_\dagger y'\right)\right|$$

$$= \left|\left(\left[E_\dagger AE^\dagger - \sum_{i=1}^{n} \Delta^i E_\dagger \theta_{i-1}\lambda_{i-1}E^\dagger\right]x', y'\right)\right|$$

$$= \left|\left(\left[E_\dagger AE^\dagger - \sum_{i=0}^{j-1} \Delta^i \theta_\dagger \Phi(\dagger, i-1)\lambda_{i-1} - \sum_{i=j}^{n} \Delta^i \theta_{i-1}\Phi(i-1, \dagger)\lambda_\dagger\right]x', y'\right)\right|$$

$$= \left|\left(\left[E_\dagger AE^\dagger - \sum_{i=0}^{j-1} \Delta^i \theta_\dagger \lambda_\dagger E^{i-1} - \sum_{i=j}^{n} \Delta^i E_{i-1}\theta_\dagger \lambda_\dagger\right]x', y'\right)\right|$$

$$= \left|\left(\left[E_\dagger AE^\dagger - \sum_{i=1}^{n} \Delta^i \theta_\dagger \lambda_\dagger\right]x', y'\right)\right| = |([E_\dagger AE^\dagger - \theta_\dagger \lambda_\dagger]x', y')|$$

showing that $\theta_\dagger \lambda_\dagger = E_\dagger AE^\dagger$.

Finally, assume that A and A' are causal operators such that

$$E_\dagger AE^\dagger = \theta_\dagger \lambda_\dagger = E_\dagger A'E^\dagger, \qquad \dagger \in \$$$

Thus $E_\dagger[A - A']E^\dagger = 0$, which implies that

$$\begin{aligned}[A - A']E^\dagger &= [E^\dagger + E_\dagger][A - A']E^\dagger \\ &= E^\dagger[A - A']E^\dagger + E_\dagger[A - A']E^\dagger = E^\dagger[A - A']E^\dagger\end{aligned}$$

showing that $[A - A'] \in \mathscr{C}^*$, which together with $[A - A'] \in \mathscr{C}$ implies that $[A - A'] \in \mathscr{M}$, as required. //

3. Example: Consider the state decomposition on $l_2[0, \infty)$ with its usual resolution structure, constructed in Example A.5. Here $S_n = R^{n+1}$,

$$\lambda_n a = \text{col}\,[(a, e_0), (a, e_1), \ldots, (a, e_n)]$$

and

$$\theta_n(\text{col}\,[x_0, x_1, \ldots, x_x]) = \sum_{i=0}^{n} x_i \hat{e}_{n+1+i}$$

where $\{e_i\}_0^\infty$ is the standard basis for $l_2[0, \infty)$ and $\hat{e}_i = e_i/e^i$. Conditions (i) and (ii) for Property 2 were verified in Example B.4, whereas condition (iii) was verified in Example B.18. It thus remains to show that there exists $M > 0$ such that

$$\left\|\sum_{i=1}^{n} \Delta^i \theta_{i-1}\lambda_{i-1}\right\| < M$$

Indeed, for a partition $\mathscr{P} = \{E^{\dagger}i : i = 0, 1, \ldots, n\}$

$$\sum_{i=1}^{n} \Delta^i \theta_{i-1} \lambda_{i-1} = \sum_{i=1}^{n} \sum_{j=0}^{\dagger_i - \dagger_{i-1} - 1} (a, e_j) \hat{e}_{\dagger_{i-1}+1+j}$$

Hence $M = e/(e-1)$ will suffice. Moreover, on taking the limit over the above net we obtain

$$Aa = \lim_{\mathscr{P}} \left[\sum_{i=1}^{n} \Delta^i \theta_{i-1} \lambda_{i-1} \right] a = \sum_{i=1}^{\infty} (a, e_0) \hat{e}_i$$

where A is characterized by the matrix representation

$$A = \begin{bmatrix} 0 & 0 & 0 & 0 & \cdots \\ e^{-1} & 0 & 0 & 0 & \cdots \\ e^{-2} & 0 & 0 & 0 & \cdots \\ e^{-3} & 0 & 0 & 0 & \cdots \\ \vdots & \vdots & \vdots & \vdots & \end{bmatrix}$$

PROBLEMS

1. Formulate a state decomposition for the difference operator

$$x_{k+1} = Ax_k + Ba_k, \qquad x_0 = 0$$
$$b_k = Cx_k,$$

defined on the Hilbert resolution space $l_2[0, m]$ with its usual resolution structure. Determine conditions under which such a state decomposition is minimal.

2. Formulate a costate decomposition for the anticausal operator on $L_2[0, 1]$ with its usual resolution structure defined by the differential operator

$$\dot{x}(\dagger) = Ax(\dagger) + Bf(\dagger), \qquad x(1) = 0$$
$$g(\dagger) = Cx(\dagger),$$

3. Assume that $\{(S_\dagger, \lambda_\dagger, \theta_\dagger) : \dagger \in \$\}$ is a completely observable state decomposition for $A \in \mathscr{B}$ and show that

$$x(\dagger) = [\theta_\dagger^* \theta_\dagger]^{-1} \theta_\dagger^* [E_\dagger y - E_\dagger A E_\dagger u]$$

where $y = Au$.

4. A state decomposition $\{(S_\dagger, \lambda_\dagger, \theta_\dagger) : \dagger \in \$\}$ is said to be *uniformly observable* if there exists $\varepsilon > 0$ such that $\|\theta_\dagger x\| \geq \varepsilon \|x\|$ for all $x \in S_\dagger$ and $\dagger \in \$$. Show that every $A \in \mathscr{B}$ admits a state decomposition which is completely controllable and uniformly observable.

5. Given two differential operators

$$\begin{aligned} \dot{x}(\dagger) &= Ax(\dagger) + Bf(\dagger), \\ & \qquad\qquad\qquad\qquad x(0) = 0 \\ g(\dagger) &= Cx(\dagger), \end{aligned}$$

and

$$\begin{aligned} \dot{x}'(\dagger) &= A'x'(\dagger) + B'f(\dagger), \\ & \qquad\qquad\qquad\qquad x'(0) = 0 \\ g(\dagger) &= C'x'(\dagger), \end{aligned}$$

show that they define the same operator on $L_2[0, 1]$ with its usual resolution structure if and only if

$$CA^iB = C'A'^iB', \qquad i = 0, 1, \ldots, n - 1$$

Furthermore, if both differential operators define minimal state decompositions as in Example A.4, show that there exists an $n \times n$ matrix T such that

$$C' = CT^{-1}, \qquad B' = TB, \qquad A' = TAT^{-1}$$

6. Let $\{(S_\dagger, \lambda_\dagger, \theta_\dagger) : \dagger \in \$\}$ be a state decomposition for $A \in \mathscr{B}$ and show that

$$E_\dagger \theta_\ddagger \lambda_\ddagger = \theta_\dagger \lambda_\dagger E^\ddagger$$

whenever $\ddagger \leq \dagger$.

7. Verify that the $\Phi(m, n)$ matrix of Example B.18 is indeed the state transition matrix for the decomposition of Example A.5.

REFERENCES

1. Nerode, A., Linear automation transformations, *Proc. Amer. Math. Soc.* **9**, 441–444 (1958).
2. Saeks, R., State in Hilbert space, *SIAM Rev.* **15**, 283–308 (1973).
3. Schnure, W., Ph.D. dissertation, Univ. of Michigan, Ann Arbor, Michigan (1974).

7

State Trajectory Factorization

A. STATE TRAJECTORIES

1. Remark: In Chapter 6 we constructed a factorization of $E_\dagger A E^\dagger = \theta_\dagger \lambda_\dagger$ through an appropriate state space $S_\dagger$. Here we would like to extend that factorization to a factorization of A through an appropriate *state trajectory space*, $A = \theta\lambda$, where for any input $u \in H$, λu is the state-valued function defined by $[\lambda u](\dagger) = x(\dagger)$. Since $S_\dagger$ is dependent on $\dagger$, λu is actually a section of $\pi[S_\dagger]$ rather than a function, though the distinction does not prove to be significant in the present context. Consistent with Property 6.A.7 we restrict consideration to strongly strictly causal $A \in \mathscr{S}$, and we assume that a minimal state decomposition $\{(S_\dagger, \theta_\dagger, \lambda_\dagger) : \dagger \in \$\}$ for A has been specified. In that case λ is uniquely defined by

$$[\lambda u](\dagger) = \lambda_\dagger u = x(\dagger)$$

In fact, one can construct a more convenient *integral representation* for λ as follows.

2. Property: Let $A \in \mathscr{S}$ and $\{(S_\dagger, \lambda_\dagger, \theta_\dagger) : \dagger \in \$\}$ be a minimal state decomposition for A. Then

$$\lambda = \mathrm{s}\,(M) \int_{\dagger_0}^{\dagger} \Phi(\dagger, \ddagger)\lambda_\ddagger \, dE(\ddagger)$$

Proof: For any $u \in H$ and partition $\{E^i : i = 0, 1, \ldots, n\}$,

$$[\lambda u](\dagger) = \lambda_\dagger u = \lambda_\dagger E^\dagger u = \lambda_\dagger \left[\sum_{i=1}^{n} \Delta^i\right] E^\dagger u = \left[\sum_{i=1}^{n} \lambda_\dagger E^i \Delta^i\right] E^\dagger u$$

$$= \left[\sum_{i=1}^{n} \Phi(\dagger, i)\lambda_i \Delta^i\right] E^\dagger u.$$

Since this holds for all u and is independent of the choice of partition, these partial sums converge strongly to

$$\lambda = \mathrm{s}\,(M) \int_{\dagger_0}^{\dagger} \Phi(\dagger, \ddagger)\lambda_\ddagger \, dE(\ddagger)$$

as was to be shown. //

3. Remark: If one expresses u in the form $u = E^{\text{✗}}u + E_{\text{✗}}u$ and applies Property 2, an alternative integral representation is obtained:

$$x(\dagger) = [\lambda u](\dagger) = \lambda_\dagger u = \lambda_\dagger E^{\text{✗}} u + \lambda_\dagger E_{\text{✗}} u = \Phi(\dagger, \text{✗})\lambda_{\text{✗}} u + \lambda_\dagger[E_{\text{✗}} u]$$

$$= \Phi(\dagger, \text{✗})x(\text{✗}) + \mathrm{s}\,(M) \int_{\text{✗}}^{\dagger} \Phi(\dagger, \ddagger)\lambda_\ddagger \, dE(\ddagger)u$$

whenever $\text{✗} \leq \dagger$. Hence the state at time $\dagger$ is entirely determined by the state at time ✗ and the input applied in the interval $[\text{✗}, \dagger]$. Also note the similarity between this representation and the classical representation for the state of the differential operator

$$\dot{x}(\dagger) = Ax(\dagger) + Bf(\dagger), \qquad x(0) = 0$$

$$g(\dagger) = Cx(\dagger)$$

of Example 6.A.4. Indeed, in that case

$$x(\dagger) = e^{A(\dagger - \text{✗})}\, x(\text{✗}) + \int_{\text{✗}}^{\dagger} e^{A(\dagger - \ddagger)} Bf(\ddagger)\, d\ddagger$$

where $e^{A(\dagger - \text{✗})}$ is the appropriate matrix representation for $\Phi(\dagger, \text{✗})$.

4. Property: Let $A \in \mathscr{S}$ have minimal state decomposition $\{(S_\dagger, \lambda_\dagger, \theta_\dagger) : \dagger \in \$\}$ and let $x(\dagger) = [\lambda u](\dagger)$ and $y = Au$. Then

$$y = \mathrm{s}\,(m) \int_{\dagger_0}^{\dagger_\infty} dE(\dagger)\, \theta_\dagger x(\dagger)$$

Proof: If $x(\dagger) = [\lambda u](\dagger) = \lambda_\dagger u$, then

$$\begin{aligned} \mathrm{s}\,(m) \int_{\dagger_0}^{\dagger_\infty} dE(\dagger)\, \theta_\dagger x(\dagger) &= \mathrm{s}\,(m) \int_{\dagger_0}^{\dagger_\infty} dE(\dagger)\, \theta_\dagger \lambda_\dagger u \\ &= \mathrm{s}\,(m) \int_{\dagger_0}^{\dagger_\infty} dE(\dagger)\, E_\dagger A E^\dagger u \\ &= \mathrm{s}\,(m) \int_{\dagger_0}^{\dagger_\infty} dE(\dagger)\, A E^\dagger u = Au = y \end{aligned}$$

by Corollary 2.B.5. //

5. Remark: Consistent with Property 4 we may now formulate our desired stable trajectory factorization of A in the form $A = \theta\lambda$, where λ is defined in Remark 1 and θ is defined by

$$\theta x = \mathrm{s}\,(m) \int_{\dagger_0}^{\dagger_\infty} dE(\dagger)\, \theta_\dagger x(\dagger)$$

It is important, however, to note that θ is only defined on the trajectories of A while x takes its values in $\pi[S_\dagger]$. Thus even though we have constructed the desired factorization it remains to formulate an appropriate state trajectory space that includes the range of λ but is small enough to permit θ to be extended to the entire space.

6. Example: For the differential operator of Example 6.A.4, $\theta_\dagger$: $R^n \to L_2[0, 1]$ is defined by

$$\theta_\dagger x(\dagger) = \begin{cases} Ce^{A(\ddagger - \dagger)}x(\dagger), & \ddagger \geq \dagger \\ 0, & \ddagger < \dagger \end{cases}$$

Hence

$$[\theta x](\ddagger) = \lim_{\mathscr{P}} \left[\sum_{i=1}^{n} \Delta^i C e^{A(\ddagger - \dagger_{i-1})} x(\dagger_{i-1}) \right]$$

Now with the usual resolution structure on $L_2[0, 1]$, Δ^i is the operator that multiplies a function by the characteristic function of the interval $(\dagger_{i-1}, \dagger_i]$. Hence if $\ddagger \in (\dagger_{j-1}, \dagger_j]$,

$$\left[\sum_{i=1}^{n} \Delta^i C e^{A(\ddagger - \dagger_{i-1})} x(\dagger_{i-1})\right] = C e^{A(\ddagger - \dagger_{j-1})} x(\dagger_{j-1})$$

Moreover, for a sufficiently fine partition $e^{A(\ddagger - \dagger)} \simeq I$ for $\ddagger \in (\dagger_{j-1}, \dagger_j]$; hence

$$[\theta x](\ddagger) = \lim_{\mathscr{P}} [C e^{A(\ddagger - \dagger_{j-1})} x(\dagger_{j-1}) = \lim_{\mathscr{P}} [C x(\dagger_{j-1})] = C x(\ddagger)$$

where we have used the continuity of the state trajectory to justify the final limiting process. As such, θ is just the multiplication operator defined by C, mapping a section of $\pi[R^n]$ to $L_2[0, 1]$. In this case it is clear that θ admits an extension to all of $\pi[R^n]$, thereby permitting any subspace of $\pi[R^n]$ containing all state trajectories to serve as the state trajectory space. This is, however, not true in general.

7. Remark: The problem of constructing a state trajectory space would be greatly simplified if one had some type of bound on $\|x(\dagger)\|$. Although this is not the case in general, if one works with the minimal state decomposition constructed in Property 6.B.8 in which $\theta_\dagger$ is an isometry, then

$$\|x(\dagger)\| = \|\lambda_\dagger u\| = \|\theta_\dagger \lambda_\dagger u\| = \|E_\dagger A E^\dagger u\| \le \|A\| \, \|u\|$$

showing that $x(\dagger)$ is bounded. Therefore every state trajectory will be contained in the subspace composed of sections of $\pi[R^n]$ that are square integrable with respect to any finite Borel measure on $ (Borel with respect to the order topology on $).

B. THE STATE TRAJECTORY SPACE

1. Remark: An initial requirement in the formulation of a state trajectory space for $\{(S_\dagger, \lambda_\dagger, \theta_\dagger) : \dagger \in \$\}$ is that it includes all state trajectories λu, $u \in H$. Furthermore, since the trajectory space must have a resolution structure, we include the truncates of the trajectories

$$[F^\dagger x](\ddagger) = \begin{cases} x(\ddagger), & \ddagger \le \dagger \\ 0, & \ddagger > \dagger \end{cases}$$

and all possible linear combinations thereof to guarantee that the trajectory space will be linear.

2. Definition: For an operator $A \in \mathscr{S}$ with state decomposition $\{(S_\dagger, \lambda_\dagger, \theta_\dagger) : \dagger \in \$\}$ its state trajectory space is the pair $(S, \mathscr{F})$ where

$$S = \left\{ \sum_{i=1}^{n} a_i F^{\dagger_i} \lambda u_i \in \pi[S] : a_i \in R,\ \dagger_i \in \$,\ u_i \in H;\ i = 1, 2, \ldots, n \right\}$$

and

$$F^\dagger \left[\sum_{i=1}^{n} a_i F^{\dagger_i} \lambda u_i \right] = \left[\sum_{i=1}^{n} a_i F^{\ddagger_i} u_i \right]$$

where $\ddagger_i = \min\{\dagger, \dagger_i\}$.

3. Remark: It is important to note that $(S, \mathscr{F})$ is an algebraic resolution space rather than a Hilbert resolution space. Although a number of researchers have attempted to artificially impose a Hilbert space structure on $(S, \mathscr{F})$, there is no natural topology on $(S, \mathscr{F})$ and we therefore prefer to work with the trajectory space in its "natural" algebraic form.[1]

4. Example: In the state decomposition of Example 6.A.5 on $l_2[0, \infty)$ with its usual resolution structure, $\lambda_n : l_2[0, \infty) \to R^{n+1}$:

$$\lambda_n a = \operatorname{col}[(a, e_0), (a, e_1), (a, e_2), \ldots, (a, e_n)]$$

Thus λa is the semi-infinite triangular matrix

$$\lambda a = \begin{bmatrix} (a, e_0) & 0 & 0 & 0 & \cdots \\ (a, e_0) & (a, e_1) & 0 & 0 & \cdots \\ (a, e_0) & (a, e_1) & (a, e_2) & 0 & \cdots \\ (a, e_0) & (a, e_1) & (a, e_2) & (a, e_3) & \cdots \\ \vdots & \vdots & \vdots & \vdots & \end{bmatrix}$$

and the set of state trajectories is just the set of lower triangular matrices with constant columns. Now, if one truncates this trajectory using F^n, the resultant truncated trajectory is zero below the nth row. Linear combinations of such trajectories and truncated trajectories will thus result in a lower triangular semi-infinite matrix whose columns are eventually constant. The set of such matrices thus constitutes S.

5. Remark: Given the above formulation of the state trajectory space with $\lambda : H \to S$ we would like to extend θ to a mapping defined on all of S, $\theta : S \to H$. It will be necessary to deal both with causality concepts formulated in an algebraic resolution space and with operators mapping one such space to another. Since the concepts of memorylessness and causality are algebraic in nature, they extend without modification to an algebraic resolution space setting and in the obvious manner to mappings from one such space to another. That is, if $A: H \to H'$ where $(H, \mathscr{E})$ and $(H', \mathscr{E}')$ are algebraic resolution spaces (which may also have a Hilbert space structure) then A is memoryless if and only if

$$E'^{\dagger}A = AE^{\dagger}, \qquad \dagger \in \$$$

and causal if and only if

$$E'^{\dagger}A = E'^{\dagger}AE^{\dagger}, \qquad \dagger \in \$$$

Unfortunately, most of the hypercausality concepts of Chapter 2 are topological in nature and thus do not extend to an algebraic resolution space setting. We can, however, define a class of strictly causal operators in an algebraic resolution space by working with the radical of the causals whenever $(H, \mathscr{E}) = (H', \mathscr{E}')$. Alternatively, we may work with the algebraic strict causality concept of Definition 2.D.12. Indeed, the latter seems to be best suited to our application and is employed throughout the chapter.

6. Property: Let $A \in \mathscr{S}$ admit a minimal state decomposition $\{(S_{\dagger}, \lambda_{\dagger}, \theta_{\dagger}): \dagger \in \$\}$ such that $\Phi(\dagger, \ddagger)$ is one-to-one for all $\ddagger \geq \dagger$. Then $\lambda: H \to S$ is algebraically strictly causal.

Proof: If $F_{\dagger}\lambda E^{\dagger}u = 0$ for all $\dagger \leq \ddagger$, then

$$\lambda_{\ddagger} E^{\dagger}u = [\lambda E^{\dagger}u](\ddagger) = [F_{\dagger}\lambda E^{\dagger}u](\ddagger) = 0, \qquad \dagger \leq \ddagger$$

Now

$$0 = \lambda_{\ddagger} E^{\dagger}u = 0(\ddagger, \dagger)\lambda_{\dagger} E^{\dagger}u = 0(\ddagger, \dagger)\lambda_{\dagger} u, \qquad \dagger \leq \ddagger$$

which implies that $x(\dagger) = \lambda_{\dagger} u = 0$, $\dagger \leq \ddagger$, since $\Phi(\ddagger, \dagger)$ is one-to-one. Therefore $F^{\ddagger}x = 0$ and

$$F^{\ddagger}\lambda u = F^{\ddagger}x = 0$$

showing that $\lambda \in \mathscr{V}$. //

7. Remark: The requirement that $\Phi(\ddagger, \dagger)$ be one-to-one in effect represents a strengthening of the minimality condition, and we say that a minimal state decomposition is *strongly minimal* if $\Phi(\ddagger, \dagger)$ is one-to-one, $\dagger \leq \ddagger$. This condition will follow upon strengthening either the complete controllability or the complete observability condition for a minimal state decomposition. For instance, we may require that

$$\text{Ker}[\lambda_{\dagger}] = \text{Ker}[\lambda_{\ddagger} E^{\dagger}], \qquad \dagger \leq \ddagger$$

or equivalently that $E_{\ddagger}\theta_{\dagger}$ be one-to-one for $\dagger \leq \ddagger$.

8. Property: Let $A \in \mathscr{S}$ admit a minimal state decomposition $\{(S_\dagger, \lambda_\dagger, \theta_\dagger) : \dagger \in \$\}$. Then θ admits a linear memoryless extension defined on S such that $A = \theta\lambda$.

Proof: From Definition 2 a typical element of S takes the form $[\sum_{i=1}^n a_i F^{\dagger_i} \lambda u_i]$, and hence we define $\theta[\sum_{i=1}^n a_i F^{\dagger_i} \lambda u_i]$ by

$$\theta\left[\sum_{i=1}^{n} a_i F^{\dagger_i} \lambda u_i\right] = \left[\sum_{i=1}^{n} a_i E^{\dagger_i} A u_i\right]$$

To show that this is well defined, it suffices to show that $E^\ddagger A u = 0$ whenever $F^\ddagger \lambda u = 0$. The latter condition is, however, equivalent to the requirement that $\lambda_\dagger u = 0$, $\dagger \leq \ddagger$, so

$$E_\dagger A E^\dagger u = \theta_\dagger \lambda_\dagger u = 0, \qquad \dagger \leq \ddagger$$

and hence by Property 2.D.14 the strong strict causality of A implies that $E^\ddagger A u = 0$. Furthermore, for the state trajectory $x = \lambda u$, $\theta x = A u = \theta \lambda u$ by Remark A.5, showing that the above definition coincides with that of the θ operator on the state trajectories with Property A.4 and justifying the continued use of the θ notation.

Finally, for $s = [\sum_{i=1}^n a_i F^{\dagger_i} \lambda u_i]$

$$E^\dagger \theta s = E^\dagger \theta\left[\sum_{i=1}^{n} a_i F^{\dagger_i} \lambda u_i\right] = E^\dagger\left[\sum_{i=1}^{n} a_i E^{\dagger_i} A u_i\right] = \left[\sum_{i=1}^{n} a_i E^{\dagger_i} A u_i\right]$$

$$= \left[\sum_{i=1}^{n} a_i F^{\ddagger_i} \lambda u_i\right] = \theta F^\dagger\left[\sum_{i=1}^{n} a_i F^{\dagger_i} \lambda u_i\right] = \theta F^\dagger s$$

where $\ddagger_i = \min\{\dagger, \dagger_i\}$, verifying that θ is memoryless. //

9. Remark: Consistent with the above development, if $A \in \mathscr{S}$ and $\{(S_\dagger, \lambda_\dagger, \theta_\dagger) : \dagger \in \$\}$ is a strong minimal state decomposition for A, then A admits the state trajectory factorization, $A = \theta\lambda$, where $\lambda : H \to S$ is algebraically strictly causal while $\theta : S \to H$ and is memoryless. Finally, we note that the complete observability condition implies that λu can be uniquely determined from $\theta \lambda u = A u$. Note that this does not imply that θ is one-to-one, only that θ is one-to-one on the range of λ, the state trajectories.

C. LIFTING THEORY

1. Remark: Given $A \in \mathscr{S}$ with a strong minimal state decomposition $\{(S_\dagger, \lambda_\dagger, \theta_\dagger) : \dagger \in \$\}$ and the associated state trajectory factorization $A = \theta\lambda$, where $\lambda : (H, \mathscr{E}) \to (S, \mathscr{F})$ is algebraically strictly causal and

$\theta : (S, \mathscr{F}) \to (H, \mathscr{E})$ is memoryless, we would like to investigate the possibility of lifting an arbitrary strongly strictly causal operator $V \in \mathscr{S}$ through the state trajectory space of A. In particular, when does there exist a memoryless operator $F : (S, \mathscr{F}) \to (H, \mathscr{E})$ such that $V = F\lambda$? Although this might at first seem to be an anomalous question, it proves to be inextricably intertwined with the spectral factorization problem and as such plays a fundamental role in the system theoretic applications to come. The *lifting problem* is described diagramatically as follows:

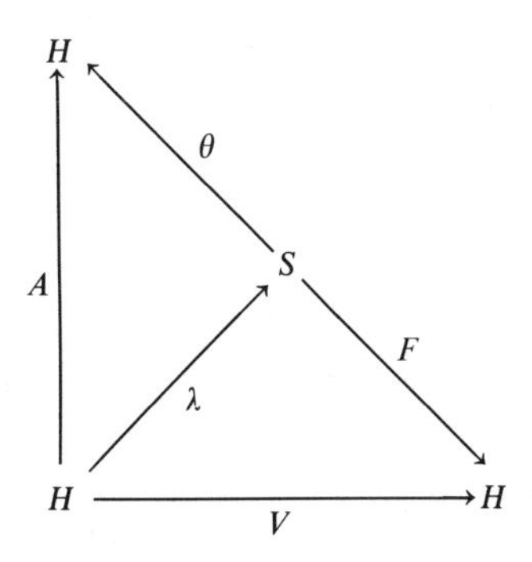

2. Lemma: Assume that $A \in \mathscr{S}$ admits a strong minimal state decomposition $\{(S_\dagger, \lambda_\dagger, \theta_\dagger) : \dagger \in \$\}$ and let $V \in \mathscr{V}$. Then there exists a memoryless operator $F : S \to H$ such that $V = F\lambda$ if and only if

$$\operatorname{Ker}[\lambda_\dagger] \subset \operatorname{Ker}[E_\dagger V E^\dagger], \qquad \dagger \in \$$$

Proof: Assume that $V = F\lambda$, $F \in \mathscr{M}$, and that $\lambda_\dagger u = 0$. Then

$$\lambda_\ddagger E^\dagger u = \Phi(\ddagger, \dagger)\lambda_\dagger E^\dagger u = \Phi(\ddagger, \dagger)\lambda_\dagger u = 0, \qquad \dagger \le \ddagger$$

which in turn implies that $F_\dagger \lambda E^\dagger u = 0$ and

$$E_\dagger V E^\dagger u = E_\dagger F \lambda E^\dagger u = F F_\dagger \lambda E^\dagger u = 0$$

Conversely, if

$$\operatorname{Ker}[\lambda_\dagger] \subset \operatorname{Ker}[E_\dagger V E^\dagger], \qquad \dagger \in \$$$

we may define F on a vector $[\sum_{i=1}^n a_i F^{\dagger_i} \lambda u_i] \in S$ by

$$F\left[\sum_{i=1}^n a_i F^{\dagger_i} \lambda u_i\right] = \left[\sum_{i=1}^n a_i F^{\dagger_i} V u_i\right]$$

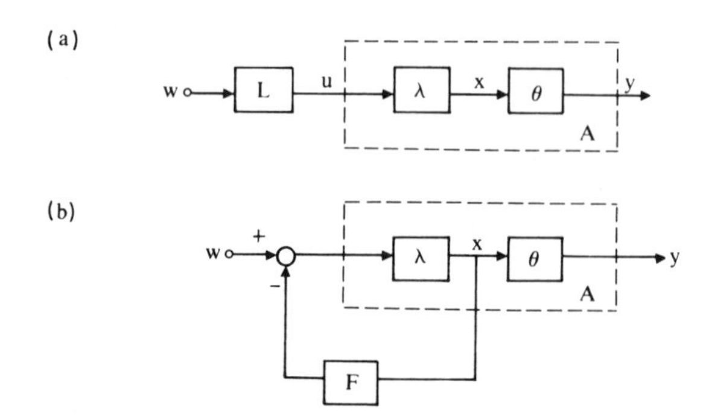

Fig. 1. (a) Open loop system and (b) equivalent memoryless state feedback system.

which is a well-defined linear memoryless map $F: S \to H$ such that $V = F\lambda$ by exactly the same argument used for θ in the proof of Property B.8, completing the proof. //

3. Remark: Although we propose to employ the above lemma in an operator setting, in the present context the result has significant system theoretic implication. Indeed, if one is given an *open loop system*, as shown in Fig. 1a, then the system has an equivalent *memoryless state feedback* implementation, as shown in Fig. 1b, if and only if

(i) L^{-1} exists,
(ii) $L^{-1} - I = V \in \mathscr{V}$, and
(iii) $\mathrm{Ker}[\lambda_\dagger] \subset \mathrm{Ker}[E_\dagger V E^\dagger]$.

Indeed, the equivalence of the block diagrams of Figs. 1a and 1b corresponds to the equality $L = (I + F\lambda)^{-1}$. Thus $L^{-1} = (I + F\lambda)$ exists, $L^{-1} - I = F\lambda \in \mathscr{V}$ by Property 2.D.15, and if $\lambda_\dagger u = 0$, then

$$\lambda_\ddagger E^\dagger u = \Phi(\ddagger, \dagger)\lambda_\dagger E^\dagger u = \Phi(\ddagger, \dagger)\lambda_\dagger u = 0, \qquad \dagger \leq \ddagger$$

which implies that $F_\dagger \lambda E^\dagger u = 0$. Hence

$$E_\dagger F\lambda E^\dagger u = FF_\dagger \lambda E^\dagger u = 0$$

showing that (i)–(iii) are satisfied. Conversely, if L satisfies (i)–(iii), then Lemma 2 implies that there exists a memoryless F such that

$$L^{-1} - I = F\lambda,$$

and hence $L = (I + F\lambda)^{-1}$ as required.

4. Remark: In essence, the lemma implies that if it is possible to lift V through the state trajectory space of A, then the dynamics of V are identical to those of A since $A = \theta\lambda$ and $V = F\lambda$ with both θ and F memoryless. Hence $\lambda \in \mathscr{V}$ characterizes all of the dynamics associated with both operators.

5. Lemma: Assume that $A \in \mathscr{S}$ admits a strong minimal state decomposition $\{(S_\dagger, \lambda_\dagger, \theta_\dagger) : \dagger \in \$\}$ and that $I + A^*A$ admits a regular factorization

$$I + A^*A = B^*B$$

where $B = [M + V]$ is decomposable into the sum of a memoryless operator and an algebraically strictly causal operator. Then

$$\mathrm{Ker}[\lambda_\dagger] \subset \mathrm{Ker}[E_\dagger V E^\dagger]$$

Proof: If $\lambda_\dagger u = 0$, then

$$E_\dagger A E^\dagger u = \theta_\dagger \lambda_\dagger u = 0$$

and

$$\begin{aligned} E_\dagger V E^\dagger u &= [E_\dagger M E^\dagger + E_\dagger V E^\dagger]u = E_\dagger B E^\dagger u = E_\dagger B^{*-1} B^* B E^\dagger u \\ &= E_\dagger B^{*-1} E_\dagger B^* B E^\dagger u = E_\dagger B^{*-1} E_\dagger [I + A^*A] E^\dagger u \\ &= E_\dagger B^{*-1} E_\dagger I E^\dagger u + E_\dagger B^{*-1} E_\dagger A^* A E^\dagger u = E_\dagger B^{*-1} E_\dagger A^* A E^\dagger u \\ &= E_\dagger B^{*-1} E_\dagger A^* E_\dagger A E^\dagger u = [E_\dagger B^{*-1} E_\dagger A^*](0) = 0 \end{aligned}$$

where we have invoked the anticausality of A^* and B^{*-1}. //

6. Remark: The condition that B be decomposable as $B = [M + V]$ is extremely weak. Indeed, for B to admit a decomposition in the form $B = [B]_{\mathscr{M}} + [B]_{\mathscr{S}}$ all that is required is the existence of the strongly convergent integral

$$[B]_{\mathscr{M}} = \mathrm{s}\,(M) \int dE(\dagger)\, B\, dE(\dagger)$$

owing to Property 3.A.4. In fact, since it suffices for V to be algebraically strictly causal, the existence conditions are even less stringent. We now proceed to our main theorem.[1]

Lifting Theorem: Assume that $A \in \mathscr{S}$ admits a strong minimal state decomposition $\{(S_\dagger, \lambda_\dagger, \theta_\dagger) : \dagger \in \$$ and that $I + A^*A$ admits a regular factorization

$$I + A^*A = B^*B$$

where $B = [M + V]$ is decomposable into the sum of a memoryless operator and an algebraically strictly causal operator. Then there exists a memoryless operator $F: S \to H$ such that $V = F\lambda$.

Proof: By Lemma 6 $\mathrm{Ker}[\lambda_\dagger] \subset \mathrm{Ker}[E_\dagger V E^\dagger]$, which is just the hypothesis required by Lemma 2 for the existence of the required memoryless operator. //

7. Example: Consider the differential operator P,

$$\dot{x}(\dagger) = Ax(\dagger) + f(\dagger), \qquad x(0) = 0$$
$$g(\dagger) = Cx(\dagger),$$

where f, g, and x are n vectors of square integrable functions in $L_2[0, 1]^n$ and A and C are $n \times n$ matrices. Adopting an operational notation with $D = d/dt\, I$, this operator takes the form

$$P = C(D - A)^{-1}$$

which has the state trajector factorization $P = \theta\lambda$ where $\theta = C$ by Example A.6, $\lambda = (D - A)^{-1}$, and $S = L_2[0, 1]^n$.

Now let us consider the possibility of constructing a spectral factorization for $I + P^*P$ in the form

$$I + P^*P = (I + V)^*(I + V)$$

From the lifting theorem, $V = F\lambda$ for some memoryless F. Hence we look for a V that takes the form

$$V = F(D - A)^{-1}$$

In our operational notation $P^* = -(D + A^t)^{-1}C^t$ and $V^* = -(D + A^t)^{-1}F^t$, and thus F must be chosen to satisfy

$$\begin{aligned} I + (D + A^t)^{-1}C^tC(D - A)^{-1} &= I + P^*P = (I + V)^*(I + V) \\ &= I - (D + A^t)^{-1}F^t + F(D - A)^{-1} \\ &\quad - (D + A^t)^{-1}F^tF(D - A)^{-1} \end{aligned}$$

Equivalently, we must solve

$$(D + A^t)^{-1}[C^tC + F^tF](D - A)^{-1} = F(D - A)^{-1} - (D + A^t)^{-1}F^t$$

which reduces to

$$[C^tC + F^tF] = (D + A^t)F - F^t(D - A)$$

Furthermore, if we require that F be symmetric, this differential equation reduces to the *algebraic Riccati equation*

$$FA + A^{t}F - FF = -C^{t}C$$

which must be solved for $F = F^{t}$ to compute the required spectral factorization. Indeed, given appropriate controllability assumptions this equation is known to admit a positive definite symmetric solution.

8. Remark: The example illustrates the essential relationship between the lifting theorem in our Hilbert resolution space approach to system theory and the omnipresent matrix Riccati equation of classical system theory. Indeed, our memoryless F operator is a natural generalization of the solution of this equation in the resolution space setting.

PROBLEMS

1. Let $\{(S_{\dagger}, \lambda_{\dagger}, \theta_{\dagger}) : \dagger \in \$\}$ be a uniformly observable (see Problem 6.D.4) state decomposition for $A \in \mathscr{S}$ and show that the resultant state trajectories are bounded.

2. Show that $\mathscr{V}$ is a linear space.

3. Let $\mathscr{R}$ denote the radical of the causal operators in an algebraic resolution space. Determine the relationship between $\mathscr{R}$ and $\mathscr{V}$.

4. Let $[A]^{\mathscr{M}}$ be the expectation operator defined in Chapter 3, and let $A \in \mathscr{C}$. Then determine whether $A - [A]^{\mathscr{M}} \in \mathscr{V}$.

5. For a minimal state decomposition $\{(S_{\dagger}, \lambda_{\dagger}, \theta_{\dagger}) : \dagger \in \$\}$ show that

$$\operatorname{Ker}[\lambda_{\lambda}] \subset \operatorname{Ker}[\lambda_{\ddagger} E^{\dagger}], \qquad \dagger < \ddagger$$

If, moreover, this containment holds with equality show that the transition operators are one-to-one.

6. For a minimal state decomposition $\{(S_{\dagger}, \lambda_{\dagger}, \theta_{\dagger}) : \dagger \in \$\}$ show that $\Phi(\dagger, \ddagger)$, $\ddagger \leq \dagger$, is one-to-one if $E_{\dagger}\theta_{\dagger}$ is one-to-one.

7. Assume $A \in \mathscr{S}$ and $V \in \mathscr{V}$ satisfy the axioms of Lemma C.2 and show that V admits a strong minimal state decomposition in the form $\{(S_{\dagger}, \lambda_{\dagger}, \pi_{\dagger}) : \dagger \in \$\}$. Furthermore, show that this state decomposition admits the same family of transition operators as that used for A.

8. Define a strong minimal costate decomposition.

9. Assume that $A \in \mathscr{S}^*$ admits a minimal costate decomposition. Show that A can be factored through a *costate trajectory space* in the form $A = \tilde{\theta}\tilde{\lambda}$ where $\tilde{\theta}$ is memoryless and $\tilde{\lambda}$ is algebraically strictly anticausal.

10. Formulate a dual lifting theorem for strongly strictly anticausal operators.

REFERENCE

1. Schumitzky, A., State feedback control for general linear systems, *Proc. Internat. Symp. Math. Networks and Systems* pp. 194–200. T. H. Delft (1979).

Notes for Part II

1. Historical Background: The state concept has its origins in automata theory, where it is used as an abstraction of computer memory.[5] The concept came to the fore in system theory in the late 1950s and early 1960s, when it was recognized that the intermediate variable $x(\dagger)$ associated with the first-order differential operator

$$\begin{aligned} \dot{x}(\dagger) &= Ax(\ddagger) + Bf(\dagger), \\ & \qquad\qquad\qquad x(0) = 0 \\ g(\dagger) &= Cx(\dagger), \end{aligned}$$

on $L_2[0, 1]$ was, in fact, the state trajectory for the system.[2,3]

In a resolution space setting the concept was first introduced by Saeks[7] and developed in the dissertations of Schnure[9] and Steinberger.[10] Although a state decomposition was employed through this work, state trajectory space did not appear or appeared only in its most rudimentary form. Indeed, state trajectory space has been the subject of much controversy. The first serious attempt at formulating a state trajectory space appeared in the dissertation by Steinberger,[10] in 1977, which was summarized in a (now classical) unpublished paper by Steinberger *et al.*[11] Although unpublished, this work set the stage for everything to follow and established the ground rules for the state trajectory concept. Since 1977 three alternative approaches to the state trajectory concept have been proposed. The algebraic formulation presented here is due to Schumitzky[8] and yields a viable lifting theory. It is, however, cast in an algebraic resolution space setting rather than a Hilbert resolution space and, as such, does not admit a viable dual theory. Although two alternative Hilbert space formulations have been proposed, both have deficiencies. In 1980 Feintuch[1] showed how to construct a Hilbert space of state trajectories in terms of an arbitrarily chosen measure. Unfortunately, no natural measure with which to carry out this construction exists except in the case where the underlying Hilbert space admits a cyclic vector. An alternative approach, due to Feintuch and Saeks, is described in Section 4 of these notes. Although this theory yields a viable Hilbert resolution space of state trajectories and a full-fledged dual theory, the state trajectory space employed is too small to permit the formulation of an interesting lifting theory.

As with the state trajectory space the lifting theory has its origin in Steinberger's dissertation[10] and the later work of Steinberger *et al.*,[11] while our presentation follows the algebraic formulation of Schumitzky.[8] Indeed, the lifting theory is the key tool that makes the state concept useful. Although it has many manifestations in various system theoretic applications its power is most readily illustrated by the transformation from an open loop system to a memoryless state feedback system of Remark 7.C.3. Moreover, the lifting theorem and its variations allows us to formulate a viable regulator theory, the Kalman filter, etc. In essence, the lifting theory plays the same role in a Hilbert resolution space setting as the matrix Riccati equation in classical system theory. Many system theoretic problems that are classically reduced to the solution of a matrix Riccati equation may therefore be resolved by the lifting theory in a resolution space setting.

2. Time-Invariant State Decompositions: If A is defined on a uniform resolution space $(H, \mathscr{E}, \mathscr{U})$, defined on an ordered locally compact group $\$$, and time invariant, then it is natural to attempt to construct a time-invariant state decomposition for A. Following the construction outlined in Property 6.B.8 it can be shown that every time-invariant A admits a time-invariant state decomposition $\{(S_\dagger, \lambda_\dagger, \theta_\dagger) : \dagger \in \$\}$ where

$$S_\dagger = S_0, \qquad \lambda_\dagger = \lambda_0 U^{-\dagger}, \qquad \text{and} \qquad \theta_\dagger = U^\dagger \theta_0$$

for all $\dagger \in \$$ while any such state decomposition defines a time-invariant strongly strictly causal operator.[1,9] Of course, a state decomposition may be time varying but defined in such a way that the "internal" time variation cancels to define a time-invariant operator. If $\{(S'_\dagger, \lambda'_\dagger, \theta'_\dagger): \dagger \in \$\}$ is such a state decomposition and is also minimal, then it is equivalent to the minimal time-invariant state decomposition constructed by the argument of Property 6.B.8. As such, it is equivalent to a time-invariant state decomposition in the sense of Definition 6.B.12 from which it follows that

$$U^\dagger \theta_\dagger \lambda_\dagger = \theta_0 \lambda_0 U^\dagger$$

Conversely, a state decomposition satisfying the above equality defines a time-invariant causal operator.

A careful analysis of the definition of a time-invariant state decomposition will reveal that $\Phi(\dagger, \ddagger)$ is of the form $\Phi(\dagger - \ddagger)$. Indeed, there exists a *strongly continuous semigroup* of operators $V(\dagger)$, $\dagger \in \$^\dagger$, such that

$$\Phi(\dagger, \ddagger) = V(\dagger - \ddagger), \qquad \ddagger \leq \dagger$$

A such, for a time-invariant state decomposition

$$x(\dagger) = V(\dagger)x(0) + \mathrm{s}\,(m) \int_0^\dagger V(\dagger - \ddagger)\lambda_0 U^{-\ddagger}\, dE(\ddagger)\, u$$

which is of exactly the same form as the representation for the state trajectory of a differential operator given in Remark 7.A.3.

3. Nonlinear State Decomposition: For a linear system we make use of the equality $E_\dagger Au = E_\dagger AE^\dagger u + E_\dagger AE_\dagger u$ to represent the response of the system after time $\dagger$ as the sum of the response of the system due to the input after time $\dagger$ and the response of the system after time $\dagger$ due to the state at time $\dagger$. In a nonlinear system no such decomposition exists, and so one must carry the future input along with the state.[6] We therefore define a *nonlinear state decomposition* to be a triple $\{(S_\dagger, \lambda_\dagger, \theta_\dagger): \dagger \in \$\}$ where $S_\dagger$ is a Hilbert space, $\dagger \in \$$, and $\lambda_\dagger: H \to S_\dagger \oplus H_\dagger$ by

$$\lambda_\dagger u = (\lambda_\dagger^1 E^\dagger u, E_\dagger u)$$

where $\lambda_\dagger^1 = \lambda_\dagger^1 E^\dagger$ is a nonlinear operator mapping $H^\dagger$ to $S_\dagger$. Similarly, $\theta_\dagger: S_\dagger \oplus H_\dagger \to H$ and satisfies $\lambda_\dagger = E_\dagger \theta_\dagger$ and the state decomposition is related to the given operator by the equality

$$E_\dagger A = \theta_\dagger \lambda_\dagger$$

These conditions are described diagramatically as follows:

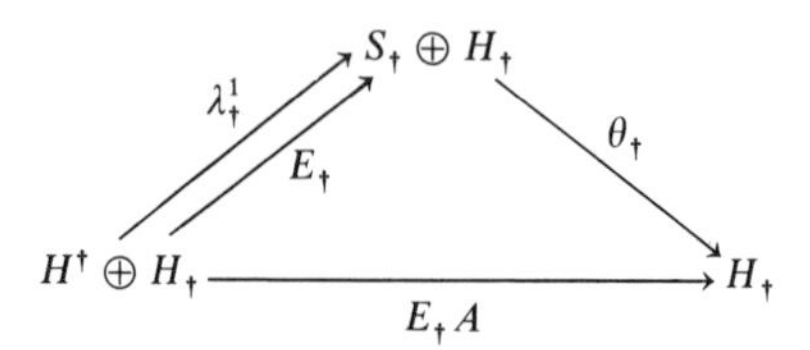

Surprisingly, with the aid of this definition, virtually the entire theory developed in Chapter 6 carries through to nonlinear operators defined on a Hilbert resolution space. Minimal state decompositions are equivalent, state transition operators exist, etc. On the other hand, because of the lack of linearity a viable state trajectory space has yet to be formulated in the nonlinear case.

One interesting class of nonlinear operators for which a more powerful state space theory can be formulated are the *polynomic operators.* These operators form a subclass of the weakly additive operators and are factorable in the form

$$P = A'\tau$$

where τ is a memoryless operator that maps H to an appropriate resolution space defined in a Hilbert scale and A' is a linear operator that maps that space back to H. Since τ is memoryless, one can define a state model for P by constructing a linear state model for A' and letting $\lambda_\dagger = \lambda'\tau$. This theory fits within the above axioms.

4. The State Trajectory Space: In Definition 7.B.2 the state trajectory space was constructed by taking linear combinations of state trajectory segments. Since this yields a space closed under truncation, the usual resolution structure for a function space is well defined. As an alternative, one can work with a space composed only of the state trajectories and define a resolution of the identity that maps trajectories to trajectories. To this end we let

$$S = \{x \in \pi[S_\dagger]: x(\dagger) = \lambda_\dagger u, u \in H\}$$

with the norm

$$\|x\|_S = \inf\{\|u\| : x = \lambda u, u \in H\}$$

and resolution of the identity $\mathscr{F}$ defined by

$$[F^{\dagger}x](\ddagger) = \begin{cases} x(\ddagger), & \ddagger \leq \dagger \\ \Phi(\ddagger, \dagger)x(\dagger), & \ddagger > \dagger \end{cases}$$

Although some manipulation is required this proves to be a well-defined Hilbert resolution space (since Ker[λ] is a closed subspace invariant under $\mathscr{E}$). Moreover, λ is a bounded memoryless operator from $(H, \mathscr{E})$ to $(S, \mathscr{F})$ and θ is a bounded strongly strictly causal operator from $(S, \mathscr{E})$ to $(H, \mathscr{F})$ that defines a state trajectory factorization for a strongly strictly causal A by

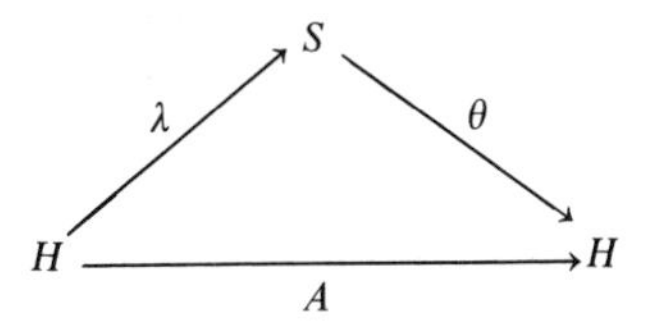

Note that relative to the results of Chapter 7 the causality properties of λ and θ have been reversed by the choice of $\mathscr{F}$.

One of the advantages of the full Hilbert resolution space structure on S is that it permits one to formulate a dual theory. To this end, if $A \in S$ we construct a *costate trajectory* factorization for A^* in the form

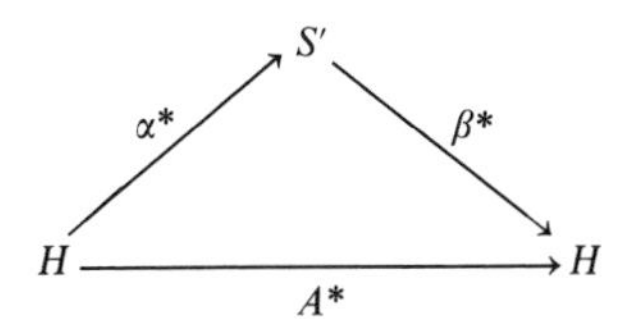

where β^* is strongly strictly anticausal and α^* is memoryless. As such, we obtain an alternative factorization of A through its costate trajectory space defined by the commutative diagram

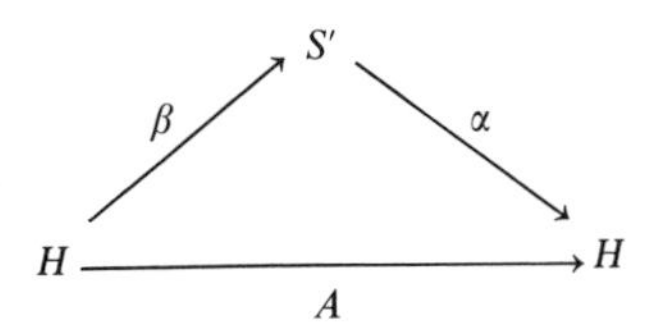

where α is memoryless and $\beta \in \mathscr{S}$. Finally, the two factorizations for A may be combined into a single diagram:

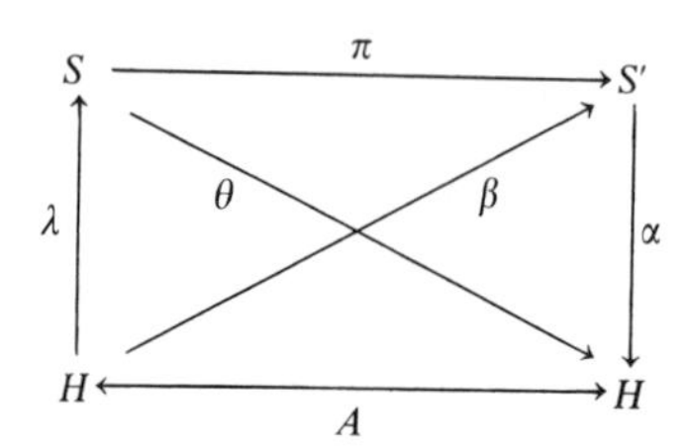

where the existence of a $\pi \in \mathscr{S}$ that makes the diagram commutative follows from the usual diagram chasing arguments.

Consistent with the above we have constructed a factorization for A in the form

$$A = \alpha\pi\lambda$$

where α and λ are memoryless and π is a strongly strictly causal mapping from the state trajectory space to the costate trajectory space that characterizes all of the dynamics in A. This factorization should be compared with the factorization of the differential operator P,

$$\dot{x}(\dagger) = Ax(\dagger) + Bf(\dagger), \qquad x(0) = 0$$
$$g(\dagger) = Cx(\dagger),$$

which takes the form

$$P = C(D - A)^{-1}B$$

in the operational notation of Example 7.C.7. Here, B and C are memoryless and $(D - A)^{-1}$ is a strongly strictly causal operator mapping the state trajectory space to itself. Although the two factorizations may at first seem to be identical, this is not the case. In particular, the costate trajectory space does not appear in the factorization of P, while the state is the input to the dynamical operator in the factorization of A and the output of the dynamical operator in the factorization of P. Finally, controllability manifests itself in the fact that λ is onto in the factorization of A, while it manifests itself in a complex relationship between A and B (see Example 6.B.7) in the factorization of P, with a similar observation for observability.

Because of the powerful duality theory associated with the state trajectory theory formulated here, it would seem to be superior to the algebraic theory of Chapter 7. Unfortunately, the state trajectory space employed is too small. Thus the lifting theorem holds only for a highly restricted class of operators. Since the lifting theorem is the main tool by which the state space theory interfaces with out system theoretic applications, to adopt this formulation of the state trajectory concept would render the theory essentially useless.

REFERENCES

1. Feintuch, A., State space theory for resolution space operators, *J. Math. Anal. Appl.* **74**, 164–191 (1980).
2. Kalman, R. E., On the general theory of control systems, *Proc. IFAC Congr., 1st, Moscow*. Butterworths, London, 1960.
3. Kalman, R. E., Canonical structure of linear dynamical systems, *Proc. Nat. Acad. Sci. USA* **48**, 596–600 (1962).
4. Krohn, K., and Rhodes, J., Algebraic theory of machines I, *Trans. Amer. Math. Soc.* **116**, 450–464 (1966).
5. Nerode, A., Linear automation transformations, *Proc. Amer. Math. Soc.* **9**, 541–544 (1958).
6. Olivier, P. D., Ph. D. Dissertation, Texas Tech. Univ., Lubbock, Texas (1980).
7. Saeks, R., State in Hilbert space, *SIAM Rev.* **15**, 283–308 (1973).
8. Schumitzky, A., State feedback control for general linear systems, *Proc. Internat. Symp. Math. Networks and Systems* pp. 194–200. T. H. Delft (1979).
9. Schnure, W., Ph.D. Dissertation, Univ. of Michigan, Ann Arbor, Michigan (1974).
10. Steinberger, M., Ph.D. Dissertation, Univ. of Southern Calif., Los Angeles, California (1977).
11. Steinberger, M., Silverman, L., and Schumitzky, A., Unpublished notes, Univ. of Southern California (1977).

PART III
Feedback Systems

A typical *feedback system* is illustrated in the block diagram of Fig. 1. Here the blocks denote operators on a Hilbert resolution space, while the signals associated with the lines denote elements of the appropriate Hilbert space and the circles denote adders. The block P is termed the *plant* and represents some physical device to be controlled. This control is implemented by measuring the output of P and processing that output in a *feedback* block F whose output modifies the *references input* r, which is applied to the plant.

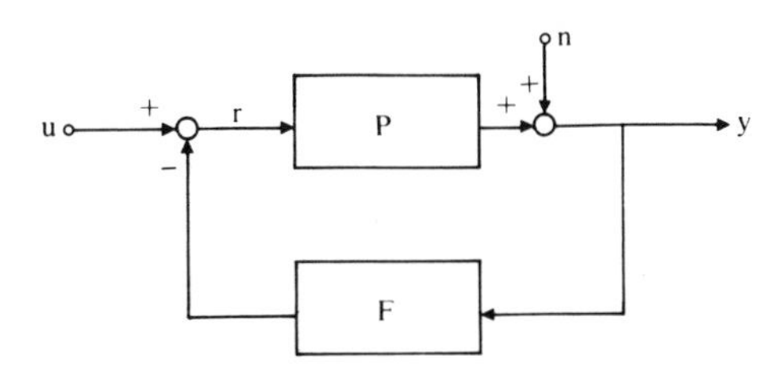

Fig. 1. Typical feedback system.

Conceivably one could control the plant by means of an *open loop configuration*, such as that illustrated in Fig. 2, wherein the reference input is processed by a *compensator* M and then directly fed to the input of the plant without reference to the plant output. In practice, the feedback system proves to be more useful since its performance may be less sensitive to modeling errors and the *noise term* n. Unlike an open loop system, however, the design of a feedback system is no simple matter. Indeed, the feedback system design equations are nonlinear while the system, itself, is prone to stability problems.

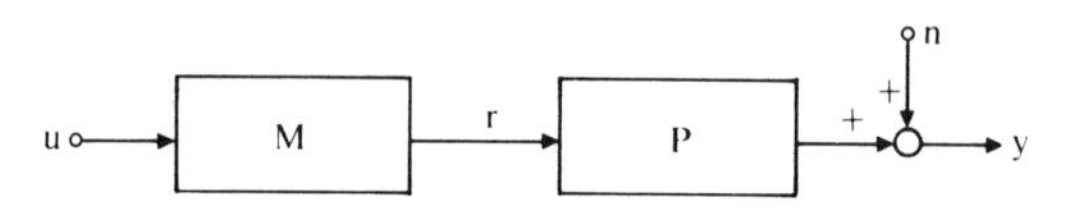

Fig. 2. Typical open loop system.

This third part of the text is, therefore, devoted to a study of the analysis and design problem for feedback systems defined on a Hilbert resolution space. In Chapter 8 the basic feedback system model is formulated. In particular, the relationship between the concepts of causality and stability is delineated and a fractional representation theory for (possibly) unstable feedback systems is formulated. A criterion for stability analysis is developed and a sensitivity theory is formulated. In Chapters 9 and 10 the feedback system design problem is introduced. In particular, Chapter 9 is devoted to the asymptotic design problem in which the stabilization, tracking, and disturbance rejection problems are considered, and Chapter 10 considers a number of deterministic optimal control problems, including the classical servomechanism and regulator problems.

8

Feedback Models

A. EXTENSION SPACES

1. Remark: Intuitively, a system is unstable if its responses blow up in some sense. Unfortunately, if the system is to be modeled by an operator on Hilbert space such instability cannot occur, since signals modeled by the elements of a Hilbert space are all bounded (in the sense that they have a finite Hilbert space norm). For the purposes of our feedback system theory these difficulties are alleviated by imbedding a Hilbert resolution space into an appropriate extension space that includes unbounded signals. Moreover, it can be shown that the causal bounded operators $\mathscr{C}$ on the given Hilbert resolution space are in one-to-one correspondence with the stable operators on the extension space. We may therefore "identify" stability and causality, thereby opening the door to the development of a feedback system theory for a Hilbert resolution space.

In the sequel we assume that $(H, \mathscr{E})$ is a given Hilbert resolution space and that A, B, and $C \in \mathscr{B}$ while the operators P, F, and M denote the plant, feedback, and compensator for a system, respectively.

2. Definition: Let $(H, \mathscr{E})$ be a Hilbert resolution space. Then the *resolution seminorms* are defined by

$$\|x\|^{\dagger} = \|E^{\dagger}x\|, \qquad \dagger \in \$$$

all for $x \in H$.

3. Remark: If $x \neq 0$ then $\|x\|^{\dagger} \neq 0$ for sufficiently large $\dagger$, and hence the family of resolution seminorms separate H. As such, the resolution seminorms define a Hausdorf topology on H, which we term the *resolution topology*. Of course, if

$$\lim_{i} \|x_i\| = 0$$

then

$$\lim_{i} \|x_i\|^{\dagger} = 0$$

Hence if the sequence $\{x_i\}$ converges in the Hilbert space topology for H, it also converges in the resolution topology, showing that the resolution topology is weaker than the Hilbert space topology.

In the following development we shall need both the Hilbert space and weak Hilbert space topologies in addition to the resolution topology. We adopt the notation lim, w lim, and r lim to denote limits in the three topologies, respectively.

4. Property: Let $A \in \mathscr{C}$ be a causal bounded operator on $(H, \mathscr{E})$. The A is continuous with respect to the resolution topology on H.

Proof: If $\text{r}\lim_i [x_i] = 0$, then

$$\lim_{i} \|x_i\|^{\dagger} = 0, \qquad \dagger \in \$$$

so

$$\lim_{i} \|Ax_i\|^{\dagger} = \lim_{i} \|E^{\dagger}Ax_i\| = \lim_{i} \| E^{\dagger}AE^{\dagger}x_i\| \leq \|E^{\dagger}A\| \left[\lim_{i} \|E^{\dagger}x_i\|\right]$$

$$\leq \|A\| \left[\lim_{i} \|x_i\|^{\dagger}\right] = 0$$

showing that $\text{r}\lim_i [Ax_i] = 0$ and that A is continuous with respect to the resolution topology on H. //

5. Example: A noncausal bounded operator may or may not be continuous with respect to the resolution topology. For instance, if $\tilde{P}$ is the *ideal predictor* on $L_2(-\infty, \infty)$ defined by

$$[\tilde{P}f](\dagger) = f(\dagger + 1)$$

then $E^{\dagger}\tilde{P} = E^{\dagger}\tilde{P}E^{\dagger+1}$, and the above argument will verify that $\tilde{P}$ is continuous in the resolution topology given the appropriate shifting of the time parameter.

Now consider the operator defined on $l_2[1, \infty)$ with its usual resolution structure via

$$b_j = \frac{1}{j}\sum_{k=j}^{\infty}\frac{a_k}{k^2}$$

and let a^i denote the sequence of input vectors $a^i = [0, 0, 0, \ldots, 0, i^2, 0, \ldots)$ where the nonzero entry lies in the ith position. Now, $E^n a^i = 0$ for $i > n$; hence

$$\lim_i \|a^i\|^n = \lim_i \|E^n a^i\| = 0$$

verifying that

$$\text{r}\lim_i [a^i] = 0$$

However, if b^i denotes the response of the specified operator to a^i, then

$$b_j^i = \begin{cases} 1/j, & j \le i \\ 0, & j > i \end{cases}$$

while

$$[E^n b^i]_j = \begin{cases} 1/j, & j \le \min\{i, n\} \\ 0, & j > \min\{i, n\} \end{cases}$$

Hence, if $i > n$,

$$\|b^i\|^n = \|E^n b^i\| = \left(1 + \frac{1}{4} + \frac{1}{9} + \cdots + \frac{1}{n^2}\right)^{1/2} > 1$$

showing that

$$\liminf_i \|b^i\|^n > 1 \qquad \text{and} \qquad \text{r}\lim_i [b^i] \ne 0$$

Thus even though the given operator is continuous in the l_2 topology it is not continuous in the resolution topology on $l_2[1, \infty)$.

6. Definition: Let $(H, \mathscr{E})$ be a Hilbert resolution space. Its completion in the resolution topology is termed its *extended resolution space.*

7. Remark: We denote the extended resolution space by $(H_e, \mathscr{E}_e)$ where $\mathscr{E}_e$ is the family of projections obtained by extending $\mathscr{E}$ to H_e by continuity. This is clearly an algebraic resolution space and may be given a topological vector space structure by extending the family of seminorms $\| \;\|^\dagger$ to seminorms $\| \;\|_e^\dagger$ defined on H_e.

8. Example: The most common example of an extended resolution space is $L_2(-\infty, \infty)_e$ with its usual resolution structure. Indeed, $L_2(-\infty, \infty)_e$ is just the space of Lebesgue integrable functions f satisfying

$$\int_{\infty}^{\dagger} |f(\ddagger)|^2 \, d\ddagger < \infty$$

for all real $\dagger$. To see that this is the case consider an arbitrary function f satisfying the above equality and let a sequence of $L_2(-\infty, \infty)$ functions f^i be defined by

$$f^i(\dagger) = \begin{cases} f(\dagger), & \dagger \le i, \\ 0, & \dagger > i, \end{cases} \qquad i = 1, 2, \ldots$$

Now, $\|f^i - f^j\|^\dagger = 0$ for $i, j > \dagger$; hence this sequence is Cauchy in the resolution topology and thus defines an element of $L_2(-\infty, \infty)_e$, which we may identify with f. Indeed, under this identification

$$\|f\|_e^\dagger = \left[\int_{-\infty}^{\dagger} |f(\ddagger)|^2 \, d\ddagger\right]^{1/2}$$

and $\|f^i - f\|_e^\dagger = 0$ for $i > \dagger$. Hence f^i converges to f in the topology defined by the extended seminorms. Moreover, since the space of functions satisfying

$$\int_{-\infty}^{\dagger} |f(\ddagger)|^2 \, d\ddagger < \infty$$

is complete in this topology, these functions represent the entirety of $L_2(-\infty, \infty)_e$. Of course, the usual resolution structure $\mathscr{E}$ for $L_2(-\infty, \infty)$ extends to the obvious family of truncation operators on $L_2(-\infty, \infty)_e$.

9. Property: $E_e^\dagger x_e \in H$ for all $\dagger \in \$$.

Proof: Since H_e is the completion of H in the resolution topology

$$x_e = r_e \lim_i [x_i]$$

for some sequence of vectors $x_i \in H$ where r_e lim denotes limits in the family of seminorms $\| \ \|_e^\dagger$ on H_e. Moreover, since $E_e^\dagger$ is continuous in this topology and extends $E^\dagger$,

$$E_e^\dagger x_e = E_e^\dagger[r_e \lim_i [x_i]] = r_e \lim_i [E_e^\dagger x_i] = r_e \lim_i [E^\dagger x_i]$$

Now consider the sequence $E^\dagger x_i$ in the Hilbert space topology of H. In particular,

$$\lim_{i,j} \|E^\dagger x_i - E^\dagger x_j\| = \lim_{i,j} \|x_i - x_j\|^\dagger = 0$$

since x_i is Cauchy in the resolution topology for H. The sequence $E^\dagger x_i \in H$ is therefore Cauchy in the Hilbert space topology and converges to some $x \in H$ (actually $H^\dagger$) in this topology. Moreover, since the resolution topology is weaker than the Hilbert space topology in H, the sequence $E^\dagger x_i$ also converges to x in the resolution topology, and

$$x = \lim_i [E^\dagger x_i] = r \lim_i [E^\dagger x_i] = r_e \lim_i [E^\dagger x_i]$$

where the last equality holds since the $\| \ \|_e^\dagger$ seminorms extend the $\| \ \|^\dagger$ seminorms. Finally, since the resolution topology is Hausdorff and the sequence $E^\dagger x_i$ has two limits in this topology, they must coincide. That is,

$$E_e^\dagger x_e = r_e \lim_i [E^\dagger x_i] = x \in H$$

as was to be shown. //

10. Corollary: $\|x_e\|_e^\dagger = \|E_e^\dagger x_e\|$ for all $x_e \in H_e$ and $\dagger \in \$$.

Proof: $\|E_e^\dagger x_e\|$ is well defined via Property 9. Now retaining the notation used in the proof of Property 9 we have

$$\|x_e\|_e^\dagger = \lim_i \|x_i\|^\dagger = \lim_i \|E^\dagger x_i\| = \|x\| = \|E_e^\dagger x_e\|$$

as was to be shown. //

11. Property: An element $x_e \in H_e$ is in H if and only if

$$\sup_\dagger \|x_e\|^\dagger < \infty$$

Proof: If $x_e \in H$,

$$\sup_{\dagger} \|x_e\|_e^{\dagger} = \sup_{\dagger} \|x_e\|^{\dagger} = \sup_{\dagger} \|E^{\dagger}x_e\| \leq \sup_{\dagger} \|E^{\dagger}\| \, \|x_e\| \leq \|x_e\| < \infty$$

Conversely, if

$$\sup_{\dagger} \|x_e\|_e^{\dagger} = \sup_{\dagger} \|E_e^{\dagger}x_e\| < \infty$$

then since a closed bounded set in Hilbert space is weakly compact there exist a sequence $\dagger_i \in \$$, $\lim_i [\dagger_i] = \dagger_\infty$, and an $x \in H$ such that

$$\text{w} \lim_i E_e^i x_e = x$$

Now, for any $\ddagger \in \$$ $E^{\ddagger}$ is weakly continuous and $\dagger_i$ is eventually greater than $\ddagger$; hence

$$E_{\ddagger}x = E^{\ddagger}[\text{w} \lim_i [E_e^{\dagger_i}x_e]] = \text{w} \lim[E^{\ddagger}E_e^{\dagger_i}x_e] = E_e^{\ddagger}x_e$$

Thus, for any $\ddagger \in \$$

$$\|x_e - x\|_e^{\ddagger} = \|E_e^{\ddagger}(x_e - x)\| = \|E_e^{\ddagger}x_e - E^{\ddagger}x\| = 0$$

verifying that $x_e = x \in H$ since the family of seminorms $\| \ \|_e^{\dagger}$ separates H_e. Thus we have shown that if the hypotheses are satisfied $x_e = x \in H$, as required. //

12. Remark: Taken together Properties 9 and 11 and Corollary 10 allow us to characterize completely the elements of the extended space in terms of their truncates $E_e^{\dagger}x_e$ in the given Hilbert resolution space. Hence the power of the normed space $(H, \mathscr{E})$ is not diluted by our excursion into the extended space, $(H_e, \mathscr{E}_e)$. As we shall see, however, the extended space permits one to formulate a physically meaningful stability concept that would not be well defined in $(H, \mathscr{E})$.

Since $\| \ \|_e^{\dagger}$ defines a natural topological vector space structure on $(H_e, \mathscr{E}_e)$, the extended space is a natural candidate for a topological vector resolution space. In fact, however, for the present purposes we only require its structure as an *algebraic resolution space* wherein the concepts of causality, memorylessness, and algebraic strict causality are well defined. In particular, an operator A on $(H_e, \mathscr{E}_e)$ is causal if

$$E_e^{\dagger}A = E_e^{\dagger}AE_e^{\dagger}, \qquad \dagger \in \$$$

13. Definition: An operator A defined on an extended resolution space $(H_e, \mathscr{E}_e)$ is *stable* if there exists an $M > 0$ such that

$$\|Ax_e\|_e^\dagger \leq M\|x_e\|_e^\dagger$$

for all $x_e \in H_e$ and $\dagger \in \$$.

14. Property: Let A map $(H_e, \mathscr{E}_e)$ to $(H_e, \mathscr{E}_e)$ and B be a bounded causal operator mapping $(H, \mathscr{E})$ to $(H, \mathscr{E})$. Then

(i) A is stable if and only if A is causal on $(H_e, \mathscr{E}_e)$ and its restriction to H is a bounded operator mapping H to H; and

(ii) B has a unique extension to a stable operator on $(H_e, \mathscr{E}_e)$.

Proof: Let A be stable and consider the operator $E_e^\dagger A[I - E_e^\dagger]$ defined on $(H_e, \mathscr{E}_e)$. Since A is stable

$$\begin{aligned}\|E_e^\dagger A[I - E_e^\dagger]x_e\| &= \|A[I - E_e^\dagger]x_e\|_e^\dagger \leq M\|[I - E_e^\dagger]x_e\|_e^\dagger \\ &= \|E_e^\dagger[I - E_e^\dagger]x_e\| = \|0\| = 0\end{aligned}$$

for each $x_e \in H_e$ and $\dagger \in \$$. As such, $E_e^\dagger A[I - E_e^\dagger] = 0$, or equivalently $E_e^\dagger A = E_e^\dagger A E_e^\dagger$, showing that A is causal on $(H_e, \mathscr{E}_e)$.

If $x \in H$ and A is stable, then

$$\sup_\dagger \|Ax\|_e^\dagger \leq \sup_\dagger [M\|x\|_e^\dagger] = \sup_\dagger [M\|x\|^\dagger] \leq M\|x\| < \infty$$

hence Property 11 implies that $Ax \in H$. Moreover,

$$\|Ax\| = \lim_\dagger \|E^\dagger Ax\| = \lim_\dagger \|Ax\|^\dagger \leq \sup_\dagger \|Ax\|^\dagger = \sup_\dagger \|Ax\|_e^\dagger \leq M\|x\|$$

showing that the restriction of A to H is bounded.

Finally, if $B \in \mathscr{C}$ then B is continuous in the resolution topology by Property 4. As such, B admits a unique continuous extension to $(H_e, \mathscr{E}_e)$, which we denote by B_e. Of course, since both B and $E^\dagger$ are extended to B_e and $E_e^\dagger$ by continuity, the equality $E^\dagger B = E^\dagger B E^\dagger$ in $(H, \mathscr{E})$ extends to the equality $E_e^\dagger B_e = E_e^\dagger B_e E_e^\dagger$ in $(H_e, \mathscr{E}_e)$, verifying that B_e is causal on $(H_e, \mathscr{E}_e)$. Moreover, since the restriction of B_e to H is B, which is bounded on H by hypothesis, (i) implies that B_e is stable on $(H_e, \mathscr{E}_e)$. //

15. Remark: Property 14 characterizes the precise relationship between the stable operators and the causal operators. They do not coincide as some have indicated. Indeed, they are not even defined on the same space. There is, however, a natural one-to-one correspondence between the causal bounded operators on $(H, \mathscr{E})$ and the stable operators $(H_e, \mathscr{E}_e)$ with each causal operator uniquely extending to a stable operator and the restriction of each satble operator causal and bounded.

This may be readily illustrated by the constant-coefficient differential operator

$$\left[\sum_{i=0}^{n} a_i D^i\right] g = \left[\sum_{i=0}^{n-1} b_i D^i\right] f$$

which is represented by the rational function

$$r(s) = \sum_{i=0}^{n-1} b_i s^i \Big/ \sum_{i=0}^{n} a_i s^i$$

If the operator is defined on $L_2(\infty, \infty)$ the Paley–Weiner condition implies that the operator is causal and bounded if and only if $r(s)$ has no poles with nonnegative real part. On the other hand, if the operator is defined on $L_2(-\infty, \infty)_e$, the Hurwitz condition implies that the differential operator is stable if and only if $r(s)$ has no poles with nonnegative real part.[2] As such, the causal bounded differential operators on $L_2(-\infty, \infty)$ and the stable differential operators on $L_2(-\infty, \infty)_e$ are in one-to-one correspondence in that they are represented by the same rational functions. The two operators are, however, different and, in fact, are defined in different spaces.

B. FRACTIONAL REPRESENTATION

1. Remark: Since stability is defined on $(H_e, \mathscr{E}_e)$ rather than $(H, \mathscr{E})$, our feedback system theory is most naturally set in the extended resolution space $(H_e, \mathscr{E}_e)$. Unfortunately, this space is not sufficiently structured to achieve the desired end. Rather, we invoke the theorem that allows us to identify the stable operators on $(H_e, \mathscr{E}_e)$ with the causal bounded operators on $(H, \mathscr{E})$, thereby permitting us to retain the full Hilbert resolution space structure in our theory with $\mathscr{C}$ serving in lieu of the stable operators. Given that such an approach is to be adopted, it remains to imbed $\mathscr{C}$ into a larger class of operators that can serve as our universe of (stable and unstable) systems. Although $\mathscr{B}$ might at first seem to be the natural choice, it proves to be too large for our purpose. Alternatively, one might choose to work with the bounded operators that are continuous in the resolution topology and thus admit an extension to $(H_e, \mathscr{E}_e)$, where stability and instability are naturally defined. Four our purposes, however, it proves to be most convenient to deal with the class of operators that admit left and right *coprime fractional representations* $A = N_{ar} D_{ar}^{-1}$ and $A =$

$D_{al}^{-1}N_{al}$ where N_{ar} and N_{al} are in $\mathscr{C}$ and D_{ar} and D_{al} are in $\mathscr{C} \cap \mathscr{B}^{-1}$. By so doing an arbitrary operator is represented as the "ratio" of stable (equivalently causal) operators, thereby simplifying the stability analysis of the feedback system in which that operator is imbedded. The fractional representation concept is formalized as follows.

2. Definition: Let $(H, \mathscr{E})$ be a Hilbert resolution space. Then an operator $A \in \mathscr{B}$ admits a *right fractional representation* if there exist $N_{ar} \in \mathscr{C}$ and $D_{ar} \in \mathscr{C} \cap \mathscr{B}^{-1}$ such that $A = N_{ar}D_{ar}^{-1}$. It admits a *left fractional representation* if there exist $N_{al} \in \mathscr{C}$ and $D_{al} \in \mathscr{C} \cap \mathscr{B}^{-1}$ such that $A = D_{al}^{-1}N_{al}$.

3. Definition: A right fractional representation is said to be *right coprime* if there exist U_{ar} and V_{ar} in $\mathscr{C}$ such that

$$U_{ar}N_{ar} + V_{ar}D_{ar} = I$$

A left fractional representation is said to be *left coprime* if there exist U_{al} and V_{al} in $\mathscr{C}$ such that

$$N_{al}U_{al} + D_{al}V_{al} = I$$

4. Remark: Although no arithmetic structure is assumed, these definitions for coprimeness are natural generalizations of the classical coprimeness criterion for polynomials, which states that polynomials $n(s)$ and $d(s)$ have no common zeros if and only if there exist polynomials $u(s)$ and $v(s)$ such that

$$u(s)n(s) + v(s)d(s) = 1$$

More to the point of our application, if $n(s)$ and $d(s)$ are rational functions representing stable constant coefficient differential operators, then they have no common zeros with nonnegative real part if and only if there exist rational functions $u(s)$ and $v(s)$ whose poles have negative real parts such that the above equality is satisfied.

The fact that these algebraic comprimeness conditions retain the character of their arithmetic counterparts in our operator theoretic setting is verified by the following properties.

5. Property: Let $A = N_{ar}D_{ar}^{-1}$ be a right comprime fractional representation for A and assume that N_{ar} and D_{ar} have a *common right factor* $R \in \mathscr{C}$. That is, $N_{ar} = \tilde{N}_{ar}R$ and $D_{ar} = \tilde{D}_{ar}R$ for some $\tilde{N}_{ar}$ and $\tilde{D}_{ar}$ in $\mathscr{C}$. Then R admits a left inverse in $\mathscr{C}$.

Proof: Since $A = N_{ar}D_{ar}^{-1}$ is right coprime there exist U_{ar} and V_{ar} in $\mathscr{C}$ such that

$$I = U_{ar}N_{ar} + V_{ar}D_{ar} = U_{ar}\tilde{N}_{ar}R + V_{ar}\tilde{D}_{ar}R = [U_{ar}\tilde{N}_{ar} + V_{ar}\tilde{D}_{ar}]R$$

verifying that $[U_{ar}\tilde{N}_{ar} + V_{ar}\tilde{D}_{ar}] \in \mathscr{C}$ is the required left inverse. //

6. Property: Let $A = N_{ar}D_{ar}^{-1}$ be right coprime and let $A = \tilde{N}_{ar}D_{ar}^{-1}$ be a second (not necessarily coprime) right fractional representation for A. Then there exists $R \in \mathscr{C}$ such that

$$\tilde{N}_{ar} = N_{ar}R \qquad \text{and} \qquad \tilde{D}_{ar} = D_{ar}R$$

Proof: Let $R = D_{ar}^{-1}\tilde{D}_{ar}$. Then $\tilde{D}_{ar} = D_{ar}R$, while

$$\tilde{N}_{ar} = A\tilde{D}_{ar} = N_{ar}D_{ar}^{-1}\tilde{D}_{ar} = N_{ar}R$$

showing that $\tilde{D}_{ar}$ and $\tilde{N}_{ar}$ have the correct form. It remains to show that $R \in \mathscr{C}$ as follows:

$$\begin{aligned} R = D_{ar}^{-1}\tilde{D}_{ar} &= [U_{ar}N_{ar} + V_{ar}D_{ar}]D_{ar}^{-1}\tilde{D}_{ar} = U_{ar}N_{ar}D_{ar}^{-1}\tilde{D}_{ar} + V_{ar}\tilde{D}_{ar} \\ &= U_{ar}\tilde{N}_{ar}\tilde{D}_{ar}^{-1}\tilde{D}_{ar} + V_{ar}\tilde{D}_{ar} = U_{ar}\tilde{N}_{ar} + V_{ar}\tilde{D}_{ar} \in \mathscr{C}. \end{aligned}$$

Indeed, we have represented R as the sum and product of elements in $\mathscr{C}$, which, since the causals form an algebra implies that $R \in \mathscr{C}$.

7. Remark: Properties 6 and 7 imply that our right fractional representation theory is similar in nature to the classical polynomial and polynomial matrix fractional representation theory with $\mathscr{C} \cap \mathscr{C}^{-1}$ serving as the *unimodular elements*.[2] Of course, similar properties hold for left fractional representations.

Although a causal A may always be represented by the trivial right fractional representation $A = AI^{-1}$, in practice more complex representations are often employed, and hence it behooves us to characterize the causality of A in terms of the constituents of an arbitrary right coprime fractional representation $A = N_{ar}D_{ar}^{-1}$ and the similar left coprime fractional representation.

8. Property:

(i) Let $A = N_{ar}D_{ar}^{-1}$ be a right coprime fractional representation. then $A \in \mathscr{C}$ if and only if $D_{ar} \in \mathscr{C} \cap \mathscr{C}^{-1}$.

(ii) Let $A = D_{al}^{-1}N_{al}$ be a left coprime fractional representation. Then $A \in \mathscr{C}$ if and only if $D_{al} \in \mathscr{C} \cap \mathscr{C}^{-1}$.

(iii) Let $A = N_{ar}D_a^{-1}N_{al}$ where $N_{ar}D_a^{-1}$ is right coprime and $D_a^{-1}N_{al}$ is left coprime. Then $A \in \mathscr{C}$ if and only if $D_a \in \mathscr{C} \cap \mathscr{C}^{-1}$.

Proof:

(i) If $D_{ar} \in \mathscr{C} \cap \mathscr{C}^{-1}$, then $D_{ar}^{-1} \in \mathscr{C}$ and hence so is $A = N_{ar}D_{ar}^{-1}$. Conversely, if $A \in \mathscr{C}$, then the fact that $N_{ar} = AD_{ar}$ and $D_{ar} = ID_{ar}$ implies that D_{ar} is a common right factor of both D_{ar} and N_{ar}, in which case Property 5 implies that D_{ar} has a left inverse in $\mathscr{C}$. On the other hand, D_{ar} is invertible in $\mathscr{B}$ by hypothesis, and hence

$$D_{ar}^{-1} = D_{ar}^{-L} \in \mathscr{C}$$

as was to be shown.

(ii) The proof is similar to that for (i) and will not be repeated.

(iii) If $D_a \in \mathscr{C} \cap \mathscr{C}^{-1}$, then $D_a^{-1} \in \mathscr{C}$ and hence so is A. Conversely, if $A \in \mathscr{C}$ then the fact that $D_a^{-1}N_{al}$ is left coprime implies the existence of U_{al} and V_{al} in $\mathscr{C}$ such that

$$N_{al}U_{al} + D_a V_{al} = I$$

Thus

$$\begin{aligned} N_{ar}D_a^{-1} &= N_{ar}D_a^{-1}[N_{al}U_{al} + D_a V_{al}] = N_{ar}D_a^{-1}N_{al}U_{al} + N_{ar}V_{al} \\ &= AU_{al} + N_{ar}V_{al} \in \mathscr{C} \end{aligned}$$

in which case the right coprimeness of $N_{ar}D_a^{-1}$ implies by (i) that $D_a \in \mathscr{C} \cap \mathscr{C}^{-1}$. //

9. Remark: Although it is obvious that $A \in \mathscr{C}$ if $D_a \in \mathscr{C} \cap \mathscr{C}^{-1}$, the converse arguments require the corprimeness conditions to guarantee that the numerators do not "cancel the noncausality" of D_a^{-1}.

10. Example: Consider the operator on $L_2(-\infty, \infty)^2$ with its usual resolution structure represented by the matrix multiplication operator

$$A(s) = \begin{bmatrix} \dfrac{e^{-1/s}}{s+1} & \dfrac{s-1}{s+1} \\ 0 & \dfrac{1}{s-1} \end{bmatrix}$$

with L_∞ entries. From the Paley–Wiener theorem the stable operators in this class are those whose entries are elements of H_∞, i.e., analytic for

Re[s] > 0. Hence we may construct a right coprime fractional representation for $A(s)$ in the form

$$A(s) = \begin{bmatrix} \dfrac{e^{-1/s}}{s+1} & \dfrac{(s-1)^2}{(s+1)^2} \\ 0 & \dfrac{1}{s+1} \end{bmatrix} \begin{bmatrix} 1 & 0 \\ 0 & \dfrac{s-1}{s+1} \end{bmatrix}^{-1}$$
$$= N_{ar}(s) D_{ar}^{-1}(s),$$

where

$$\begin{bmatrix} 0 & 0 \\ 0 & 2 \end{bmatrix} \begin{bmatrix} \dfrac{e^{-1/s}}{s+1} & \dfrac{(s-1)^2}{(s+1)^2} \\ 0 & \dfrac{1}{s+1} \end{bmatrix} + \begin{bmatrix} 1 & 0 \\ 0 & 1 \end{bmatrix} \begin{bmatrix} 0 & 0 \\ 0 & \dfrac{s-1}{s+1} \end{bmatrix}$$
$$= U_{ar}(s) N_{ar}(s) + V_{ar}(s) D_{ar}(s) = I,$$

whereas a left fractional representation takes the form

$$A(s) = \begin{bmatrix} 0 & 0 \\ 0 & \dfrac{s-1}{s+1} \end{bmatrix}^{-1} \begin{bmatrix} \dfrac{e^{-1/s}}{s+1} & \dfrac{s-1}{s+1} \\ 0 & \dfrac{1}{s+1} \end{bmatrix}$$
$$= D_{al}^{-1}(s) N_{al}(s)$$

with

$$\begin{bmatrix} \dfrac{e^{-1/s}}{s+1} & \dfrac{s-1}{s+1} \\ 0 & \dfrac{1}{s+1} \end{bmatrix} \begin{bmatrix} 0 & 0 \\ 0 & 2 \end{bmatrix} + \begin{bmatrix} 1 & 0 \\ 0 & \dfrac{s-1}{s+1} \end{bmatrix} \begin{bmatrix} 1 & \dfrac{-2(s-1)}{s+1} \\ 0 & 1 \end{bmatrix}$$
$$= N_{al}(s) U_{al}(s) + D_{al}(s) V_{al}(s) = I$$

11. Remark: In Example 10 we constructed both left and right coprime fractional representations for the given operator, typical of feedback system design, wherein we frequently assume that the plant and feedback operators are both characterized by left and right coprime representations. Indeed, given that an operator is characterized by both left and right comprime models it is possible to assume additional structure without loss of generality.

12: Property: Let $A \in \mathscr{B}$ and assume that A admits right and left coprime fractional representations $A = N_{ar}D_{ar}^{-1} = D_{al}^{-1}N_{al}$ where

$$U_{ar}N_{ar} + V_{ar}D_{ar} = I \qquad \text{and} \qquad N_{al}U_{al} + D_{al}V_{al} = I$$

for some U_{ar}, V_{ar}, U_{al}, and V_{al} in $\mathscr{C}$. Then there exists $\tilde{U}_{al}$ and $\tilde{V}_{al}$ such that

$$N_{al}\tilde{U}_{al} + D_{al}\tilde{V}_{al} = I \qquad \text{and} \qquad V_{ar}\tilde{U}_{al} = U_{ar}\tilde{V}_{al}$$

Proof: Recall that the inverse of a 2×2 matrix of operators is given by

$$\begin{bmatrix} X & Y \\ Z & W \end{bmatrix}^{-1} = \begin{bmatrix} \Delta^{-1} & -\Delta^{-1}YW^{-1} \\ -W^{-1}Z\Delta^{-1} & W^{-1} + W^{-1}Z\Delta^{-1}YW^{-1} \end{bmatrix}$$

where $\Delta = X - YW^{-1}Z$. Applying this formula to the matrix

$$\begin{bmatrix} V_{ar} & U_{ar} \\ -N_{al} & D_{al} \end{bmatrix}$$

we find

$$\begin{aligned} \Delta &= V_{ar} + U_{ar}D_{al}^{-1}N_{al} = V_{ar} + U_{ar}N_{ar}D_{ar}^{-1} = V_{ar} + [I - V_{ar}D_{ar}]D_{ar}^{-1} \\ &= V_{ar} + D_{ar}^{-1} - V_{ar} = D_{ar}^{-1} \end{aligned}$$

Therefore

$$\Delta^{-1} = D_{ar} \in \mathscr{C}$$

Similarly,

$$-W^{-1}Z\Delta^{-1} = D_{al}^{-1}N_{al}D_{ar} = N_{ar}D_{ar}^{-1}D_{ar} = N_{ar} \in \mathscr{C}$$

Now, let

$$\tilde{U}_{al} = \Delta^{-1}YW^{-1} = D_{ar}U_{ar}D_{al}^{-1}$$

and

$$\tilde{V}_{al} = W^{-1} + W^{-1}Z\Delta^{-1}YW^{-1} = D_{al}^{-1} - D_{al}^{-1}N_{al}D_{ar}U_{ar}D_{al}^{-1}$$

Although we have no simple expression for these two entries in the inverse matrix, they are both causal. Indeed,

$$\begin{aligned} \tilde{U}_{al} &= D_{ar}U_{ar}D_{al}^{-1}[N_{al}U_{al} + D_{al}V_{al}] = D_{ar}U_{ar}N_{ar}D_{ar}^{-1}U_{al} + D_{ar}U_{ar}V_{al} \\ &= D_{ar}[I - V_{ar}D_{ar}]D_{ar}^{-1}U_{al} + D_{ar}U_{ar}V_{al} \\ &= U_{al} - D_{ar}V_{ar}U_{al} + D_{ar}U_{ar}V_{al} \in \mathscr{C} \end{aligned}$$

while

$$\begin{aligned}\tilde{V}_{al} &= D_{al}^{-1} - D_{al}^{-1}N_{al}\tilde{U}_{al} = D_{al}^{-1}N_{al}[U_{al} - D_{ar}V_{ar}U_{al} + D_{ar}U_{ar}V_{al}] \\ &= D_{al}^{-1}[I - N_{al}U_{al}] + D_{al}^{-1}N_{al}[D_{ar}V_{ar}U_{al} - D_{ar}U_{ar}V_{al}] \\ &= D_{al}^{-1}[D_{al}V_{al}] + N_{ar}D_{ar}^{-1}D_{ar}[V_{ar}U_{al} - U_{ar}V_{al}] \\ &= V_{al} + N_{ar}V_{ar}U_{al} - N_{ar}U_{ar}V_{al} \in \mathscr{C}\end{aligned}$$

Thus

$$\begin{bmatrix} V_{ar} & U_{ar} \\ -N_{al} & D_{al} \end{bmatrix}^{-1} = \begin{bmatrix} D_{ar} & -\tilde{U}_{al} \\ N_{ar} & \tilde{V}_{al} \end{bmatrix}$$

is the specified matrix of causal operators, and the four equalities of the matrix equation

$$\begin{bmatrix} V_{ar} & U_{ar} \\ -N_{al} & D_{al} \end{bmatrix}\begin{bmatrix} D_{ar} & -\tilde{U}_{al} \\ N_{ar} & \tilde{V}_{al} \end{bmatrix} = \begin{bmatrix} I & 0 \\ 0 & I \end{bmatrix}$$

imply the desired result. //

13. Remark: Consistent with the result of Problem 12, if one assumes that a given operator admits both left and right coprime fractional representations, then without loss of generality one may also assume that the U's and V's satisfy the equality

$$V_{ar}U_{al} = U_{ar}V_{al}$$

This, in turn, implies that

$$\begin{bmatrix} V_{ar} & U_{ar} \\ -N_{al} & D_{al} \end{bmatrix}^{-1} = \begin{bmatrix} D_{ar} & -U_{al} \\ N_{ar} & V_{al} \end{bmatrix}$$

and thus eight equalities are satisfied among the N's, D's, U's, and V's. Indeed, the four defining equalities for the N's, D's, U's, and V's are specified by the matrix equation

$$\begin{bmatrix} V_{ar} & U_{ar} \\ -N_{al} & D_{al} \end{bmatrix}\begin{bmatrix} D_{ar} & -U_{al} \\ N_{ar} & V_{al} \end{bmatrix} = \begin{bmatrix} I & 0 \\ 0 & I \end{bmatrix}$$

while four additional equalities are given by the matrix equation

$$\begin{bmatrix} D_{ar} & -U_{al} \\ N_{ar} & V_{al} \end{bmatrix}\begin{bmatrix} V_{ar} & U_{ar} \\ -N_{al} & D_{al} \end{bmatrix} = \begin{bmatrix} I & 0 \\ 0 & I \end{bmatrix}$$

in which the order of the given matrix and its inverse have been reversed.

14. Definition: If right and left coprime fractional representations satisfying $V_{ar}U_{al} = U_{ar}V_{al}$ are given for an operator A, then A is said to admit a *doubly coprime fractional representation.*

15. Remark: A double coprime fractional representation for A is denoted by the matrix equality

$$\begin{bmatrix} V_{ar} & U_{ar} \\ -N_{al} & D_{al} \end{bmatrix}^{-1} = \begin{bmatrix} D_{ar} & -U_{al} \\ N_{ar} & V_{al} \end{bmatrix}$$

16. Example: For the system of Example 10

$$\begin{bmatrix} V_{ar} & U_{ar} \\ -N_{al} & D_{al} \end{bmatrix}^{-1} = \left[\begin{array}{cc:cc} 1 & 0 & 0 & 0 \\ 0 & 1 & 0 & 2 \\ \hdashline \dfrac{e^{-1/s}}{s+1} & -\dfrac{s-1}{s+1} & 0 & 0 \\ 0 & \dfrac{1}{s+1} & 0 & \dfrac{s-1}{s+1} \end{array}\right]^{-1}$$

$$= \left[\begin{array}{cc:cc} 1 & 0 & 0 & 0 \\ 0 & \dfrac{s-1}{s+1} & 0 & -2 \\ \hdashline \dfrac{e^{-1/s}}{s+1} & \dfrac{(s-1)^2}{(s+1)^2} & 1 & \dfrac{2(s-1)}{s+1} \\ 0 & \dfrac{1}{s+1} & 0 & 1 \end{array}\right] = \begin{bmatrix} D_{ar} & -U_{al} \\ N_{ar} & V_{al} \end{bmatrix}$$

Hence this is, in fact, a double coprime fractional representation.

17. Remark: Returning to the problem of formulating a viable model for the feedback system of Fig. 1 of Part III, we assume that P and F are characterized by fractional representations

$$P = N_{pr}D_{pr}^{-1} = D_{pl}^{-1}N_{pl}$$

and

$$F = N_{fr}D_{fr}^{-1} = D_{fl}^{-1}N_{fl}$$

such that

(i) $N_{fr}[D_{pl}D_{fr}]^{-1}$ is right comprime,
(ii) $[D_{pl}D_{fr}]^{-1}N_{pl}$ is left coprime,
(iii) $N_{pr}[D_{fl}D_{pr}]^{-1}$ is right coprime, and
(iv) $[D_{fl}D_{pr}]^{-1}N_{fl}$ is left coprime.

These conditions imply that the above fractional representations for P and F are both right and left coprime, and hence by Property 12 we may assume, without loss of generality, that U's and V's exist such that

$$\begin{bmatrix} V_{pr} & U_{pr} \\ -N_{pl} & D_{pl} \end{bmatrix}^{-1} = \begin{bmatrix} D_{pr} & -U_{pl} \\ N_{pr} & V_{pl} \end{bmatrix}$$

and

$$\begin{bmatrix} V_{fr} & U_{fr} \\ -N_{fl} & D_{fl} \end{bmatrix}^{-1} = \begin{bmatrix} D_{fr} & -U_{fl} \\ N_{fr} & V_{fl} \end{bmatrix}$$

define doubly coprime representations for P and F. Conditions (i)–(iv) are, however, stronger in that they imply a degree of coprimeness between the numerators of F and the denominators of P and vice versa. Intuitively, this is a generalization of the classical *frequency domain condition*, which states that there are no pole/zero cancellations between F and P.

Before proceeding with the explicit description of our feedback system in terms of the above described model, we state the following classical lemma in our operator theoretic setting:

18. Lemma: Let $A, B \in \mathscr{C}$. Then $(I + AB)^{-1}$ exists if and only if $(I + BA)^{-1}$ exists. Moreover, one of these inverses is causal if and only if the other is causal, and the following equalities hold:

$$(I + AB)^{-1} = I - A(I + BA)^{-1}B, \qquad (I + AB)^{-1}A = A(I + BA)^{-1}$$

Proof: Using that fact that $A(I + BA) = (I + AB)A$ we have

$$\begin{aligned} (I + AB)[I - A(I + BA)^{-1}B] &= I + AB - (I + AB)A(I + BA)^{-1}B \\ &= I + A(I + BA)(I + BA)^{-1}B \\ &\quad - (I + AB)A(I + BA)^{-1}B \\ &= I + (I + AB)A(I + BA)^{-1}B \\ &\quad - (I + AB)A(I + BA)^{-1}B = I \end{aligned}$$

which verifies the first equality and shows that $(I + AB)^{-1}$ exists whenever $(I + BA)^{-1}$ exists. Moreover, since $A, B \in \mathscr{C}$ $(I + AB)^{-1} \in \mathscr{C}$ if

$(I + BA)^{-1} \in \mathscr{C}$. Of course, the opposite implications can be obtained by reversing the role of A and B in the above manipulations. Finally, since $A(I + BA) = (I + AB)A$, multiplying on the right by $(I + BA)^{-1}$ and on the left by $(I + AB)^{-1}$ yields the remaining equality. //

19. Remark: Once we have formulated the appropriate doubly coprime plant and feedback models, the feedback system analysis problem reduces to the characterization of the gains between the various inputs and outputs associated with the feedback system of Fig. 1 of Part III. Although we are primarily interested in the *gain* H_{yu} observed between the system input and output in practice, the remaining gains H_{ru}, H_{yn}, and H_{rn} also play a significant role. Indeed, H_{ru} is used to characterize the input energy applied to the plant, which plays a significant role in optimal control theory, while H_{yn} and H_{rn} model the effect of noise and/or initial condition terms on the system responses. Moreover, for the system to be stable all four gains must be stable. Indeed, this is the case even if the input n does not explicitly appear in a given system since in any "real world" system inadvertent noise terms that might excite unstable modes will always appear at the plant output.

20. Property: Let a feedback system admit plant and feedback models as formulated in Remark 17. Then whenever the appropriate inverses exist

$$\begin{bmatrix} H_{ru} & H_{rn} \\ H_{yu} & H_{yn} \end{bmatrix} = \begin{bmatrix} I - F(I + PF)^{-1}P & -F(I + PF)^{-1} \\ (I + PF)^{-1}P & (I + PF)^{-1} \end{bmatrix}$$

$$= \begin{bmatrix} (I + FP)^{-1} & -(I + FP)^{-1}F \\ P(I + FP)^{-1} & I - P(I + FP)^{-1}F \end{bmatrix}$$

$$= \begin{bmatrix} D_{pr}(D_{fl}D_{pr} + N_{fl}N_{pr})^{-1}D_{fl} & -D_{pr}(D_{fl}D_{pr} + N_{fl}N_{pr})^{-1}N_{fl} \\ N_{pr}(D_{fl}D_{pr} + N_{fl}N_{pr})^{-1}D_{fl} & I - N_{pr}(D_{fl}D_{pr} + N_{fl}N_{pr})^{-1}N_{fl} \end{bmatrix}$$

$$= \begin{bmatrix} I - N_{fr}(D_{pl}D_{fr} + N_{pl}N_{fr})^{-1}N_{pl} & -N_{fr}(D_{pl}D_{fr} + N_{pl}N_{fr})^{-1}D_{pl} \\ D_{fr}(D_{pl}D_{fr} + N_{pl}N_{fr})^{-1}N_{pl} & D_{fr}(D_{pl}D_{fr} + N_{pl}N_{fr})^{-1}D_{pl} \end{bmatrix}$$

Proof: For the sake of brevity we derive only the formulas for H_{ru}; the remaining gains can be derived by similar arguments. By tracing through the block diagram of Fig. 1 of Part III with $n = 0$ (by linearity), we obtain

$$r = u - Fy = u - FPr$$

Hence

$$(I + FP)r = u$$

and

$$r = (I + FP)^{-1}u$$

Thus

$$H_{ru} = (I + FP)^{-1} = I - F(I + PF)^{-1}P$$

where the last equality follows from Lemma 18. Now upon substituting $FP = D_{fl}^{-1}N_{fl}N_{pr}D_{pr}^{-1}$ into $(I + FP)^{-1}$ we obtain

$$\begin{aligned} H_{ru} &= (I + FP)^{-1} = [I + D_{fl}^{-1}N_{fl}N_{pr}D_{pr}^{-1}]^{-1} \\ &= [D_{fl}^{-1}(D_{fl}D_{pr} + N_{fl}N_{pr})D_{pr}^{-1}]^{-1} \\ &= D_{pr}(D_{fl}D_{pr} + N_{fl}N_{pr})^{-1}D_{fl} \end{aligned}$$

while substituting $PF = D_{pl}^{-1}N_{pl}N_{fr}D_{fr}^{-1}$ into $I - F(I + PF)^{-1}P$ will yield

$$H_{ru} = I - N_{fr}(D_{pl}D_{fr} + N_{pl}N_{fr})^{-1}N_{pl}$$

as required. //

C. STABILITY

1. Remark: Given our identification of the stable operators on $(H_e, \mathscr{E}_e)$ with the causal bounded operators on $(H, \mathscr{E})$ it is natural to identify the stable feedback systems with those systems which have causal bounded gains.

2. Definition: A feedback system is *stable* if H_{ru}, H_{rn}, H_{yu}, and H_{yn} exist and are in $\mathscr{C}$.

3. Remark: Note that it is possible to give examples of systems with any combination of the gains H_{ru}, H_{rn}, H_{yu}, and H_{yn} causal and the remaining noncausal, and hence the requirement that all four gains lie in $\mathscr{C}$ for stability is nonredundant.[1] Note, however, that by Lemma 18 if F and P are causal and bounded, then the feedback system is stable if and only if $(I + FP)^{-1} \in \mathscr{C}$ (or $(I + PF)^{-1} \in \mathscr{C}$).

4. Property: Let a feedback system admit plant and feedback models as formulated in Remark B.17. Then the following are equivalent:

(i) The system is stable.
(ii) $(D_{f1}D_{pr} + N_{f1}N_{pr}) \in \mathscr{C} \cap \mathscr{C}^{-1}$.
(iii) $(D_{pl}D_{fr} + N_{pl}N_{fr}) \in \mathscr{C} \cap \mathscr{C}^{-1}$.

Proof: Clearly, if $(D_{f1}D_{pr} + N_{f1}N_{pr}) \in \mathscr{C} \cap \mathscr{C}^{-1}$ then it follows from Property B.20 and the fact that $\mathscr{C}$ is an algebra that the system is stable. Conversely, if the feedback system is stable $H_{yn} \in \mathscr{C}$ and hence so is $N_{pr}(D_{f1}D_{pr} + N_{f1}N_{pr})^{-1}N_{f1}$. Moreover, B.17 implies that $N_{pr}[D_{f1}D_{pr}]^{-1}$ is right coprime and $[D_{f1}D_{pr}]^{-1}N_{f1}$ is left coprime hence there exist U's and V's such that

$$U_r[N_{pr}] + V_r[D_{f1}D_{pr}] = I$$

and

$$[N_{f1}]U_1 + [D_{f1}D_{pr}]V_1 = I$$

Therefore

$$[U_r - V_rN_{f1}]N_{pr} + V_r[D_{f1}D_{pr} + N_{f1}N_{pr}] = U_rN_{pr} + V_rD_{f1}D_{pr} = I$$

showing that $N_{pr}(D_{f1}D_{pr} + N_{f1}N_{pr})^{-1}$ is right coprime. Similarly,

$$N_{f1}[U_1 - N_{pr}V_1] + [D_{f1}D_{pr} + N_{f1}N_{pr}]V_1 = N_{f1}U_1 + D_{f1}D_{pr}V_1 = I$$

implies that $(D_{f1}D_{pr} + N_{f1}N_{pr})^{-1}N_{f1}$ is left coprime. Given these coprimeness conditions and the fact that $N_{pr}(D_{f1}D_{pr} + N_{f1}N_{pr})^{-1}N_{f1} \in \mathscr{C}$, Property B.8.iii implies that $(D_{f1}D_{pr} + N_{f1}N_{pr}) \in \mathscr{C} \cap \mathscr{C}^{-1}$, as was to be shown, while a parallel argument applies to $(D_{pl}D_{fr} + N_{pl}N_{fr})$. //

5. Remark: Note that in the case of a *unity gain feedback system*, where $F = I$, we may take $D_{f1} = N_{f1} = D_{fr} = N_{fr} = I$, in which case the above result reduces to the classical system theoretic conditions[1] that $D_{pr} + N_{pr} \in \mathscr{C} \cap \mathscr{C}^{-1}$ or $D_{pl} + N_{pl} \in \mathscr{C} \cap \mathscr{C}^{-1}$.

6. Remark: Consistent with the result of Property 4, the feedback system stability problem is reduced to the causal invertibility problem. The entirety of Chapter 5 could thus be replicated at this point in the context of stability. Indeed, many of the results presented in Chapter 5 were motivated by classical stability theory. The topological approach to the causal invertibility problem represents an extension of classical Nyquist theory to our Hilbert resolution space setting, while the numerical range test for causal invertibility is motivated by the classical passivity condition for feedback system stability.

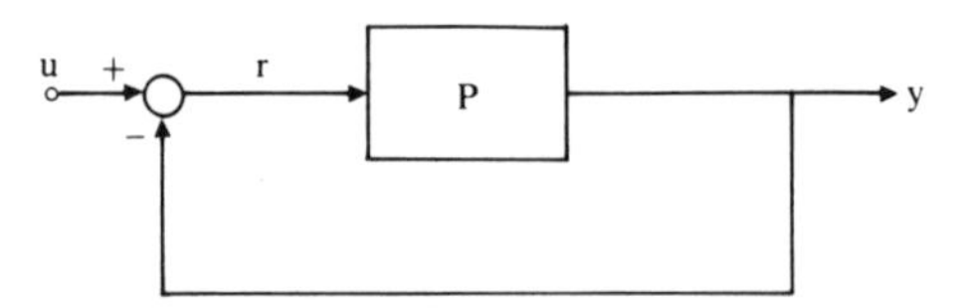

Fig.1. Unity gain feedback system.

7. Example: Consider the feedback system shown in Fig. 1 defined on $l_2(-\infty, \infty)$ with its usual resolution structure where $F = I$ and P is the *discrete convolution operator*, defined by

$$[Pr]_i = \sum_{j=-\infty}^{\infty} P_{i-j} r_j$$

where

$$P_k = \begin{cases} 1/2^k, & k \geq 0 \\ 0, & k < 0 \end{cases}$$

Since $P_k = 0$ for $k < 0$, P is causal. Moreover, since the discrete convolution operators form a commutative algebra we may adopt the doubly coprime representation

$$\begin{bmatrix} I & 0 \\ -P & I \end{bmatrix}^{-1} = \begin{bmatrix} I & 0 \\ P & I \end{bmatrix}$$

for the plant. Consistent with Remark 5 the resultant feedback system will be stable if and only if $I + P \in \mathscr{C} \cap \mathscr{C}^{-1}$. In the approach of Example 5.B.10 this will be the case if and only if $\deg_0 [1 + \hat{P}] = 0$, where $\hat{P}$ is the *discrete Fourier transform* of $\{P_k\}_{-\infty}^{\infty}$, defined by

$$\hat{P}(\theta) = \sum_{k=-\infty}^{\infty} \hat{P}_k e^{ik\theta} = \sum_{k=-\infty}^{\infty} \left[\frac{1}{2^k}\right] e^{ik\theta} = \sum_{k=-\infty}^{\infty} \left[\frac{e^{i\theta}}{2}\right]^k = \frac{2}{2 - e^{i\theta}}$$

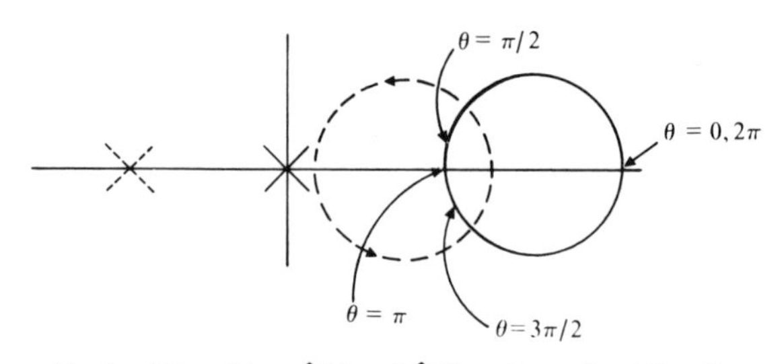

Fig. 2. Plot of $1 + \hat{P}(\theta)$ and $\hat{P}(\theta)$ as θ goes from 0 to 2π.

so that

$$1 + \hat{P}(\theta) = (4 - e^{i\theta})/(2 - e^{i\theta})$$

This function is plotted in Fig. 2 for θ from 0 to 2π. Since the plot does not encircle the origin $(I + P) \in \mathscr{C} \cap \mathscr{C}^{-1}$ and the feedback system is stable. Note, that if we plot simply $\hat{P}(\theta)$ rather than $1 + \hat{P}(\theta)$ the result is a simple shifted in the complex plane, as shown by the dashed lines in the Fig. 2, and we may test for stability by computing $\deg_{-1}[\hat{P}(\theta)]$ rather than $\deg_0[1 + \hat{P}(\theta)]$. Of course, this is the classical Nyquist test for the stability of such a feedback system.

D. SENSITIVITY

1. Remark: Intuitively, there are two reasons for adopting a feedback control strategy rather than an open loop strategy. First, by use of feedback it is possible to stabilize an otherwise unstable plant. Second, if a feedback system is properly designed it may be less sensitive to the noise input n and to deviations of the plant from its nominal model. In practice, this proves to be a major advantage of feedback since the model of the plant P_0 used in designing the system is, at best, an approximation to the actual plant P. Moreover, even if one begins with an accurate plant model, "real world" plants change their characteristics with aging, temperature, environmental factors, etc. Hence the ability of a feedback system to reduce the effect of the plant modeling error $P - P_0$ is an important feature in any feedback system design. Unlike those of the plant, the design and construction of the feedback F, or the open loop compensator M are under the control of the system designer, and we may therefore assume that the actual values for F and M coincide with their nominal values. Indeed, in many modern control systems F and M are implemented in the form of a subroutine on a control computer and are thus as accurate as the computer on which they are implemented.

Consistent with the above comments, the purpose of the present section is the formulation of an appropriate measure of the sensitivity of a feedback system. Although a number of absolute sensitivity concepts are possible, it proves most convenient to normalize our feedback system sensitivity measure to the sensitivity of a nominally equivalent open loop system. Since u represents the primary input to a system, we say that a given feedback system is *nominally equivalent* to an open loop system (as per Fig. 2 of Part III) if $H^{\mathrm{f}}_{yu} = H^{\mathrm{o}}_{yu}$ and $H^{\mathrm{f}}_{ru} = H^{\mathrm{o}}_{ru}$ when $P = P_0$. Here, the superscript f denotes the feedback system gain and the superscript o denotes the corresponding open loop system gain. We then have the following result:

2. Lemma: Let a given feedback system and a corresponding open loop system be designed so that they are nominally equivalent. Then

$$I - M - FP_0 M = 0$$

Proof: Since $y = n + P_0 r$ for both the feedback and open loop system, it suffices to consider the equality $H_{ru}^{\mathrm{f}} = H_{ru}^{\mathrm{o}}$. With the aid of Property B.20, we have $H_{ru}^{\mathrm{f}} = (1 + FP_0)^{-1}$ while $H_{ru}^{\mathrm{o}} = M$; hence

$$(1 + FP_0)^{-1} - M = 0$$

which reduces to the required equality upon multiplication on the left by $I + FP_0$. //

3. Remark: To obtain the desired sensitivity comparison we let $y \in H$ and

$$\delta^{\mathrm{f}} = y^{\mathrm{f}} - y_0^{\mathrm{f}}$$

where y^{f} is the response of the actual feedback system (with plant P) to u and y_0^{f} is the response of the nominal feedback system (with plant P_0) to u. Similarly, we let

$$\delta^{\mathrm{o}} = y^{\mathrm{o}} - y_0^{\mathrm{o}}$$

where y^{o} is the response to the actual open loop system to u and y_0^{o} is the response of the nominal open loop system to u. The ratio, in an appropriate sense, of δ^{f} and δ^{o} is then used as our normalized sensitivity measure.

4. Property: Let a given feedback system be nominally equivalent to an open loop system. Then

$$\delta^{\mathrm{f}} = H_{yn}^{\mathrm{f}} \delta^{\mathrm{o}}$$

whenever H_{yn}^{f} is well defined.

Proof: Since the two systems are nominally equivalent $y_0^{\mathrm{o}} = y_0^{\mathrm{f}}$. Hence

$$\delta^{\mathrm{o}} = \delta^{\mathrm{f}} + (y^{\mathrm{o}} - y^{\mathrm{f}})$$

Now, $y^{\rm o} = PMu$, while $y^{\rm f} = Pu + PFy^{\rm f}$. Hence

$$y^{\rm o} - y^{\rm f} = PMu - Pu + PFy^{\rm f}$$

Moreover, with the aid of the lemma and the fact that the nominal values for the plant input for the two systems $r_0^{\rm f}$ and $r_0^{\rm o}$ coincide, we have

$$PFy_0^{\rm f} = PFP_0 r_0^{\rm f} = PFP_0 r_0^{\rm o} = PFP_0 MU = -P(M - I)u$$

Combining these equalities we obtain

$$\begin{aligned}\delta^{\rm o} &= \delta^{\rm f} + (y^{\rm o} - y^{\rm f}) = \delta^{\rm f} + P(M - I)u + PFy^{\rm f} \\ &= \delta^{\rm f} - PFy_0^{\rm f} + PFy^{\rm f} + \delta^{\rm f} + PF\delta^{\rm f} = (I + PF)\delta^{\rm f}\end{aligned}$$

and $\delta^{\rm f} = (I + PF)^{-1}\delta^{\rm o} = H_{yn}^{\rm f}\delta^{\rm f}$, where the last equality follows from Property B.20. //

5. Remark: Although the formal derivation is rather complex, the equality $\delta^{\rm f} = H_{yn}\delta^{\rm o}$ is quite natural if one observes that the effect of the plant perturbation $P - P_0$ can be modeled as a noise term $n = (P - P_0)r$, from which it follows that H_{yn} is a natural measure of the effect of the plant perturbation on y.

6. Remark: Clearly our feedback system has sensitivity characteristics superior to those of the equivalent open loop system if H_{yn} is a contraction. Of course, for stability H_{yn} must also be well defined and causal. We are therefore interested in determining conditions under which the *sensitivity operator* H_{yn} is a causal contraction. Interestingly enough, such operators are commonly encountered in circuit theory, where they are termed *passive operators*.

7. Example: For the unity gain feedback system of Example C.7 $H_{yn} = (I + P)^{-1}$ is causal since the system is stable. Moreover, since H_{yn} is a discrete convolution operator,

$$\|H_{yn}\| = \sup_{\theta} |(1 + \hat{P}(\theta))^{-1}| = 1/\inf_{\theta} |(1 + \hat{P}(\theta))|$$

Now, an inspection of the plot for $(1 + \hat{P}(\theta))$ given in Fig. 2 will reveal that $|(1 + \hat{P}(\theta))| > 1$; hence

$$\|H_{yn}\| < 1$$

verifying that the sensitivity of this feedback system is superior to the nominally equivalent open loop system (with $M = (I + P)^{-1}$).

PROBLEMS

1. Show that any sequence convergent in the resolution topology has a unique limit.

2. Let $F: \$ \to \$$ and assume that $A \in \mathscr{B}$ satisfies $E^{\dagger}A = E^{\dagger}AE^{F(\dagger)}$. Then show that A is continuous in the resolution topology.

3. Give a characterization of $l_2[0, n]_e$, $l_2[0, \infty)_e$, and $l_2(-\infty, \infty)_e$, all with their usual resolution structure.

4. Determine whether every operator that admits a right (left) coprime fractional representation is continuous in the resolution topology.

5. Derive the formula for the inverse of a 2×2 matrix of operators used in the proof of Property B.12.

6. Show that conditions (i)–(iv) of Remark B.17 imply that $P = N_{pr}D_{pr}^{-1} = D_{pl}^{-1}N_{pl}$ and $F = N_{fr}D_{fr}^{-1} = D_{fl}^{-1}N_{fl}$ are both right and left coprime.

7. Give examples of feedback systems in which any three of the four gains H_{ru}, H_{yu}, H_{rn}, and H_{yn} are stable and the fourth is unstable.

8. Assuming that $A^{-1} \in \mathscr{C}$, show that

$$\begin{bmatrix} 0 & I \\ -I & A \end{bmatrix}^{-1} = \begin{bmatrix} A & -I \\ I & 0 \end{bmatrix}$$

is a doubly coprime fractional representation for A.

9. For a given nominal plant P_0 imbedded in a unity gain feedback system, derive a formula for an open loop compensator M such that the resultant open loop system will be nominally equivalent to the unity gain feedback system.

10. Show that an operator A is *passive* (i.e., a causal contraction) if and only if

$$(\|x\|^{\dagger})^2 - (\|y\|^{\dagger})^2 \geq 0, \qquad \dagger \in \$$$

whenever $y = Ax$.

11. An operator A is said to be *lossless* if and only if it is isometric and causal. Show that A is lossless if and only if it is passive and if

$$\|x\|^2 - \|y\|^2 = 0$$

whenever $y = Ax$.

12. An alternative concept of passivity requires that A be causal and have positive hermitian part $A + A^* \geq 0$. Show that this is the case if and only if

$$(E^{\dagger}x, Ax) \geq 0, \qquad \dagger \in \$, \quad x \in H$$

13. Show that the inverse of an operator that is passive in the sense of Problem 12 is also passive in that sense.

14. Let A and B be passive in the sense of Problem 12 and show that $A + B$ is passive in that sense.

15. Let A and B be passive in the sense of Problem 10 and show that AB is passive in that sense.

REFERENCES

1. Callier, F. M., and Desoer, C. A., Open loop unstable convolution feedback systems with dynamical feedback, *Automatica* **12**, 507–518 (1976).
2. Rosenbrock, H. H., "State-Space and Multivariable Theory." Wiley, New York, 1970.

9

Asymptotic Design

A. STABILIZATION

1. Remark: The feedback system design problem may be naturally subdivided into two sequential tasks:

(i) the satisfaction of design constraints and
(ii) the "optimization" of system performance within these constraints.

The most important design constraint is *stabilization*, which must be achieved in any practical design. Depending on the application one may also face additional design constraints in the form of asymptotic tracking and disturbance rejection specifications. Once all design constraints have been satisfied, any remaining design latitude may then be used to optimize the system performance. This may take the form of a formal optimization problem wherein one minimizes energy consumption, tracking error, or norm of the sensitivity operator. Alternatively, one may simple minimize a qualitative measure of system performance, such as overshoot, vibration, or reliability, etc.

The purpose of the present section is to characterize those feedback laws F stabilizing any given plant P. In the process a transformation of variables is formulated wherein the stabilizing feedback laws are expressed in terms of a new design parameter $W \in \mathscr{C}$. Moreover, the resultant feedback system gains H_{ru}, H_{rn}, H_{yu}, and H_{yn} prove to be linear (actually affine) in this new design parameter, thereby greatly simplifying the remainder of the design process!

Stabilization Theorem: Let a feedback system plant P admit a doubly coprime fractional representation

$$\begin{bmatrix} V_{pr} & U_{pr} \\ -N_{pl} & D_{pl} \end{bmatrix}^{-1} = \begin{bmatrix} D_{pr} & -U_{pl} \\ N_{pr} & V_{pl} \end{bmatrix}$$

where $P = N_{pr}D_{pr}^{-1} = D_{pl}^{-1}N_{pl}$. Then the class of feedback laws $F \in \mathscr{B}$ that stabilize the feedback system are given by

$$\begin{aligned} F &= N_{fr}D_{fr}^{-1} = (D_{pr}W + U_{pl})(N_{pr}W + V_{pl})^{-1} \\ &= (WN_{pl} + V_{pr})^{-1}(-WD_{pl} + U_{pr}) = D_{fl}^{-1}N_{fl} \end{aligned}$$

whenever the appropriate inverses exist in $\mathscr{B}$ with $W \in \mathscr{C}$. Moreover, F admits the doubly coprime fractional representation

$$\begin{aligned} \begin{bmatrix} V_{fr} & U_{pr} \\ -N_{fl} & D_{fl} \end{bmatrix} &= \begin{bmatrix} D_{pl} & N_{pl} \\ -(-WD_{pl} + U_{pr}) & WN_{pl} + V_{pr} \end{bmatrix}^{-1} \\ &= \begin{bmatrix} N_{pr}W + V_{pl} & -N_{pr} \\ -D_{pr}W + U_{pl} & D_{pr} \end{bmatrix} = \begin{bmatrix} D_{fr} & -U_{fl} \\ N_{fr} & V_{fl} \end{bmatrix} \end{aligned}$$

and the feedback systems gains resulting from the use of such a feedback law take the form

$$\begin{bmatrix} H_{ru} & H_{ru} \\ H_{yu} & H_{yn} \end{bmatrix} = \begin{bmatrix} [D_{pr}WN_{pl} + D_{pr}V_{pr}] & [D_{pr}WD_{pl} - D_{pr}U_{pr}] \\ [N_{pr}WN_{pl} + N_{pr}V_{pr}] & [N_{pr}WD_{pl} + V_{pl}D_{pl}] \end{bmatrix}$$

Proof: By Property 8.C.4 the feedback system will be stable if and only if $D_{fl}D_{pr} + N_{fl}N_{pr} \in \mathscr{C} \cap \mathscr{C}^{-1}$. Hence we begin by choosing $K \in \mathscr{C} \cap \mathscr{C}^{-1}$ such that

$$D_{fl}D_{pr} + N_{fl}N_{pr} = K$$

for D_{f1} and N_{f1} in $\mathscr{C}$ given the double coprime fractional representation for P. Now let $D^h_{f1} = LN_{pl}$ and $N^h_{f1} = -LD_{pl}$ where $L \in \mathscr{C}$ is arbitrary. Then

$$D^h_{f1}D_{pr} + N^h_{f1}N_{pr} = L[N_{pl}D_{pr} - D_{pl}N_{pr}] = 0$$

since $N_{pl}D_{pr} = D_{pl}N_{pr}$ for a doubly coprime fractional representation. D^h_{f1} and N^h_{f1} are thus homogeneous solutions to the given equation. Indeed, they represent all possible homogeneous solution in $\mathscr{C}$. To verify this fact let $\tilde{N}^h_{f1}$ and $\tilde{D}^h_{f1}$ be arbitrary homogeneous solutions in $\mathscr{C}$ to the given equation. Then

$$\tilde{D}^h_{f1}D_{pr} + \tilde{N}^h_{f1}N_{pr} = 0$$

Now let $\tilde{L} = -\tilde{N}^n_{f1}D^{-1}_{pl}$, in which case

$$\tilde{N}^h_{f1} = -\tilde{L}D_{pl}$$

while

$$\tilde{D}^h_{f1} = -\tilde{N}^h_{f1}N_{pr}D^{-1}_{pr} = -N^h_{f1}D^{-1}_{pl}N_{pl} = LN_{pl}$$

showing that $\tilde{N}^h_{f1}$ and $\tilde{D}^h_{f1}$ have the same form as the class of solutions already formulated. To show that they fall into that class it remains to show that $\tilde{L} \in \mathscr{C}$. To this end consider the equality

$$\tilde{L} = \tilde{N}^h_{f1}D^{-1}_{pl} = -\tilde{N}^h_{f1}D^{-1}_{pl}(N_{pl}U_{pl} + D_{pl}V_{pl}) = \tilde{D}^h_{f1}U_{pl} - \tilde{N}^h_{f1}V_{pl} \in \mathscr{C}$$

We have constructed all possible homogeneous solutions in $\mathscr{C}$ for the given equation. Moreover, since $U_{pr}N_{pr} + V_{pr}D_{pr} = I$,

$$[KV_{pr}]D_{pr} + [KU_{pr}]N_{pr} = K$$

showing that $D^p_{f1} = KV_{pr}$ and $N^p_{f1} = KU_{pr}$ define a particular solution to the given equation. The set of all possible feedback laws that stabilize the given plant therefore take the form

$$\begin{aligned} F &= (LN_{pl} + KV_{pr})^{-1}(-LD_{pl} + KU_{pr}) \\ &= (WN_{pl} + V_{pr})^{-1}(-WD_{pl} + U_{pr}) = D^{-1}_{f1}N_{f1} \end{aligned}$$

whenever the inverse exists. Here $W = K^{-1}L$ is an arbitrary element of $\mathscr{C}$ since $L \in \mathscr{C}$ and $K \in \mathscr{C} \cap \mathscr{C}^{-1}$.

Now, observe that with $D_{f1} = WN_{pl} + V_{pr}$ and $N_{f1} = -WD_{pl} + U_{pr}$ as defined above

$$\begin{aligned} D_{f1}D_{pr} + N_{f1}N_{pr} &= (WN_{pl} + V_{pr})D_{pr} + (-WD_{pl} + U_{pr})N_{pr} \\ &= W(N_{pl}D_{pr} - D_{pl}N_{pr}) + (V_{pr}D_{pr} + U_{pr}N_{pr}) = I \end{aligned}$$

Substituting this equality into the expressions of Property 8.B.20 now yields

$$\begin{bmatrix} H_{ru} & H_{rn} \\ H_{yu} & H_{yn} \end{bmatrix} = \begin{bmatrix} D_{pr}D_{fl} & -D_{pr}N_{fl} \\ N_{pr}D_{fl} & I - N_{pr}N_{fl} \end{bmatrix}$$

$$= \begin{bmatrix} [D_{pr}WN_{pl} + D_{pr}V_{pr}] & [D_{pr}WD_{pl} - D_{pr}U_{pr}] \\ [N_{pr}WN_{pl} + N_{pr}V_{pr}] & I + [N_{pr}WD_{pl} - N_{pr}U_{pr}] \end{bmatrix}$$

$$= \begin{bmatrix} [D_{pr}WN_{pl} + D_{pr}V_{pr}] & [D_{pr}WD_{pl} - D_{pr}U_{pr}] \\ [N_{pr}WN_{pl} + N_{pr}V_{pr}] & [N_{pr}WD_{pl} + V_{pl}D_{pl}] \end{bmatrix}$$

where the last equality follows from Remark 8.B.13. Moreover, the equality $D_{fl}D_{pr} + N_{fl}N_{pr} = I$ implies that the coprimeness conditions (iii) and (iv) of Remark 8.B.17, required for any feedback system to be well defined, are satisfied. Before proceeding to the verification of the coprimeness conditions (i) and (ii) of Remark 8.B.17 we need to construct an appropriate right fractional representation for our feedback law $F = D_{fl}^{-1}N_{fl}$. To this end consider the equality

$$\begin{aligned}
&(-WD_{pl} + U_{pr})(N_{pr}W + V_{pl}) \\
&\quad = -WD_{pl}N_{pr}W - WD_{pl}V_{pl} + U_{pr}N_{pr}W + U_{pr}V_{pl} \\
&\quad = -WN_{pl}D_{pr}W - W[I - N_{pl}U_{pl}] + [I - V_{pr}N_{pr}]W + V_{pr}U_{pl} \\
&\quad = -WN_{pl}D_{pr}W + WN_{pl}U_{pl} - V_{pr}N_{pr}W + V_{pr}U_{pl} \\
&\quad = (WN_{pl} + V_{pr})(-D_{pr}W + U_{pl})
\end{aligned}$$

Multiplying this equality on the right by $(WN_{pl} + V_{pr})^{-1}$ and on the left by $(N_{pr}W + V_{pl})^{-1}$, we obtain a right fractional representation for F in the form

$$\begin{aligned}
F = D_{fl}^{-1}N_{fl} &= (WN_{pl} + V_{pr})^{-1}(-WD_{pl} + U_{pr}) \\
&= (-D_{pr}W + U_{pl})(N_{pr}W + V_{pl})^{-1} = N_{fr}D_{fr}^{-1}
\end{aligned}$$

Furthermore, with the aid of this choice for N_{fr} and D_{fr} we obtain

$$\begin{aligned}
D_{pl}D_{fr} + N_{pl}N_{fr} &= D_{pl}(N_{pr}W + V_{pl}) + N_{pl}(-D_{pr}W + U_{pl}) \\
&= (D_{pl}N_{pr} - N_{pl}D_{pr})W + (D_{pl}V_{pl} + N_{pl}U_{pl}) = I
\end{aligned}$$

Hence coprimeness conditions (i) and (ii) of Remark 8.B.17 are satisfied. Moreover, the above equalities imply that

$$\begin{bmatrix} D_{pl} & N_{pl} \\ -(WD_{pl} + U_{pr}) & WN_{pl} + V_{pr} \end{bmatrix}^{-1} = \begin{bmatrix} N_{pr}W + V_{pl} & -N_{pr} \\ -D_{pr}W + V_{pl} & D_{pr} \end{bmatrix}$$

defines a doubly coprime fractional representation for F, thereby completing the proof. //

2. Remark: The significance of the theorem is twofold. First, it gives a complete characterization of the stabilizing feedback control laws for a given plant parameterized by $W \in \mathscr{C}$. Additionally, however, if one uses W as the system design parameter rather than F, all four system gains are linear (actually affine) in the new design parameter. As we shall see in the sequel, this greatly simplifies the remainder of the design problem by effectively converting it from a feedback problem in terms of F to an open loop problem in terms of W.

3. Example: Let P be defined by the ordinary differential operator

$$P = \{[D^2 - 4]^{-1}[D + 1]\}$$

defined on $L_2(-\infty, \infty)$ with its usual resolution structure. Here, D denotes the derivative operator, and the differential operator defines an element of $\mathscr{C}$ if and only if it is proper and the roots of its characteristic polynomials have negative real parts. In this case the operator is not in $\mathscr{C}$ though it admits the fractional representation

$$\begin{aligned} P &= \{[D^2 - 4]^{-1}[D + 1]\} \\ &= \{[D^2 + 4D + 4]^{-1}[D + 1]\}\{[D + 2]^{-1}[D - 2]\}^{-1} = N_p D_p^{-1} \end{aligned}$$

Since such ordinary differential operators are commutative, this fractional representation may serve as both a right and a left fractional representation, for which reason we have deleted the subscript l or r from the notation. Similarly, commutativity allows us to restrict consideration to a single pair of operators U_p and V_p that characterize the coprimeness of this representation by

$$\begin{aligned} &\{[3]^{-1}[16]\}\{[D^2 + 4D + 4]^{-1}[D + 1]\} \\ &\qquad + \{[D + 2]^{-1}[D + \tfrac{2}{3}]\}\{[D + 2]^{-1}[D - 2]\} \\ &= U_p N_p + V_p D_p = I \end{aligned}$$

Indeed, because of the commutativity U_p, N_p, V_p, and D_p may be used in both right and left fractional representations and, in fact, define a doubly coprime fractional representation. Hence the stabilizing feedback laws for this plant take the form

$$\begin{aligned} F &= \{WN_p + V_p\}^{-1}\{-WD_p + U_p\} \\ &= \{W[D^2 + 4D + 4]^{-1}[D + 1] + [D + 2]^{-1}[D + \tfrac{2}{3}]\}^{-1} \\ &\quad \times \{-W[S + 2]^{-1}[D - 2] + [3]^{-1}[16]\} \end{aligned}$$

where W is an arbitrary causal operator that can be chosen to satisfy additional design objectives over and above stabilization.

For example, it follows from Property 8.D.4 that the feedback system will be insensitive to plant perturbations if

$$\begin{aligned} H_{yn} &= \{N_p W D_p + V_p D_p\} \\ &= \{W[D+1] + [D+\tfrac{2}{3}][D+2]\} \\ &\quad \times \{[D^3 + 6D^2 + 12D + 8]^{-1}[D-2]\} \end{aligned}$$

is small in an appropriate sense. Now, for any causal W this system will have a gain of 1 at high frequencies, hence it is impossible simultaneously to stabilize the feedback system and to make $H_{yn} = 0$. By proper choice of W it is, however, possible simultaneously to stabilize the system and to make H_{yn} small in some desired frequency band. For instance, if we desire to make H_{yn} small in the frequency band below 1 rad/sec, we may choose W so that H_{yn} has a factor of the form $D[D^2 + 1]$, which will cause the system to have a null reponse to dc and 1-rad noise inputs (and, one hopes, small responses to other low-frequency noise inputs). Moreover, it follows from Property 8.D.4 that the effect of plant perturbations at these frequencies will also be null. To achieve this end, let us assume that W takes the form

$$W = \{[D+c]^{-1}[aD+b]\}$$

where $c > 0$ for causality. We then desire to solve the operator equation

$$\begin{aligned} &\{[D+c]^{-1}[aD+b][D+1] + [D+\tfrac{2}{3}][D+2]\} \\ &\quad = \{[D+c]^{-1}[D^3 + D]\} \end{aligned}$$

A little algebra will now reveal that this operator equation has the unique solution $a = -\frac{29}{3}$, $b = -\frac{28}{3}$, and $c = 7$. With this solution, if we let W be the causal differential operator $W = [-\frac{1}{3}]\{[D+7]^{-1}[29D + 28]\}$, then the corresponding feedback law F yields the sensitivity operator

$$H_{yn} = \{[D^4 + 13D^3 + 54D^2 + 92D + 56]^{-1}[D^4 - 2D^3 + D^2 - 2D]\}$$

which has the required small low-frequency response.

4. Example: Although our theory is simplified in the commutative case, it remains valid in the general case. For instance, consider the plant on $L_2(-\infty, \infty)$ with its usual resolution structure characterized by the operator $[f][\tilde{P}]$, where $[f]$ is an L_∞ function with L_∞ reciprocal $[f^{-1}]$ and $\tilde{P}$ is the *ideal predictor*, defined by

$$[\tilde{P}g](\dagger) = g(\dagger + 1)$$

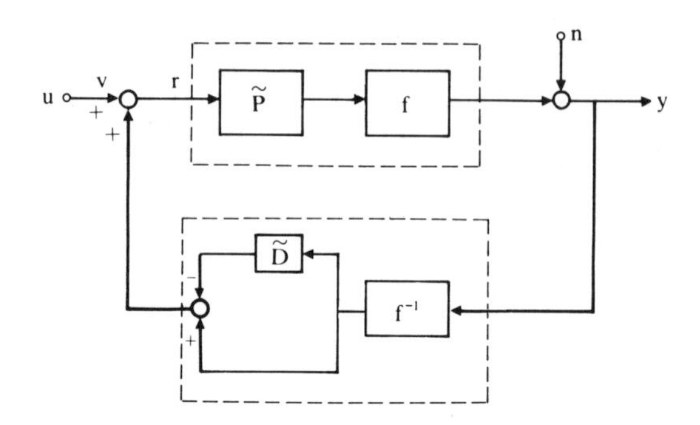

Fig. 1. Time-varying distributed feedback system.

Of course, $\tilde{P}^{-1}$ is the *ideal delay* $\tilde{D}$, defined by

$$[\tilde{D}g](\dagger) = g(\dagger - 1)$$

For this plant we have

$$[f][\tilde{P}] = [f][\tilde{D}]^{-1} = [\tilde{D}f^{-1}]^{-1}[I] = N_{pr}D_{pr}^{-1} = D_{pl}^{-1}N_{pl}$$

while a doubly coprime fractional representation takes the form

$$\begin{bmatrix} V_{pr} & U_{pr} \\ -N_{pl} & D_{pl} \end{bmatrix}^{-1} = \begin{bmatrix} I & [f^{-1}] \\ -I & [\tilde{D}f^{-1}] \end{bmatrix}^{-1} = \begin{bmatrix} [\tilde{D}] & -I \\ [f] & 0 \end{bmatrix} = \begin{bmatrix} D_{pr} & -U_{pl} \\ N_{pr} & V_{pl} \end{bmatrix}$$

so that the general form of the stabilizing feedback law is

$$F = [W]^{-1}(-W[\tilde{D}][f^{-1}] + [f^{-1}])$$

In particular, if we take $W = I$, we obtain the feedback law

$$F = [I - \tilde{D}][f^{-1}]$$

which is implemented in the feedback system of Fig. 1.

B. TRACKING

1. Remark: Although stabilization is the first and foremost constraint in any feedback system design, one must also design a feedback law to control the system in some prescribed manner. Often this takes the form of a requirement that the response of the system track (or follow the input) in an appropriate sense. Indeed, this is the essence of the so-called *servomechanism problem.* Typically, one deals with a restricted class of inputs,

$x \in D$ and requires that the response $H_{uy}x$ to such inputs, be asymptotic to x. Unfortunately, all $x \in H$ are asymptotic to 0 in the sense that

$$\lim_{\dagger} \|E_\dagger x\| = 0$$

and hence asymptotic tracking is trivial in a Hilbert space setting. On the other hand one may say that two elements x_e and y_e in an extended resolution space $(H_e, \mathscr{E}_e)$ are asymptotic if $y_e - x_e \in H$, i.e. if their difference is asymptotic to 0. With our identification of stable systems defined on $(H_e, \mathscr{E}_e)$ with causal bounded operators on $(H, \mathscr{E})$, the required definition takes the following form:

2. Definition: Let $T \in \mathscr{B}$ admit a left coprime fractional representation $T = D_{tl}^{-1}N_{tl}$. Then we say that a feedback system *tracks* $T_e(H)$ *if* $[I - H_{uy}]T \in \mathscr{C}$.

3. Remark: This definition is compatible with the intuition of Remark 1 if one lifts the operator $[I - H_{yu}]T$ to the extended space. Since T may fail to be stable, $T_e(H)$ may include unbounded elements in H_e, which, by Property 8.A.14, are mapped to H by $[I_e - H_{yue}]$. Hence if $r_e = T_e x$, $x \in H$, and $z_e = H_{yue} r_e$, then

$$z_e - r_e \in H$$

verifying that the output of our system tracks its input as we have described.

4. Property: Let a stable feedback system have input/output gain H_{yu}. Then the system tracks $T_e(H)$ if and only if D_{tl} is a right divisor of $[I - H_{yu}]$, i.e., $[I - H_{yu}] = XD_{tl}$ for some $X \in \mathscr{C}$.

Proof: If $[I - H_{yu}] = XD_{tl}$ for some $X \in \mathscr{C}$, then

$$[I - H_{yu}]T = [I - H_{yu}]D_{tl}^{-1}N_{tl} = XN_{tl} \in \mathscr{C}$$

Conversely, if $[I - H_{yu}]T \in \mathscr{C}$, then

$$\begin{aligned} X &= [I - H_{yu}]D_{tl}^{-1} = [I - H_{yu}]D_{tl}^{-1}[N_{tl}U_{tl} + D_{tl}V_{tl}] \\ &= [I - H_{yu}]TU_{tl} + [I - H_{yu}]V_{tl} \in \mathscr{C} \end{aligned}$$

Hence $[I - H_{yu}] = XD_{tl}$ for some $X \in \mathscr{C}$. //

5. Example: Consider a stable feedback system with

$$\begin{aligned} I - H_{uy} = {} & [D^5 + 15D^4 + 80D^3 + 200D^2 + 240D + 112]^{-1} \\ & \times [D^4 + D^3 + D^2 + D] \end{aligned}$$

where the characteristic roots of this proper differential operator are -2 (with multiplicity 4) and -7. Now, to track the response of an integrator, $T = [D]^{-1}$ we write

$$T = \{[D+1]^{-1}[D]\}^{-1}\{[D+1]^{-1}[1]\} = D_t^{-1}N_t$$

where

$$\{[D+1]^{-1}[1]\}\{[1]\} + \{[D+1]^{-1}[D]\}\{[1]\} = N_t U_t + D_t V_t = I$$

Since these differential operators are all commutative, we have dropped the left/right notation in the above fractional representation as in Example A.3. Now a little algebra will reveal that

$$[I - H_{yu}]D_t^{-1} = \{[D^5 + 15D^4 + 80D^3 + 200D^2 + 240D + 112]^{-1} \times [D^4 + 2D^3 + 2D^2 + 2D + 1]\}$$

which has the same characteristic roots as $[I - H_{yu}]$ and therefore also represents a causal operator. Accordingly, the feedback system tracks $T_e(H)$. In particular, $T_e(H)$ includes functions that are asymptotic to the step function and are commonly employed in control theory.

If one were to replace T by the coprime fractional representation

$$\{[D^3 + D]^{-1}[I]\} = \{[D^3 + 3D^2 + 3D + 1]^{-1}[D^3 + D]\}^{-1} \times \{[D^3 + 3D^2 + 3D + 1]^{-1}[I]\}$$

then

$$[I - H_{yu}]D_t^{-1} = \{[D^5 + 15D^4 + 80D^3 + 200D^2 + 240D + 112]^{-1} \times [D^4 + 4D^3 + 6D^2 + 4D + 1]\} \in \mathscr{C}$$

and hence the system also tracks the responses of this more general T. Indeed, $T_e(H)$ includes functions asymptotic both to step functions and to 1-rad/sec oscillations.

6. Property: Let a feedback system with plant $P = N_{pr}D_{pr}^{-1} = D_{pl}^{-1}N_{pl}$ be characterized by the doubly coprime fractional representation

$$\begin{bmatrix} V_{pr} & U_{pr} \\ -N_{pl} & D_{pl} \end{bmatrix}^{-1} \begin{bmatrix} D_{pr} & -U_{pl} \\ N_{pr} & V_{pl} \end{bmatrix}$$

and let T admit a left coprime fractional representation $T = D_{tl}^{-1}N_{tl}$. Then there exists a feedback law that simultaneously stabilizes the system and causes it to tract $T_e(H)$ if and only if the equation

$$N_{pr}WN_{pl} + XD_{tl} = [I - N_{pr}V_{pr}]$$

admits solutions W and X in $\mathscr{C}$. Moreover, when such solutions exist the required feedback law is given by

$$F = (-D_{pr}W + U_{pl})(N_{pr}W + V_{pl})^{-1}$$

$$= (WN_{pl} + V_{pr})^{-1}(-WD_{pl} + U_{pr})$$

Proof: Since every stabilizing feedback law for the system is characterized by the stabilization theorem of Section A, the corresponding H_{yu} takes the form $H_{yu} = N_{pr}WN_{pl} + N_{pr}V_{pr}$ for some $W \in \mathscr{C}$. Thus if a stabilizing feedback law also causes the system to track $T_e(H)$, Property 4 implies that there exists $X \in \mathscr{C}$ such that

$$XD_{tl} = I - H_{yu} = I - N_{pr}WN_{pl} - N_{pr}V_{pr}$$

which upon rearrangement of terms reduces to the desired design equation. Conversely, if the given design equation admits solutions X and $W \in \mathscr{C}$, then the stabilization theorem implies that the specified F will stabilize the system and satisfy $I - H_{yu} = XD_{tl}$. Thus Property 4 implies that the resultant system will track $T_e(H)$ and the proof is complete. //

7. Corollary: Under the hypotheses of Property 6 a necessary condition for the existence of a stabilizing feedback law that causes the system to track $T_e(H)$ is that $N_{pl}D_{tl}^{-1}$ define a right coprime fractional representation.

Proof: By Property 6 the required feedback law exists if and only if there exist W and $X \in \mathscr{C}$ such that

$$N_{pr}WN_{pl} + XD_{tl} = I - N_{pr}V_{pr}$$

Now, Remark 8.B.13 implies that $N_{pr}V_{pr} = V_{pl}N_{pl}$, which upon substitution into the above yields

$$[N_{pr}W + V_{pl}]N_{pl} + [X]D_{tl} = I$$

as required. //

8. Remark: If one restricts consideration to an algebra of commutative operators (say, the constant coefficient linear differential operators), the coprimeness condition of Corollary 7 can be shown to be both necessary and sufficient for the existence of a stabilizing feedback law that also causes the system to track $T_e(H)$. Indeed, in that case one can give a complete parametrization of the solution space for the design equations. In fact, such results extend to a class of multivariate systems.[2] No such result is, however, known in our general operator theoretic setting.

9. Example: Consider the feedback system with $P = \tilde{P}$, the ideal predictor on $L_2(-\infty, \infty)$ with its usual resolution structure. Here $P = [\tilde{D}]^{-1}[I] = [I][\tilde{D}]^{-1}$ defines a doubly coprime fractional representation where

$$\begin{bmatrix} V_{pr} & U_{pr} \\ -N_{pl} & D_{pl} \end{bmatrix}^{-1} = \begin{bmatrix} 0 & I \\ -I & \tilde{D} \end{bmatrix}^{-1} = \begin{bmatrix} \tilde{D} & -I \\ I & 0 \end{bmatrix} = \begin{bmatrix} D_{pr} & -U_{pl} \\ N_{pr} & V_{pl} \end{bmatrix}$$

with this, the design equation of Property 6 takes the form

$$W = I - XD_{tl}$$

and hence the required $W \in \mathscr{C}$ is given explicitly in terms of D_{tl} and an arbitrary design parameter $X \in \mathscr{C}$. As an example consider the tracking model

$$T = \{[D^3 + D]^{-1}[1]\} = \{[D^3 3D^2 + 3D + 1]^{-1}[D^3 + D]\}^{-1} \times \{[D^3 + 3D^2 + 3D + 1]^{-1}[1]\}$$

of Example 5 where D is the differential operator. Recall that $T_e(H)$ includes functions that are asymptotic to step functions and 1-rad/sec oscillations. With $X = I$ we obtain

$$\begin{aligned} W &= \{[D^3 + 3D^2 + 3D + 1]^{-1}[3D^2 + 2D + 1]\} \\ F &= -\tilde{D} + W^{-1} = \{-\tilde{D} + [3D^2 + 2D + 1]^{-1}[D^3 + 3D^2 + 3D + 1]\} \\ &= \{[3D^3 + 2D +]^{-1}[D^2 + 3D + 3D + 1 \\ &\quad - D^3\tilde{D} - 3D^2\tilde{D} - 3D\tilde{D} - D]\} \end{aligned}$$

and

$$[I - H_{yu}] = \{[D^3 + 3D^2 + 3D + 1]^{-1}[D^3 + D]\}$$

Alternatively, with $X = 0$ we have $W = I$, in which case $F = I - \tilde{D}$ and $[I - H_{yu}] = 0$. Indeed, in this degenerate case H_{yu} is the identity and the system tracks everything.

C. DISTURBANCE REJECTION

1. Remark: In addition to requiring that a feedback system be stable and asymptotically track $T_e(H)$, one may also require that it asymptotically reject a prescribed set of disturbances $R_e(H)$ where $R = D_{rl}^{-1}N_{rl}$ admits a left coprime fractional representation.

2. Definition: Let $R \in \mathscr{B}$ admit a left coprime fractional representation $R = D_{rl}^{-1}N_{rl}$. Then we say that a feedback system *rejects* $R_e(H)$ if $[H_{yn}]R \in \mathscr{C}$.

3. Remark: As in the case of the tracking concept, since R may not be in $\mathscr{C}$, $R_e(H)$ may include unbounded elements $n_e \in H_e$. However, since $[H_{yn}]R \in \mathscr{C}$, Property 8.A.14 implies that $H_{yn}n_e \in H$ and hence asymptotic to zero in the sense of Remark B.1. Hence the requirement that $[H_{yn}]R$ lie in $\mathscr{C}$ in our Hilbert resolution space setting is equivalent to an *asymptotic disturbance rejection* criterion when the feedback system model is lifted to the extension space. The theory surrounding the disturbance rejection concept closely parallels that formulated for the tracking problem in the previous section, and the following results are stated without proof.

4. Property: Let a stable feedback system have noise/output gain H_{yn}. Then the system rejects $R_e(H)$ if and only if D_{rl} is a *right divisor* of $[H_{yn}]$, i.e., $[H_{yn}] = -YD_{rl}$ for some $Y \in \mathscr{C}$.

5. Property: Let a feedback system with plant $P = N_{pr}D_{pr}^{-1} = D_{pl}^{-1}N_{pl}$ be characterized by the doubly coprime fractional representation

$$\begin{bmatrix} V_{pr} & U_{pr} \\ -N_{pl} & D_{pl} \end{bmatrix}^{-1} \begin{bmatrix} D_{pr} & -U_{pl} \\ N_{pr} & V_{pl} \end{bmatrix}$$

and let R admit a left coprime fractional representation $R = D_{rl}^{-1}N_{rl}$. Then there exists a feedback law that simultaneously stabilizes the system and causes it to reject $R_e(H)$ if and only if the equation

$$N_{pr}WD_{pl} + YD_{rl} = -[I - N_{pr}U_{pr}]$$

admits solutions W and Y in $\mathscr{C}$. Moreover, when such solutions exist the required feedback law is given by

$$\begin{aligned} F &= (-D_{pr}W + U_{pl})(N_{pr}W + V_{pl})^{-1} \\ &= (WN_{pl} + V_{pr})^{-1}(-WD_{pl} + U_{pr}) \end{aligned}$$

6. Remark: If the above design equation is satisfied by some W and $Y \in \mathscr{C}$, then

$$N_{pr}[-WD_{pl} - U_{pr}] + [-Y]D_{rl} = I$$

so that N_{pr} and D_{rl} are *skew coprime* in the obvious sense. Of course, if we restrict consideration to a commutative subalgebra of operators, this reduces to a classical coprimeness condition that, like the corresponding condition for the solution of the tracking problem, is in fact necessary and sufficient. In our general setting, however, this "cross coprimeness" condition proves to be of little value.

7. Example: For the feedback system of Example B.9 the design equation of Property 5 reduces to

$$W\tilde{D} + YD_{rl} = N_{pr}WD_{pl} + YD_{rl} = -[I - N_{pr}U_{pr}] = 0$$

Thus $Y = \tilde{D}$ and $W = D_{rl}$ is always a solution for any specified $R = D_{rl}^{-1}N_{rl}$.

8. Remark: As a final variation on the asymptotic design problem, consider the case where one is required to construct a feedback law that will simultaneously stabilize the system, cause it to track $T_e(H)$, and cause it to reject $R_e(H)$. Needless to say, this will be achieved if and only if we can solve the design equations for the tracking and rejection problem simultaneously with the same W (and hence the same feedback law). We thus obtain the following:

9. Property: Let a feedback system with plant $P = N_{pr}D_{pr}^{-1} = D_{pl}^{-1}N_{pl}$ be characterized by the doubly coprime fractional presentation

$$\begin{bmatrix} V_{pr} & U_{pr} \\ -N_{pl} & D_{pl} \end{bmatrix}^{-1} \begin{bmatrix} D_{pr} & -U_{pl} \\ N_{pr} & V_{pl} \end{bmatrix}$$

and let T and R admit left coprime fractional representations $T = D_{tl}^{-1}N_{tl}$ and $R = D_{rl}^{-1}N_{rl}$. Then there exists a feedback law that simultaneously stabilizes the system, causes it to track $T_e(H)$, and causes it to reject $R_e(H)$ if and only if the equations

$$N_{pr}WN_{pl} + XD_{tl} = [I - N_{pr}V_{pr}]$$

and

$$N_{pr}WD_{pl} + YD_{rl} = -[I - N_{pr}U_{pr}]$$

admit simultaneous solutions X, Y, and W in $\mathscr{C}$. Moreover, when such solutions exist the required feedback law is given by

$$F = (-D_{pr}W + U_{pl})(N_{pr}W + V_{pl})^{-1} = (WN_{pl} + V_{pr})^{-1}(-WD_{pl} + U_{pl})$$

10. Example: For the feedback system of Example B.9 the design equations of Property 9 reduce to

$$W = I - XD_{tl} \qquad \text{and} \qquad \tilde{W}D + YD_{rl} = 0$$

Substituting the former into the latter, we obtain

$$XD_{t1}\tilde{D} - YD_{r1} = \tilde{D}$$

To simplify the remainder of the analysis let us restrict consideration to the case of *differential-delay systems*, wherein our operators are all generated by the differential operator D and the delay operator $\tilde{D}$. Since $D\tilde{D} = \tilde{D}D$ this algebra is commutative. Hence if D_{t1} and D_{r1} are coprime, there exist differential-delay operators P and $Q \in \mathscr{C}$ such that

$$PD_{t1} + QD_{r1} = I$$

or equivalently, $PD_{t1}\tilde{D} + Q\tilde{D}D_{r1} = D$. Thus if we take $X = P$ and $Y = -Q\tilde{D}$ we have

$$XD_{t1}\tilde{D} - YD_{r1} = PD_{t1}\tilde{D} + Q\tilde{D}D_{r1} = \tilde{D}$$

as required, while the corresponding W is given by

$$W = I - XD_{t1} = I - PD_{t1}$$

Now assume that we desire to track the "steplike" functions in $T_e(H)$ characterized by

$$T = [D]^{-1}[1] = \{[D + 1]^{-1}[D]\}^{-1}\{[D + 1]^{-1}[1]\}$$

and reject the 1-rad/sec oscillations in $R_e(H)$ characterized by

$$R = \{[D^2 + 1]^{-1}[1]\} = \{[D^2 + 2D + 1]^{-1}[D^2 + 1]\}^{-1} \times \{[D^2 + 2D + 1]^{-1}[1]\}$$

We obtain

$$\{[D + 1]^{-1}[2]\}\{[D + 1]^{-1}[D]\} + \{[1]][[D^2 + 2D + 1]^{-1}[D^2 + 1]\} = PD_{t1} + QD_{r1} = I$$

and we may take

$$W = I - PD_{t1} = \{[D^2 + 2D + 1]^{-1}[D^2 + 1]\}$$

which yields

$$F = \{-\tilde{D} + [D^2 + 1]^{-1}[D^2 + 2D + 1]\}$$

$$I - H_{yu} = \{[D^2 + 2D + 1]^{-1}[2D]\}$$

and

$$H_{yn} = \{[D^2 + 2D + 1]^{-1}[D^2\tilde{D} + \tilde{D}]\}$$

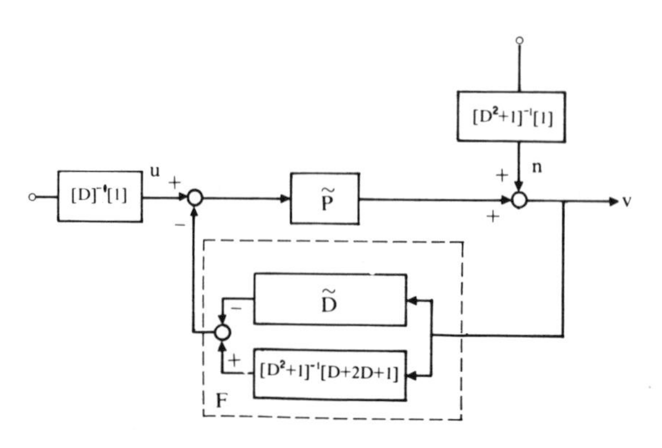

Fig. 2. Feedback system that tracks $[D]^{-1}[1]$ and rejects $[D^2 + 1]^{-1}[1]$.

Note, here that the D term in the numerator of $I - H_{yu}$ indicates that the desired tracking property is achieved, while the $[D^2 + 1]$ factor in the numerator of H_{yn} indicates that the required rejection property is realized. The final system is illustrated in Fig. 2.

PROBLEMS

1. Show that if $A = N_a D_a^{-1}$ where $U_a N_a + V_a D_a = I$ with N_a, D_a, U_a, and V_a all in $\mathscr{C}$ and pairwise commutative, then these operators may serve as both left and right fractional representations in a doubly coprime fractional representation of A.

2. Let $\mathscr{A}$ denote a subalgebra closed under inversion, and assume that $A \in \mathscr{A}$ admits a fractional representation $A = N_a D_a^{-1}$ where $N_a \in \mathscr{A} \cap \mathscr{C}$, $D_a \in \mathscr{A} \cap \mathscr{C} \cap \mathscr{B}^{-1}$. Furthermore, assume that D_a commutes with $\mathscr{A}$. Then show that there exist U_{ar}, V_{ar}, U_{al}, and V_{al} in $\mathscr{A} \cap \mathscr{C}$ such that

$$\begin{bmatrix} V_{ar} & U_{ar} \\ -N_a & D_a \end{bmatrix}^{-1} = \begin{bmatrix} D_a & -U_{pl} \\ N_a & V_{pl} \end{bmatrix}$$

defines a doubly coprime fractional representation for A. Can any of the operators U_{ar}, V_{ar}, U_{al}, and V_{al} be chosen to commute with $\mathscr{A}$?

3. In Example A.4 what feedback law is obtained if we take $W = \tilde{D}$? What are the advantages and disadvantages of this feedback law compared with that obtained using $W = I$? What about $W = 0$?

4. Will the system of Example B.5 track $T_e(H)$ where $T = [D^2]^{-1}[1]$?

5. For the system of Example B.9 choose a feedback law that simultaneously stabilizes the system and causes it to track $T_e(H)$ where $T = [D^2]^{-1}[1]$.

6. What H_{yn} results from the use of $W = D_{r1}$ in the feedback system Example B.9? Is D_{r1} a right divisor of H_{yn}?

REFERENCES

1. Desoer, C. A., Liu, R.-W., Murray, J., and Saeks, R., Feedback system design: The fractional representation approach to analysis and synthesis, *IEEE Trans. Automat. Control* **AC-25** (1980) (to appear).
2. Saeks, R., and Murray, J., Feedback system design: The tracking and disturbance rejection problems, *IEEE Trans. Automat. Control* **AC-26**, 203–217 (1981).

10

Optimal Control

A. A BASIC OPTIMIZATION PROBLEM

1. Remark: Once the asymptotic design constraints for a feedback system have been achieved, additional design latitude, which can be used to optimize some measure of system performance, may still remain. In the present chapter, we investigate the formulation and solution of some of the more commonly encountered deterministic optimal control problems. We begin with a basic optimization problem, the solution of which will point the way to the open and closed loop servomechanism problems and the regulator problem.

Let $A \in \mathscr{B}$ where $(A^*A)^{-1}$ exists, and let w and z be fixed vectors in H. We then term the problem of minimizing

$$J(B) = \|z - ABw\|^2$$

over $B \in \mathscr{C}$ the *basic optimization problem*. Without the causality constraint this is a standard generalized inverse problem with solution any B_0

satisfying

$$B_0 w = (A^*A)^{-1}A^*z$$

This may be verified by a projection theorem argument. A more elementary, if somewhat tedious, *completion of the squares* argument is as follows. With this choice of B_0 and any $B \in \mathscr{C}$,

$$(z, ABw) = (AB_0 w, ABw) \qquad \text{and} \qquad (z, AB_0 w) = (AB_0 w, AB_0 w)$$

Hence

$$\begin{aligned} J(B) &= \|z - ABw\|^2 = (z, z) - (z, ABw) - (ABw, z) + (ABw, ABw) \\ &= (z, z) - (AB_0 w, ABw) - (ABw, AB_0 w) + (ABw, ABw) \\ &\quad + (AB_0 w, AB_0 w) - (AB_0 w, AB_0 w) \\ &= \|A(B - B_0)w\|^2 + (z, z) - (AB_0 w, AB_0 w) \\ &= \|A(B - B_0)w\|^2 + (z, z) - (z, AB_0 w) \\ &\quad + [(AB_0 w, AB_0 w) - (AB_0 w, z)] \\ &= J(B_0) + \|A(B - B_0)w\|^2 \geq J(B_0) \end{aligned}$$

which verifies that B_0 minimizes $J(B)$ over $\mathscr{B}$. Unfortunately, even if $A \in \mathscr{C}$, $(A^*A)^{-1}A^*$ may fail to be in $\mathscr{C}$ (except for memoryless A). Moreover, if $z \in H_\dagger$ $(A^*A)^{-1}A^*z$ may fail to be in $H_\dagger$, in which case it will be impossible to satisfy

$$B_0 w = (A^*A)^{-1}A^*z$$

with $B_0 \in \mathscr{C}$ even for fixed w and z in $H_\dagger$. That is, since $H_\dagger$ is an invariant subspace for every causal operator if $w \in H_\dagger$, a necessary condition for a causal operator B_0 to *interpolate* the points w and $(A^*A)^{-1}A^*z$ is that $(A^*A)^{-1}A^*z \in H_\dagger$. Fortunately, by appropriately modifying the optimization criteria, an optimal causal B_0 can be constructed.[2]

Basic Optimization Theorem: Let $A \in B$, assume that A^*A admits a spectral factorization $A^*A = C^*C$ where $C \in \mathscr{C} \cap \mathscr{C}^{-1}$, and let $w \in H_\dagger$ and $z \in H$. Then

$$J(B) = \|z - ABw\|^2$$

is minimized over $B \in \mathscr{C}$ by any $B_0 \in \mathscr{C}$ such that

$$B_0 w = C^{-1}E_\dagger C^{*-1}A^*z$$

Proof: Since $w \in H_\dagger$, for any $B \in \mathscr{C}$

$$CBw = BCE_\dagger w = E_\dagger CBE_\dagger w = E_\dagger CBw$$

Hence

$$
\begin{aligned}
(z, ABw) &= (A^*z, Bw) = (C^*C^{*-1}A^*z, Bw) = (C^{*-1}A^*z, CBw) \\
&= (C^{*-1}A^*z, E_\dagger CBw) = (E_\dagger C^{*-1}A^*z, CBw) \\
&= (C^{*-1}C^*CC^{-1}E_\dagger C^{*-1}A^*z, CBw) = (C^{*-1}A^*AB_0, CBw) \\
&= (AB_0 w, AC^{-1}CBw) = (AB_0 w, ABw)
\end{aligned}
$$

while a similar argument implies that

$$(z, AB_0 w) = (AB_0 w, AB_0 w)$$

With these equalities in hand, precisely the argument used in Remark 1 reveals that for any $B \in \mathscr{C}$

$$J(B) = J(B_0) + \|A(B - B_0)y\|^2 \geq J(B_0)$$

and hence we have verified that B_0 minimizes the given performance measure over $\mathscr{C}$ whenever it exists. //

2. Remark: It is important to note the fundamental difference between the generalized inverse "solution" to our optimization problem and the solution of the theorem. Like $(A^*A)^{-1}A^*$, $C^{-1}E_\dagger C^{*-1}A^*z$ is not causal, however, since C^{-1} is causal,

$$C^{-1}E_\dagger C^{*-1}A^*z = E_\dagger C^{-1}E_\dagger C^{*-1}A^*z \in H_\dagger$$

and hence the pair $\{w, C^{-1}E_\dagger C^{*-1}A^*z\}$ satisfies the necessary condition for causal interpolation. It may then be possible to interpolate this pair with a causal operator B_0. Although no general construction for such an interpolating operator exists,[2] a necessary and sufficient condition for the existence of a causal interpolating operator will be derived in the sequel; indeed, in a number of special cases it is even possible to interpolate an infinite set of input/output pairs $\{w_\pi, C^{-1}E_{\dagger_\pi} C^{*-1}A^*z_\pi\}$.[3]

3. Example: Consider the Hilbert resolution space $L_2(-\infty, \infty)$ with its usual resolution structure and let $A = \tilde{D}$ be the ideal delay on $L_2(-\infty, \infty)$. Then since $\tilde{D}^* = \tilde{P}$, the ideal predictor

$$[\tilde{D}^*\tilde{D}] = [\tilde{P}\tilde{D}] = I = [I^*I]$$

allowing us to take $C = I$. Now if $w = z = \chi_{(0,1]}$, i.e., multiplication by the characteristic function of the interval $(0, 1]$, then $C^{*-1}A^*z = \tilde{P}z = \chi_{(-1,0]}$ and hence

$$E_0 C^{*-1}A^*z = E_0 \chi_{(-1,0]} = 0$$

We must therefore choose B_0 in $\mathscr{C}$ so that

$$B_0 \chi_{(0,1]} = B_0 w = C^{-1} E_0 C^{*-1} A^* z = 0$$

Clearly $B_0 = 0$ suffices.

Now, let us consider the case where $w = \chi_{(0,1]}$ but $z = \chi_{(0,2]}$. Then

$$C^{*-1} A^* z = \tilde{P} z = \chi_{(-1,1]}$$

Hence

$$E_0 C^{*-1} A^* z = E_0 \tilde{P} z = \chi_{(0,1]} = w$$

and

$$B_0 w = C^{-1} E_0 C^{*-1} A^* z = w$$

allowing us to take $B_0 = I$ as our causal interpolating operator. Note that with this optimal choice of B_0 we have

$$z - A B_0 w = \chi_{(0,2]} - \chi_{(1,2]} = \chi_{(0,1]}$$

so that $\|z - A B_0 w\|^2 = 1$. On the other hand, without the causality constraint, one could take $B_0 = \tilde{P} = \tilde{D}^{-1}$, in which case $z - A B_0 w = 0$ and so is its norm. Accordingly, the causality constraint has strictly reduced the performance of our estimator.

4. Example: Now consider the Hilbert resolution space $l_2(-\infty, \infty)$ with its usual resolution structure and assume that $w \in H_0$. This space turns out to be especially well suited for our interpolation problem since one can identify a signal $x \in l_2(-\infty, \infty)$ with an operator $X \in \mathscr{B}$ by means of the equality $Xu = x * u$ where $*$ denotes convolution. In particular, $X \in \mathscr{C}$ if and only if $x \in H_0$ (1.B.11), and so if we let $x = E_0 C^{*-1} A^* z$, then the operator $C^{-1} X W^{-1}$ interpolates the pair $\{w, C^{-1} E_0 C^{*-1} A^* z\}$. To verify this recall that $w = W\delta$ and $x = X\delta$ where the δ sequence in $l_2(-\infty, \infty)$ is defined by

$$\delta_i = \begin{cases} 1, & i = 0 \\ 0, & i \neq 0 \end{cases}$$

Thus

$$\begin{aligned}[C^{-1} X W^{-1}](w) &= [C^{-1} X W^{-1} W](\delta) = [C^{-1} X](\delta) = [C^{-1}](x) \\ &= C^{-1} E_0 C^{*-1} A^* z\end{aligned}$$

verifying the interpolation property. Moreover, $C^{-1} \in \mathscr{C}$ by construction, while $X \in \mathscr{C}$ since $x = E_0 C^{*-1} A^* z \in H_0$. Hence for $C^{-1} X W^{-1}$ to be causal it suffices that W^{-1} be causal. To this end the assumption that $w \in H_0$ implies that $W \in \mathscr{C}$ and hence the only restriction required to obtain a solution of our optimization problem is that $W \in \mathscr{C} \cap \mathscr{C}^{-1}$.

Although this process for obtaining a causal interpolating operator might at first seem to be somewhat roundabout, it in fact reduces to the classical frequency domain solution to the optimal servomechanism problem. Here, however, one typically lets $v = C^{*-1}A^*z$ and then recognizes that $X = [V]_C$, obtaining the more familiar solution $B_0 = C^{-1}[V]_C W^{-1}$ where $[V]_C$ denotes the *causal part* of V.

5. Remark: Although the above examples indicate that the causal interpolating operator needed for the solution of the basic optimization problem can often be computed by ad hoc means, it remains to formulate a criterion for its existence.[1] We begin in the finite-dimensional case.

6. Lemma: Let $x = \text{col}(x_i)$ and $y = \text{col}(y_i)$ be n vectors and suppose that

$$\sum_{i=1}^{r} |y_i|^2 \leq K^2 \sum_{i=1}^{r} |x_i|^2$$

Then there exists a lower triangular $n \times n$ matrix such that $y = Ax$ and $\|A\| \leq K$.

Proof: Since the case for $n = 1$ is trivial, we give an inductive proof: We assume that the lemma holds for $n = 1, 2, \ldots, p-1$ and show that it also holds for $n = p$. To this end we let $x' = \text{col}(x_1, x_2, \ldots, x_{p-1})$ and $y' = \text{col}(y_1, y_2, \ldots, y_{p-1})$, in which case the inductive hypotheses implies that there exists a lower triangular $(p-1) \times (p-1)$ matrix A' such that $y' = A'x'$ and $\|A'\| \leq K$. Taking the entries of this matrix to be the first $p-1$ rows and columns of the desired matrix, we have constructed a_{ij}, $i, j = 1, 2, \ldots, p-1$, such that

$$y_i = \sum_{j=1}^{p-1} a_{ij}x_j, \qquad i = 1, 2, \ldots, p-1$$

$$a_{ij} = 0, \qquad j > i$$

and

$$\sum_{i=1}^{p-1} \left| \sum_{j=1}^{i} a_{ij}z_j \right|^2 \leq K^2 \sum_{i=1}^{p-1} |z_i|^2$$

for any p vector $z = \mathrm{col}(z_i)$. Now, since

$$\sum_{i=1}^{p} |y_i|^2 \leq K^2 \sum_{i=1}^{p} |x_i|^2$$

$$|y_p|^2 \leq K^2 \sum_{i=1}^{p-1} |x_i|^2 - \sum_{i=1}^{p-1} \left| \sum_{j=1}^{i} a_{ij} x_j \right|^2 + K^2 |x_n|^2$$

Therefore we may define a linear functional on the one-dimensional subspace spanned by the vector x:

$$\phi(cx) = cy_n$$

for any scalar c. Moreover

$$|\phi(cx)| \leq \|z\|_S$$

where $\| \ \|_S$ is the seminorm defined on the p vector $z = \mathrm{col}(z_i)$ by

$$\|z\|_S^2 = K^2 |z_n|^2 + K^2 \sum_{i=1}^{p-1} |z_i|^2 - \sum_{i=1}^{p-1} \left| \sum_{j=1}^{i} a_{ij} z_j \right|^2$$

As such, $\phi(cx)$ by the Hahn–Banach theorem admits an extension to p space $\phi(z)$ satisfying the inequality $\phi(z) \leq \|z\|_S$.

Finally, since every linear functional on n space can be represented by a row vector, there exist complex scalars a_{pj} such that

$$\phi(z) = \sum_{j=1}^{p} a_{pj} z_j$$

that define the remainder of the nonzero entries in the desired A matrix. //

7. Remark: In finite dimensions the inequality required by the hypotheses of Lemma 6 is always satisfied for some K if $x_1 \neq 0$, in which case an interpolating matrix exists. The inequality, however, characterizes the norm of the resultant interpolating matrix, thereby allowing the theory to be extended to the infinite-dimensional case.

8. Lemma: Let $\mathscr{P} = \{E^i: i = 1, 2, \ldots, n\}$ be a partition of $\mathscr{E}$ and assume that

$$\|E^i y\| \leq K \|E^i x\|, \qquad i = 1, 2, \ldots, n$$

for specified $x, y \in H$. Then there exists $A \in \mathscr{B}$ such that

(i) $E^i A = E^i A E^i$, $i = 1, 2, \ldots, n$;
(ii) $y = Ax$; and
(iii) $\|A\| \leq K$

Proof: Let $x_i = \Delta^i x$ and $y_i = \Delta^i y$ and define two n vectors, x' and y' by $x' = \text{col}(\|x_i\|)$ and $y' = \text{col}(\|y_i\|)$. Now

$$\sum_{i=1}^{r} |y_i'|^2 = \sum_{i=1}^{r} \|\Delta^i y\|^2 = \|E^r y\|^2$$

$$\leq K^2 \|E^r x\|^2 = K^2 \sum_{i=1}^{r} \|\Delta^i x\|^2 = K^2 \sum_{i=1}^{r} |x_i'|^2$$

The hypotheses of Lemma 6 are satisfied, and there exists a lower triangular $n \times n$ matrix B with $\|B\| \leq K$ such that $y' = Bx'$; i.e.,

$$\sum_{i=1}^{n} b_{ij} \|x_j\| = \|y_i\|$$

Moreover, without loss of generality we may assume that $b_{ij} = 0$ when either $x_i = 0$ or $y_i = 0$, allowing us to define a matrix C by

$$c_{ij} = \begin{cases} b_{ij}/\|x_j\| \, \|y_i\|, & b_{ij} \neq 0 \\ 0, & b_{ij} = 0 \end{cases}$$

Finally, the required A is constructed:

$$A = \sum_{i=1}^{n} \sum_{j=1}^{n} c_{ij}[x_j \otimes y_i]$$

Now,

$$Ax = \sum_{i=1}^{n} \sum_{j=1}^{n} c_{ij}(x, x_j) y_i = \sum_{i=1}^{n} \sum_{j=1}^{n} c_{ij}(x, \Delta^j x) y_i$$

$$= \sum_{i=1}^{n} \sum_{j=1}^{n} c_{ij} \|x_j\|^2 y_i = \sum_{i=1}^{n} \sum_{j=1}^{n} b_{ij} \frac{\|x_j\|^2}{\|x_j\| \, \|y_i\|} y_i$$

$$= \sum_{i=1}^{n} \left[\sum_{j=1}^{n} b_{ij} \|x_j\| \right] \frac{y_i}{\|y_i\|} = \sum_{i=1}^{n} \frac{[\|y_i\|] y_i}{\|y_i\|} = \sum_{i=1}^{n} y_i = y$$

where we have used the fact that C is lower triangular to limit the range of summation over j. Now, for any $k = 0, 1, 2, \ldots, n$ $E^k y_i = E^k \Delta^i y = y_i$ if $i \leq k$ while $E^k y_i = 0$ for $i > k$, and similarly for $E^k x_j$. Accordingly,

$$E^k Az = \sum_{i=1}^{n} \sum_{j=1}^{i} c_{ij}(z, x_j) E^k y_i = \sum_{i=1}^{k} \sum_{j=1}^{i} c_{ij}(z, x^j) E^k y_i$$

$$= \sum_{i=1}^{k} \sum_{j-1}^{i} c_{ij}(z, E^k x_j) E^k y_i = \sum_{i=1}^{k} \sum_{j=1}^{i} c_{ij}(E^k z, x_j) E^k y_i$$

$$= \sum_{i=1}^{n} \sum_{j=1}^{i} c_{ij}(E^k z, x_j) E^k y_i = E^k A E^k z$$

verifying property (i) of the lemma.

Finally, to show that $\|A\| \le K$ we denote by H_x the finite-dimensional subspace space

$$H_x = V_i\{x_i\}$$

in which case

$$\|A\| = \sup_{z \neq 0} \|Az\|/\|z\| = \sup_{z \neq 0,\, z \in H_x} \|Az\|/\|z\|$$

since $Az = 0$ if z is orthogonal to H_x. Moreover, since the x_is are mutually orthogonal if $z \in H_x$,

$$z = \sum_{i=1}^{n} a_i x_i, \qquad \text{where} \quad \|z\|^2 = \sum_{i=1}^{n} |a_i|^2 \|x_i\|^2$$

Thus

$$\begin{aligned}
\|Az\|^2 &= \sup_{\|z\| \neq 0,\, z \in H_x} \left\| \sum_{i=1}^{n} \sum_{j=1}^{i} c_{ij}(z, x_j) y_i \right\|^2 \\
&= \sup_{\|z\| \neq 0,\, z \in H_x} \left\| \sum_{i=1}^{n} \sum_{j=1}^{i} c_{ij} \left(\sum_{k=1}^{n} a_i x_i, x_j \right) y_i \right\|^2 \\
&= \sup_{\|z\| \neq 0,\, z \in H_x} \left\| \sum_{i=1}^{n} \left[\sum_{j=1}^{i} c_{ij} a_j \|x_j\|^2 \right] y_i \right\|^2 \\
&= \sup_{\|z\| \neq 0,\, z \in H_x} \sum_{i=1}^{n} \sum_{j=1}^{i} c_{ij} a_j \|x_j\|^2 |^2 \, \|y_i\|^2 \\
&= \sup_{\|z\| \neq 0,\, z \in H_x} \sum_{i=1}^{n} \left| \sum_{j=1}^{n} c_{ij} a_j \|x_j\|^2 \, \|y_i\| \right|^2 \\
&= \sup_{\|z\| \neq 0,\, z \in H_x} \sum_{i=1}^{n} \left| \sum_{j=1}^{n} b_{ij} a_j \|x_j\| \right|^2 = \|Bz'\|^2 \le K^2 \|z'\|^2
\end{aligned}$$

where $z' = \operatorname{col}(a_i \|x_i\|)$ and we have used the fact that $\|B\| \le K$. Now,

$$\|z'\|^2 = \sum_{i=1}^{n} |a_i \|x_i\| |^2 = \sum_{i=1}^{n} |a_i|^2 \|x_i\|^2 = \|z\|^2$$

We have verified that

$$\|Az\|^2 \le K^2 \|z\|^2$$

thereby completing the proof. //

9. Property: Let $x, y \in H$. Then there exists $A \in \mathscr{C}$ such that $y = Ax$ if and only if

$$K = \sup_{\dagger \in \$} \|E^{\dagger}y\|/\|E^{\dagger}x\| < \infty$$

where $\|E^{\dagger}y\|/\|E^{\dagger}x\|$ is taken to be zero if $E^{\dagger}y = E^{\dagger}x = 0$ and infinite if $E^{\dagger}x = 0$ but $E^{\dagger}y \neq 0$. Moreover, when this condition is satisfied $\|A\| = K$, and if $B \in \mathscr{C}$ also satisfies $y = Bx$, then $\|B\| \geq \|A\| = K$.

Proof: Assume that $B \in \mathscr{C}$ satisfies $y = Bx$. Then

$$\|E^{\dagger}y\| = \|E^{\dagger}Bx\| = \|E^{\dagger}BE^{\dagger}x\| \leq \|B\|\,\|E^{\dagger}x\|$$

showing that $\|B\| \geq \|E^{\dagger}y\|/\|E^{\dagger}x\|$ for all $\dagger \in \$$. Hence for any $\varepsilon > 0$ there exists $\dagger \in \$$ such that

$$\|B\| \geq \|E^{\dagger}y\|/\|E^{\dagger}x\| > K - \varepsilon$$

verifying that $\|B\| \geq K$ and $K < \infty$ if there exists $B \in \mathscr{C}$ such that $y = Bx$.

Conversely, if $K < \infty$ Lemma 8 implies that for each partition $\mathscr{P} = \{E^i : i = 0, 1, 2, \ldots, n\}$ of $\mathscr{E}$ there exists an operator $A_{\mathscr{P}}$ such that $y = A_{\mathscr{P}}x$, $\|A_{\mathscr{P}}\| \leq K$, and

$$E^i A_{\mathscr{P}} = E^i A_{\mathscr{P}} E^i, \qquad i = 1, 2, \ldots, n$$

Since the partitions of $\mathscr{E}$ form a directed set, the family of operators $A_{\mathscr{P}}$ defines a net of operators taking values in the ball of radius K in $\mathscr{B}$. Moreover, since this ball is weakly compact, the net $A_{\mathscr{P}}$ admits a convergent subsequence with limit A in the ball. Clearly $\|A\| \leq K$ whereas for any $z \in H$

$$(Ax, z) = \lim_{\mathscr{P}} (A_{\mathscr{P}}x, z) = (y, z)$$

and hence $Ax = y$. Finally, to verify that $A \in \mathscr{C}$ we let $\mathscr{Q}^{\dagger}$ be the partition $\{0, E^{\dagger}, I\}$ and we consider the subnet of operators $A_{\mathscr{P}^{\dagger}}$ composed of operators in the convergent subnet which also refine $\mathscr{Q}^{\dagger}$. Now, by construction $E^{\dagger}A_{\mathscr{P}^{\dagger}} = E^{\dagger}A_{\mathscr{P}^{\dagger}}E^{\dagger}$ for each element of this subnet. Moreover, since the subnet $A_{\mathscr{P}^{\dagger}}$ converges to A it follows that $E^{\dagger}A = E^{\dagger}AE^{\dagger}$, as was to be shown. //

10. Remark: Although no constructive algorithm for computing A is known in general, an interpolating operator is given by the formula

$$A = \mathrm{s}(M) \int [E^{\dagger}x \otimes dE(\dagger)\, y]/\|E^{\dagger}x\|^2$$

which is convergent if

$$s(M) \int \|dE(\dagger)\, y\|^2/\|E^\dagger x\|^2 < \infty$$

Indeed,

$$Ax = s(M) \int \frac{(x, E^\dagger x)}{\|E^\dagger x\|^2}\, dE(\dagger)\, y = s(M) \int dE(\dagger)\, y = y$$

verifying the interpolation condition, whereas for any $z \in H$

$$E^\ddagger Az = s(M) \int_{-\infty}^{\ddagger} \frac{(z, E^\dagger x)}{\|E^\dagger x\|^2}\, dE(\dagger)\, y = s(M) \int_{-\infty}^{\ddagger} \frac{(z, E^\ddagger E^\dagger x)}{\|E^\dagger x\|^2}\, dE(\dagger)\, y$$

$$= s(M) \int_{-\infty}^{\ddagger} \frac{(E^\dagger z, E^\dagger x)}{\|E^\dagger x\|^2}\, dE(\dagger)\, y = E^\ddagger A E^\ddagger z$$

verifying that $A \in \mathscr{C}$. Here, we have used the fact that $E^\dagger = E^\ddagger E^\dagger$ in the interval of integration.

B. THE SERVOMECHANISM PROBLEM

1. Remark: The most elementary *servomechanism problem* is illustrated in Fig. 1. Here, we have an open loop system with plant P and compensator M that is disturbed at the output by a signal $n \in H_+$. Our problem is to operate on n with a causal compensator that minimizes

$$J(M) = \|y\|^2 + \|r\|^2$$

over $M \in \mathscr{C}$. Physically the requirement that we minimize $\|y\|^2$ means that the output of the plant must follow $-n$, so that our problem is really an optimal tracking problem. Of course, since P is linear, if one could apply arbitrarily large inputs r to P, this tracking problem would be straightforward. In practice, however, the norm of the input must be limited, and hence we minimize the sum of $\|y\|^2$ and $\|r\|^2$ to obtain a compromise between the tracking requirement and input energy.

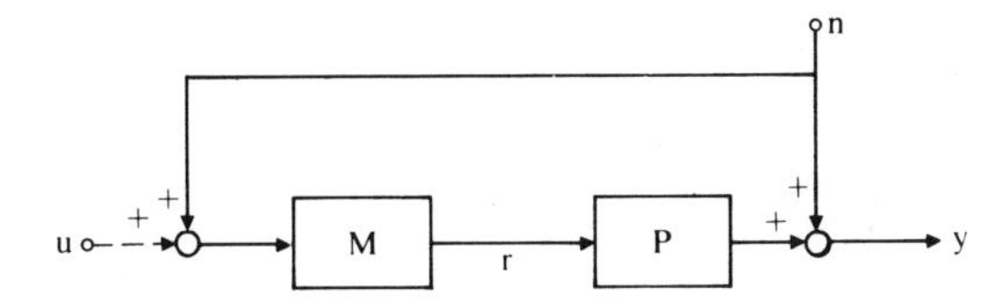

Fig. 1. Open loop servomechanism problem.

2. Property: Let $P \in \mathscr{C}$ and assume that $[I + P^*P]$ admits a spectral factorization $[I + P^*P] = C^*C$, $C \in \mathscr{C} \cap \mathscr{C}^{-1}$. Then if y and r are defined as in Remark 1 and $n \in H_\dagger$, then

$$J(M) = \|y\|^2 + \|r\|^2$$

is minimized over $M \in \mathscr{C}$ by any $M_0 \in \mathscr{C}$ satisfying

$$M_0 n = -C^{-1}E_\dagger C^{*-1}P^*n$$

Proof: This optimal servomechanism problem is a special case of the result of the basic optimization theorem. Here

$$z = (n, 0) \in H^2, \qquad w = n \in H_\dagger,$$
$$B: H \to H, \quad B \in \mathscr{C}$$

where $B = M$. Finally $A: H \to H^2$ is characterized by the 2×1 matrix of operators

$$A = \begin{bmatrix} -P \\ \hdashline -I \end{bmatrix}$$

Indeed, this yields

$$z - ABw = \begin{bmatrix} n + PMn \\ \hdashline Mn \end{bmatrix} = \begin{bmatrix} y \\ \hdashline r \end{bmatrix}$$

so that the minimization of $\|z - ABy\|^2$ over $B \in \mathscr{C}$ is equivalent to the minimization of $\|y\|^2 + \|r\|^2$ over $M \in \mathscr{C}$.

Substituting these terms into the result of the theorem now yields

$$[A^*A] = [I + P^*P] = [C^*C]$$

and

$$\begin{aligned} M_0 n = B_0 y &= C^{-1}E_\dagger C^{*-1}A^*z \\ &= C^{-1}E_\dagger C^{*-1}[-P^* \mid -I] \begin{bmatrix} n \\ \hdashline 0 \end{bmatrix} = -C^{-1}E_\dagger C^{*-1}P^*n \end{aligned}$$

as was to be shown. //

3. Example: Consider the space $L_2(-\infty, \infty)$ with its usual resolution structure, let $P = \tilde{D}$ be the ideal delay, and let $n = \chi_{(0,2]} \in H_0$. Now

$$[I + P^*P] = [I + \tilde{D}^*\tilde{D}] = [2I] = [(\sqrt{2}I)^*(\sqrt{2}I)]$$

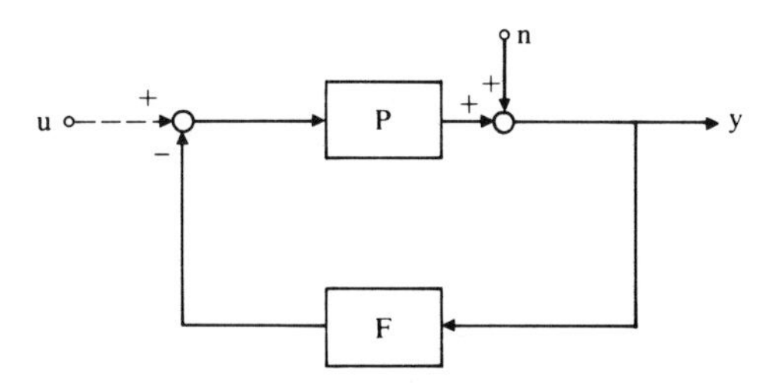

Fig. 2. Closed loop servomechanism problem.

showing that $C = \sqrt{2}I$ and that

$$M_0 \chi_{(0,2]} = -\tfrac{1}{2}E_0 \tilde{P} \chi_{(0,2]} = -\tfrac{1}{2}\chi_{(0,1]}.$$

Our compensator must take the form of a causal operator that interpolates the vectors $\chi_{(0,2]}$ and $-\frac{1}{2}\chi_{(0,1]} \in H$. One such M_0 that will achieve this goal is

$$M_0 = -\tfrac{1}{2}E^1$$

4. Remark: Although our servomechanism theory can be formulated for the open loop system of Fig. 1, in practice a feedback formulation is preferred. First, the feedback formulation allows us to employ an unstable plant so long as the plant is stabilized by the loop, as illustrated in Fig. 2. Second, in the feedback configuration one senses the system output rather than directly sensing the disturbance n, which may be physically inaccessible.

5. Remark: Since we are interested only in feedback laws that stabilize the closed loop system, we adopt the formulation used in the stabilization theorem. In particular, we assume that P admits a doubly coprime fractional representation

$$\begin{bmatrix} V_{pr} & U_{pr} \\ -N_{pl} & D_{pl} \end{bmatrix}^{-1} = \begin{bmatrix} D_{pr} & -U_{pl} \\ N_{pr} & V_{pl} \end{bmatrix}$$

and we optimize the design over the choice of $W \in \mathscr{C}$. Since our only input to the system is the disturbance n, the stabilization theorem implies that

$$y = H_{yn}n = [N_{pr}WD_{pl} + V_{pl}D_{pl}]n$$

and

$$r = H_{rn}n = [D_{pr}WD_{pl} - D_{pr}U_{pr}]n$$

To obtain the optimal stabilizing feedback law we then minimize

$$J(F(W)) = \|y\|^2 + \|r\|^2$$

over $W \in \mathscr{C}$ and use this W_0 to define the optimal feedback law

$$F_0 = (W_0 N_{pl} + V_{pr})^{-1}(-W_0 D_{pl} + U_{pr})$$

6. Property: Assume that P admits a doubly coprime fractional representation

$$\begin{bmatrix} V_{pr} & U_{pr} \\ -N_{pl} & D_{pl} \end{bmatrix}^{-1} = \begin{bmatrix} D_{pr} & -U_{pl} \\ N_{pr} & V_{pl} \end{bmatrix}$$

and assume that $[N_{pr}^* N_{pr} + D_{pr}^* D_{pr}]$ admits a spectral factorization $[N_{pr}^* N_{pr} + D_{pr}^* D_{pr}] = [C^* C]$, $C \in \mathscr{C} \cap \mathscr{C}^{-1}$. Then if y and r are defined as in Remark 5 and $n \in H_\dagger$, then

$$J(F) = \|y\|^2 + \|r\|^2$$

is minimized over the set of stabilizing feedback laws by

$$F_0 = (W_0 N_{pl} + V_{pr})^{-1}(-W_0 D_{pl} + U_{pr})$$

where W_0 is any causal operator such that $(W_0 N_{pl} + V_{pr})^{-1} \in \mathscr{B}$ and

$$W_0 D_{pl} n = C^{-1} E_\dagger C^{*-1}[-N_{pr}^* V_{pl} D_{pl} + D_{pr}^* D_{pr} U_{pr}]n$$

Proof: As with the open loop servomechanism problem the solution of the closed loop problem follows from the basic optimization theorem of Section A. In particular, we let

$$z = (V_{pl} D_{pl} n, -D_{pr} U_{pr} n) \in H^2, \qquad w = D_{pl} n \in H_\dagger,$$

$$B: H \to H, \qquad B \in C$$

where $B = W$. Finally, $A: H \to H^2$ is characterized by the 2×1 matrix of operators

$$A = \begin{bmatrix} -N_{pr} \\ \hdashline -D_{pr} \end{bmatrix}$$

This yields

$$z - ABw = \begin{bmatrix} [N_{pr} W D_{pl} + V_{pl} D_{pl}]n \\ \hdashline [D_{pr} W D_{pl} + D_{pr} U_{pr}]n \end{bmatrix} = \begin{bmatrix} H_{yn} n \\ \hdashline H_{rn} n \end{bmatrix} = \begin{bmatrix} y \\ \hdashline r \end{bmatrix}$$

Hence the minimization of $\|z - ABw\|^2$ over $B \in \mathscr{C}$ is equivalent to the minimization of $\|y\|^2 + \|r\|^2$ over W, which in turn parametrizes the stabilizing feedback laws.

To solve the minimization problem we have

$$[A^*A] = [N_{pr}^* N_{pr} + D_{pr}^* D_{pr}] = [C^*C]$$

and

$$\begin{aligned} W_0 D_{pl} n &= B_0 \tilde{y} = C^{-1} E_\dagger C^{*-1} A^* z \\ &= C^{-1} E_\dagger C^{*-1} [-N_{pr}^* \; \vdots \; -D_{pr}^*] \begin{bmatrix} V_{pl} D_{pl} n \\ \hdashline -D_{pl} U_{pr} n \end{bmatrix} \\ &= C^{-1} E_\dagger C^{*-1} [-N_{pr}^* V_{pl} D_{pl} + D_{pr}^* D_{pr} U_{pr}] n \end{aligned}$$

Finally, if W_0 is any causal solution of this equation for which $(W_0 N_{pl} + V_{pr})^{-1} \in \mathscr{B}$, the stabilization theorem implies that

$$F_0 = (W_0 N_{pl} + V_{pr})^{-1} (-W_0 D_{pl} + U_{pr})$$

is a stabilizing feedback law minimizing $J(F) = \|y\|^2 + \|r\|^2$ over the set of stabilizing feedback laws as required. //

7. Example: Let us repeat Example 3 in the case of a closed loop design. As before, $P = \tilde{D}$, which admits the doubly coprime fractional representation

$$\begin{bmatrix} I & 0 \\ -\tilde{D} & I \end{bmatrix}^{-1} = \begin{bmatrix} I & 0 \\ \tilde{D} & I \end{bmatrix}$$

Hence

$$[N_{pr}^* N_{pr} + D_{pr}^* D_{pr}] = [\tilde{D}^* \tilde{D} + I] = [2I] = [(\sqrt{2}I)^* (\sqrt{2}I)]$$

with $C = (\sqrt{2}I)$. As in Example 3, we take $n = \chi_{(0,2]} \in H_0$, in which case W_0 is defined by

$$W_0 \chi_{(0,2]} = -\tfrac{1}{2} E_0 \tilde{P} \chi_{(0,2]} = -\tfrac{1}{2} \chi_{(0,1]}$$

and if we take $W_0 = -\frac{1}{2} E^1$,

$$F_0 = [-\tfrac{1}{2} E^1 \tilde{D} + I]^{-1} [\tfrac{1}{2} E^1]$$

is the desired feedback law. For the purpose of comparison, the open and closed loop realizations for our optimal servomechanism are illustrated in Fig. 3.

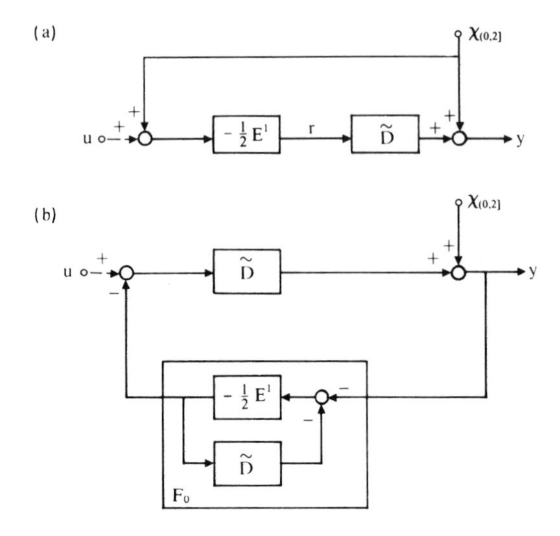

Fig. 3. (a) Open and (b) closed loop realization of an optimal servomechanism.

8. Remark: Although the open and closed loop configurations of Fig. 3 are quite different, a little algebra will reveal that the two systems have identical responses y and r to any disturbance n. This similarity, as well as the identity between M_0 and W_0, is a manifestation of the simple doubly coprime structure applicable to a stable plant. Of course, if P is unstable, the open loop problem is not well defined and a comparison between the two approaches becomes meaningless.

C. THE REGULATOR PROBLEM

1. Remark: The *regulator problem* is actually a special case of the servomechanism problem wherein the disturbance $n = \theta_\dagger x(\dagger)$, represents the system's response at time † to a nonzero initial state, which we assume represents a perturbation from nominal (which we take to be zero). A regulator may thus be viewed as a device that drives a system back to its nominal trajectory whenever it deviates therefrom (presumably the effect of some outside force not included in the model). Unlike the general servomechanism problem, however, we can exploit the structure of the state model and the special nature of the disturbance $n = \theta_\dagger x(\dagger)$ to obtain an especially powerful solution to the regular problem. First, the

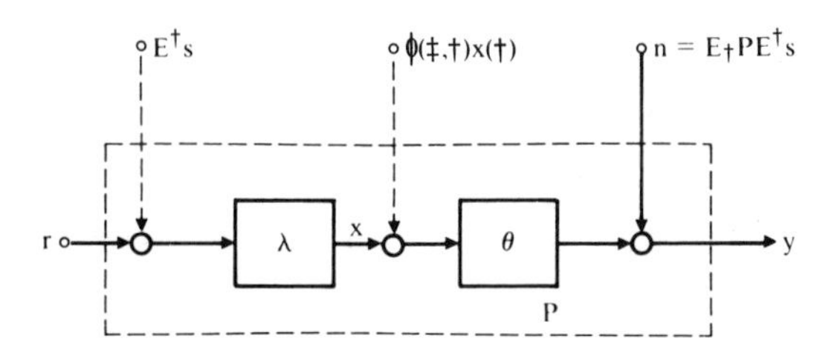

Fig. 4. Plant with disturbance applied at the output, state, and input.

solution is valid independently of the initial state, and second, the solution can be implemented by memoryless state feedback.[4,5]

In the sequel we assume that $P \in \mathscr{S}$ and admits a strong minimal state decomposition $\{(S_\dagger, \lambda_\dagger, \theta_\dagger) : \dagger \in \$\}$. It then follows from the results of Section 7.B that P admits a state trajectory decomposition $P = \theta\lambda$ where λ is algebraically strictly causal and θ is memoryless. The resultant plant is illustrated in Fig. 4, where the disturbance is shown in three equivalent forms. Here we have assumed that $x(\dagger) = \lambda_\dagger s$ for some $s \in H^\dagger$ so that $n = \theta_\dagger x(\dagger) = \theta_\dagger \lambda_\dagger s = E_\dagger P E^\dagger s$. Equivalently, however, we can view the disturbance as the term $\Phi(\ddagger, \dagger)x(\dagger)$ added to the state at time $\ddagger \geq \dagger$. Finally, the disturbance can be viewed as the application of the term $E^\dagger s$ at the input (if it is understood that we are interested only in controlling its effects after time $\dagger$).

Regulator Theorem: Let $P \varepsilon S$ admit a strong minimal state decomposition $\{(S_\dagger, \lambda_\dagger, \theta_\dagger) : \dagger \in \$\}$ and assume that $[I + P^*P]$ admits a spectral factorization $[I + P^*P] = [(I + V)^*(I + V)]$ where $C = I + V$ with V algebraically strictly causal. Then with the notation of Remark 1 the $r_0 \in H$ minimizing

$$J(r) = \|E_\dagger y\|^2 + \|E_\dagger r\|^2$$

over $r \in H_\dagger$ satisfies $E_\dagger r_0 = -E_\dagger V[E_\dagger r_0 + E^\dagger s]$. Furthermore, if $x_0 = \lambda[E_\dagger r_0 + E^\dagger s]$ is the state trajectory resulting from the application of r_0, then there exists a memoryless operator F independent of s such that $E_\dagger r_0 = E_\dagger F x_0$.

2. Remark: The fundamental point of the theorem is the equality $E_\dagger r_0 = E_\dagger F x_0$, which permits the optimal regulator to be implemented by means of memoryless state feedbacks, as illustrated in Fig. 5. Note further that F is independent of s so that this implementation is optimal for any disturbing initial state. Indeed, it is this simple implementation that

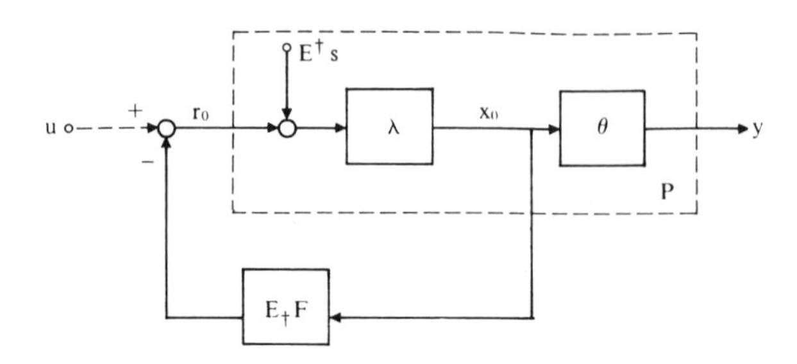

Fig. 5. Memoryless state feedback implementation of the optimal regulator.

makes the regulator especially important in control theory and, more generally, it is the memoryless state feedback concept that makes the state model useful.

3. Proof of The Theorem: From Property B.3 the optimal solution of the regulator problem satisfies

$$E_{\dagger}r_0 = M_0 n = -C^{-1}E_{\dagger}C^{*-1}P^{*}n = -C^{-1}E_{\dagger}C^{*-1}P^{*}E_{\dagger}PE^{\dagger}s$$

where $n = E_{\dagger}PE^{\dagger}s$. Thus

$$\begin{aligned} E_{\dagger}C^{*}CE_{\dagger}r_0 &= -E_{\dagger}C^{*}CC^{-1}E_{\dagger}C^{*-1}P^{*}E_{\dagger}PE^{\dagger}s \\ &= -E_{\dagger}C^{*}E_{\dagger}C^{*-1}P^{*}E_{\dagger}PE^{\dagger}s \\ &= -E_{\dagger}C^{*}C^{*-1}P^{*}PEs = -E_{\dagger}P^{*}PE^{\dagger}s \end{aligned}$$

where we have invoked the anticausality of C^{*}, C^{*-1}, and P^{*}. Recalling that $[I + P^{*}P] = [C^{*}C] = [(I + V)^{*}(I + V)]$, this becomes

$$E_{\dagger}[I + P]E_{\dagger}r_0 = -E_{\dagger}[P^{*}P]E^{\dagger}s$$

Moreover, $P^{*}P = V^{*} + V + V^{*}V = V^{*} + (I + V)^{*}V$, which upon substitution into the preceding equation yields

$$\begin{aligned} E_{\dagger}(I + V)^{*}(I + V)E_{\dagger}r_0 &= -E_{\dagger}[V^{*} + (I + V)^{*}V]E^{\dagger}s \\ &= -E_{\dagger}(I + V)^{*}VE^{\dagger}s \end{aligned}$$

where we have used the fact that $E_{\dagger}V^{*}E^{\dagger} = 0$ since $V^{*} \in \mathscr{C}^{*}$. Moreover, multiplying both sides of the above equation by $E_{\dagger}(I + V)^{*-1}$ yields

$$E_{\dagger}(I + V)E_{\dagger}r_0 = -E_{\dagger}VE^{\dagger}s$$

or equivalently,

$$E_{\dagger}r_0 = -E_{\dagger}V[E_{\dagger}r_0 + E^{\dagger}s]$$

as was to be shown.

Finally, from the lifting theorem it follows that there exists a memoryless operator F such that $V = F\lambda$

$$E_\dagger r_0 = -E_\dagger F\lambda[E_\dagger r_0 + E^\dagger s] = -E_\dagger F x_0$$

completing the proof. //

3. Example: Consider the plant modeled by the differential operator

$$\dot{x}(\dagger) = [-1]x(\dagger) + [\tfrac{1}{2}]r(\dagger), \qquad x(0) = 0$$
$$y(\dagger) = [2]x(\dagger),$$

defined on $L_2[0, \infty)$ with its usual resolution structure. This plant has the equivalent differential operator model

$$P = [D + 1]^{-1}[1]$$

while its input to state trajectory model is characterized by the differential operator

$$\lambda = [D + 1]^{-1}[\tfrac{1}{2}]$$

Now, using the differential operator $Q[-D]$ as a representation of $Q[D]^*$ we have

$$\begin{aligned} I + P^*P &= [D^2 - 1]^{-1}[D^2 - 2] \\ &= [(D + 1)(D - 1)]^{-1}[(D + \sqrt{2})(D - \sqrt{2})] \end{aligned}$$

and so

$$C = [D + 1]^{-1}[D + \sqrt{2}] = I + [D + 1]^{-1}[\sqrt{2} - 1] = I + V$$

We thus have $V = [D + 1]^{-1}[\sqrt{2} - 1]$ and $\lambda = [D + 1]^{-1}[\tfrac{1}{2}]$. Hence if $V = F\lambda$,

$$F = [2\sqrt{2} - 2]$$

is our required memoryless state feedback operator.

PROBLEMS

1. Use the projection theorem to show that $B_0 w = (A^*A)^{-1}A^*z$ minimizes $\|z - ABw\|^2$ over $B \in \mathscr{B}$ whenever $(A^*A)^{-1}$ exists.

2. Show that $(z, ABw) = (AB_0 w, ABw)$ and $(z, AB_0 w) = (AB_0 w, AB_0 w)$ whenever $B_0 w = (A^*A)^{-1}A^*z$.

3. On the space $L_2(-\infty, \infty)$ with its usual resolution structure let $A = \tilde{P}$ be the ideal predictor, let $w = \chi_{(0,1]}$, and let $z = \chi_{(0,2]}$. Determine the causal operator B_0 that minimizes $\|z - ABw\|^2$ over $B \in \mathscr{C}$.

4. Repeat Example B.3 with disturbance $n = \chi_{(0,T]}$.

5. Using the formulation of Property B.6 show that $[N_{pr}^* N_{pr} + D_{pr}^* D_{pr}]$ is always invertible.

6. Repeat Example B.7 with the plant taken to be the ideal predictor $\tilde{P}$.

7. Determine the memoryless state feedback gain that optimally regulates initial condition disturbances for the plant with state trajectory decomposition $P = \theta\lambda$ where

$$\lambda = [D + 2]^{-1}[2] \qquad \text{and} \qquad \theta = [2]$$

REFERENCES

1. Lance, E., Some properties of nest algebraes, *Proc. London Math. Soc.* (3) **19**, 45–68 (1969).
2. Porter, W. A., A basic optimization problem in linear systems, *Math. Systems Theory* **5**, 20–44 (1971).
3. Porter, W. A., Data interpolation, causality structure, and system identification, *Inform. and Control* **29**, 217–229 (1975).
4. Schumitzky, A., State feedback control for general linear systems, *Proc. Internat. Symp. Math. Networks and Systems* pp. 194–200. T. H. Delft (1979).
5. Steinberger, M., Ph.D. Dissertation, Univ. of Southern California, Los Angeles, California (1977).

Notes for Part III

1. Historical Background: From its beginnings the feedback system analysis and design problem has been the primary motivation for the development of the theory of operators defined on a Hilbert resolution space. The earliest work in the field was centered around the feedback system stability problem[21,24,28] which, in turn, motivated much of the research on the causal invertibility problem.[5,24,25] Furthermore, the derivation of the optimal causal solution to the servomechanism problem attracted interest from researchers in the field from the early 1960s, though a viable solution thereto was not achieved until the 1970s.[1,15]

The extended space concept has its origins in the classical paper of Youla *et al.*[27] on passive network theory in which a space of functions on compact support is embedded into a space of functions with support bounded on the left ($D \subset D_+$). The first application of the concept in a Hilbert space setting appears to be due to Sandberg[21] who used the embedding of $L_2(-\infty, \infty)$ into $L_2(-\infty, \infty)_e$ and applied it to a problem in stability analysis. During the 1960s there were considerable controversy over the relative merits of the extended space formulation and simply defining a stable system to be a system characterized by a causal bounded operator.[3,4] Fortunately, this controversy was settled in a classical paper of Willems,[24] wherein the relationship between causality and stability was made precise. Interestingly, the result was implicit in the original work of Youla *et al.*[27]

The use of a fractional representation model for unstable systems has become quite common in multivariate system theory during the late 1970s, through usually in the form of a polynomial matrix fraction representation for a rational matrix.[16] The extension of the concept to an operator theoretic setting using stable factors was introduced only recently in a paper by Desoer *et al.*[7] that forms the basis of the present approach. The stabilization theorem, although also based on the results of these authors, in turn extends a stabilization theory developed by Youla *et al.*[26] Similarly, the tracking and disturbance rejection theory represents a natural extension of multivariate theory to Hilbert resolution space.[9,20]

The servomechanism problem and its abstraction to the basic optimization problem was one of the first "unsolved" problems in resolution space theory and was often used as an example to motivate the formulation of the resolution space concept.[12,13] The solution presented here is due to Porter[15] as modified by DeSantis.[6] The key to this solution, however, was the observation that a global optimal controller did not exist so that one had to interpolate a noncausal operator with a causal operator at an appropriate set of points. Indeed, such an interpolation is the key to the classical frequency domain solution of the servomechanism problem.[1] As indicated in Example 10.A.4, however, the servomechanism problem is classically formulated in a space where one can identify vectors and operators. Hence the interpolation problem is usually so intermingled with the remainder of the derivation of the optimal servomechanism that its presence is not at all clear to a casual reader.[1]

The optimal regulator problem is fundamental to state space system theory. Indeed, the original solution of the regulator problem in the context of vector first-order ordinary differential equations "made" state space theory.[10,11] The fundamental point here is not the solution of the problem, which is a standard quadratic optimization problem, but the implementation of the solution in the form of memoryless state feedback.[22] Indeed, the only real justification for studying the state of a system is that direct access to the state may greatly simplify a system design, though the same information is (implicitly) available from the system output if it is observable. The general solution of the regulator problem had its origin

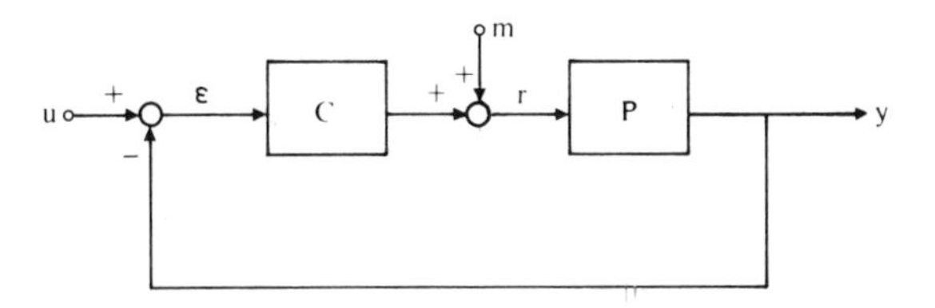

Fig. 1. Alternative feedback system model.

in the thesis of Steinberger[23] and was formalized in a paper by Schumitzky.[22]

2. Algebraic Systems: It should be noted that the last three sections of Chapter 8 and the entirety of Chapter 9 are totally algebraic and make no use whatsoever of the topological structure on our Hilbert resolution space. Hence many researchers believe that an entirely algebraic feedback system theory can be formulated. Although this would open the door to the study of a number of important classes of system that are not topologically closed, it rules out many of the deep topologically based ideas presented here. Indeed, even though one can formulate a feedback system stability criteria using a purely algebraic causal invertibility test, the more powerful causal invertibility theorems formulated in Chapter 5 are highly topological in nature. Similarly, our optimization theory is inherently topological in nature. Moreover, most of the algebraic results of Chapters 8 and 9 were originally discovered in a topological setting and only later reformulated in purely algebraic terms.[2,7,9,14,20]

3. Compensator Design: As an alternative to the feedback model used in the introduction to Part III one can work with a model such as that shown in Fig. 1. Here, the plant is the same as before, but rather than designing a feedback law one designs a *compensator* C that controls the plant input on the basis of observations of the difference between the plant output y and the reference input u. Interestingly, from a mathematical point of view this formulation of the feedback system model is virtually identical to the feedback law formulation. For instance, if one assumes that the plant is characterized by a doubly coprime fractional representation

$$\begin{bmatrix} V_{pr} & -U_{pr} \\ -N_{pl} & D_{pl} \end{bmatrix}^{-1} = \begin{bmatrix} D_{pr} & -U_{pl} \\ N_{pr} & V_{pl} \end{bmatrix}$$

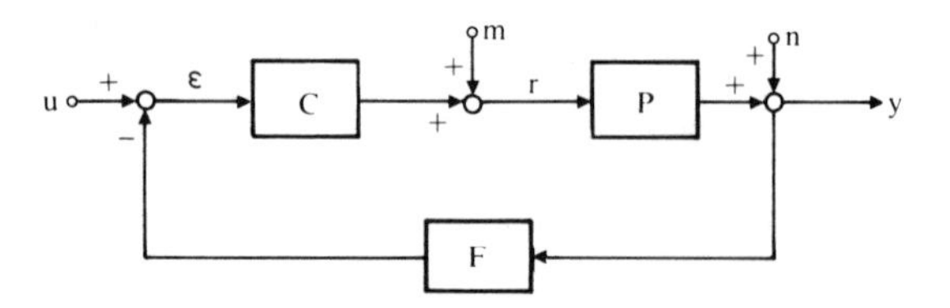

Fig. 2. Feedback system model with both compensator and feedback law.

then the set of stabilizing compensators for the feedback system is given by exactly the same formula as the set of stabilizing feedback laws derived in Chapter 9. That is,

$$C = (WN_{pl} + V_{pr})^{-1}(-WD_{pl} + U_{pr})$$

while similar stability, sensitivity, tracking, and disturbance rejection theories can be formulated.[7,20]

A final, and possibly most realistic, feedback system model is illustrated in Fig. 2. Here, we have both a compensator and a feedback law, though in this case the feedback law is usually fixed and represents a model of the instrumentation used to measure the plant output. For a relistic model this is required since the actual output in many systems (say, the temperature at the core of a nuclear reactor) is not directly measurable and sophisticated measurement devices may be required. Since such devices have only a finite bandwidth and are certainly far from the identity, $F \neq I$ may be required to model the system. The output of our measurement device is then compared with a reference input as in the compensator model, and the compensator is designed to control the plant by operating on the difference between the reference input and measured plant output. Interestingly, the theory for this most realistic feedback model is far less well developed than the single compensator or feedback law theory. Two papers of Youla *et al.*,[26] however, give a stabilization theory in the multivariate case and are indicative of the techniques required.

4. Cascade Loading: In mathematical network theory,[8,17] rather than dealing with a feedback system model one often works with a *cascade load model*, as illustrated in Fig. 3. Here, we have a $2n$ port Σ loaded by an n-port network S_L. Although the model makes sense using the standard voltage and current variables, it is most naturally studied with the *scattering*

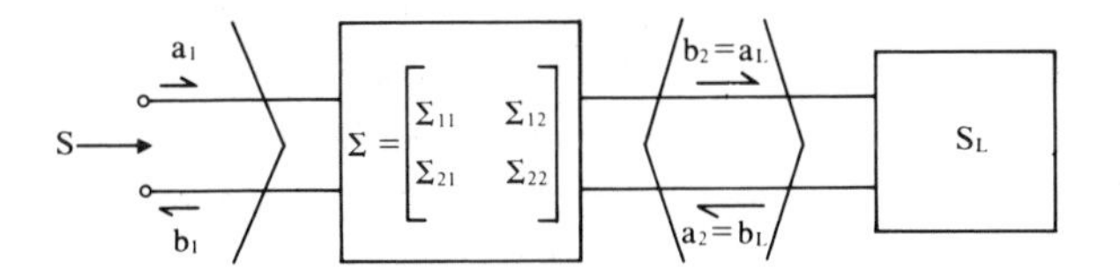

Fig. 3. Cascade load configuration.

variables $a = v + i$ and $b = v - i$. The *coupling network* Σ is then modeled by the scattering equations

$$\begin{bmatrix} b_1 \\ b_2 \end{bmatrix} = \begin{bmatrix} \Sigma_{11} & \Sigma_{12} \\ \Sigma_{21} & \Sigma_{22} \end{bmatrix} \begin{bmatrix} a_1 \\ a_2 \end{bmatrix}$$

while the load is modeled by

$$b_L = S_L a_L$$

Physically, the *incident wave a* represents the power entering a port, while the *reflective wave b* represents the power leaving a port. Our cascade connection is thus constrained by the conservation laws

$$b_2 = a_L \qquad \text{and} \qquad a_2 = b_L$$

while the n port observed at the 1 terminals of the coupling network satisfies

$$b_1 = Sa_1 = [\Sigma_{11} + \Sigma_{12}(I - S_L\Sigma_{22})^{-1}S_L\Sigma_{21}]a_1$$

Although the physical interpretation is different, this is identical to the equation arising from the feedback system illustrated in Fig. 4. Accordingly, the cascade loading model of network theory is completely equivalent to the feedback model developed here. However, one usually asks different

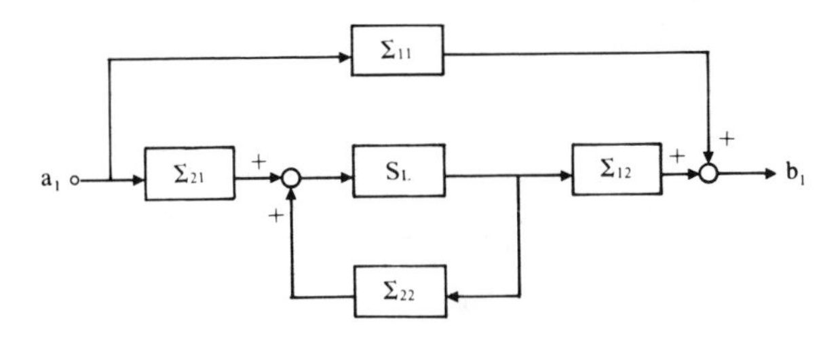

Fig. 4. Feedback system that is equivalent to a cascade load model.

questions about the cascade load model from those studied in a feedback system context. For instance, the stronger passivity condition replaces the stability condition, and one deals with questions motivated by the problems of network synthesis rather than feedback system design.

5. Passivity: The requirement that the operator H_{yn} be a causal contraction was encountered in Remark 8.D.6 in a sensitivity theory. The concept, however, arises much more naturally in a network theory as the requirement for a scattering operator to model a system containing no internal energy sources. Indeed, if

$$b = Sa$$

then we say that S is *passive*[27] if

$$\|E^{\dagger}a\|^2 - \|E^{\dagger}b\|^2 \geq 0, \qquad \dagger \in \$$$

That is, the energy that has entered the network before time † exceeds the energy leaving the network before time †. (Hence no internal energy sources are required to account for the observed energy transations.) In fact, this condition is completely equivalent to the requirement that S be a causal contraction.[18,27] Related to passivity is the concept of *losslessness*. A lossless S is required to be passive and also satisfy

$$\|a\|^2 - \|b\|^2 = 0$$

Equivalently, S is lossless if and only if it is a causal isometry.

Using these concepts and the cascade loading configuration, one may now formulate the network synthesis problem in our Hilbert resolution space setting.[19] For the active synthesis problem one assumes that the load is characterized by the property $a_L = 0$ (corresponding to -1-Ω resistors) and an $S \in \mathscr{C}$ has been specified. We then ask if it is possible to find a passive Σ such that the specified S is realized by the cascade load configuration; i.e., $b_1 = Sa_1$ when the appropriate passive Σ is used in the cascade load configuration under the constraint $a_L = 0$ imposed by our load of -1-Ω resistors. Similarly, one may consider a passive network synthesis problem, where one is given a passive S and desires to realize it in the cascade load configuration with a lossless Σ and a load $S_L = 0$ (corresponding to 1-Ω resistors). Interestingly, this problem reduces to a causal isometric extension problem:[18] Given a causal contraction on a Hilbert resolution space $(H, \mathscr{E})$, find a Hilbert resolution space $(H', \mathscr{E}')$ and a causal isometry Σ on $(H \oplus H', E \oplus E')$ such that

$$\Sigma = \begin{bmatrix} S & \Sigma_{12} \\ \Sigma_{21} & \Sigma_{22} \end{bmatrix}$$

Both the active and passive network synthesis problems have been solved in the affirmative though many open questions remain.[18] In particular, conditions under which one can take $(H', \mathscr{E}') = (H, \mathscr{E})$ are unknown.

6. Nonlinear Feedback Systems: From the inception of their work most researchers in Hilbert resolution space topics have been interested in nonlinear, as well as linear, feedback systems. Indeed, from the point of view of feedback system analysis is a rather powerful theory of nonlinear systems exists.[5,21,24,25,28,29] As in the linear case the stability problem can be reduced to a causal invertibility problem for which a rich nonlinear theory exists. Interestingly, while the right fractional representation theory holds for nonlinear operators, the left fractional representation theory fails, and hence so does the doubly coprime theory.[7] Little of the linear feedback system design theory carries over to the nonlinear case.

7. Tracking and Disturbance Rejection: Although explicit conditions for the existence of a solution to the tracking and disturbance rejection problems of Sections 9.B and 9.C are not known in the general case, if one restricts consideration to operators that lie in a commutative subalgebra of $\mathscr{B}$ an explicit solution to these problems can be obtained.[20] Since we are working within a commutative algebra we assume a doubly coprime fractional representation for our plant in the form

$$\begin{bmatrix} V_p & U_p \\ -N_p & D_p \end{bmatrix}^{-1} = \begin{bmatrix} D_p & -U_p \\ N_p & V_p \end{bmatrix}$$

where the subscripts for left and right have been deleted since all of the required operators lie in a commutative subalgebra and thus may serve as both left and right factors. Similarly, we assume that the tracking generator is modeled by $T = D_t^{-1}N_t$ while the disturbance generator is modeled by the equality $R = D_r N_r^{-1}$. Then a solution to the tracking problem exists if and only if N_p and D_t are coprime, a solution to the disturbance rejection problem exists if and only if N_p and D_r are coprime, and a solution to the simultaneous tracking and disturbance rejection problem exists if and only if N_p and D_t are coprime, N_p and D_r are coprime, and D_t and D_r are coprime. Moreover, in any of these cases an explicit description of the $W \in \mathscr{C}$ satisfying the appropriate design equation can be given in the form

$$W = AX + B$$

where $A, B \in \mathscr{C}$ and $X \in \mathscr{C}$ becomes our new design parameter. Finally, we note that these concepts can be extended to a multivariable setting at the cost of some additional complexity[20] though, to our knowledge, no extension to a full Hilbert resolution space setting is known.

8. Robust Design: Unlike our theoretical investigations, in any "real world" feedback system design problem one does not have an exact model for the plant, because of modeling errors, aging, environmental changes, etc. Even though we design around a nominal plant model P_0, our design constraints must be met by the actual plant, which is known only to be in some neighborhood of P_0. We say that a design is *robust* if it satisfies the design constraints for all plants in some neighborhood of P_0 (on appropriately defined topology). Fortunately, the stabilization problem is automatically robust since $\mathscr{C} \cap \mathscr{C}^{-1}$ is an open set in $\mathscr{B}$. On the other hand a solution to the tracking or disturbance rejection problem may fail to be robust.[9] Interestingly, however, in the case where one restricts consideration to operators in a commutative subalgebra of $\mathscr{B}$, robust solutions to the tracking and disturbance rejection problems exist whenever the coprimeness conditions described in Note 7 are satisfied, though the set of robust solutions is strictly smaller than the set of all solutions. Indeed, the $W \in \mathscr{C}$ satisfying the design equations robustly take the form

$$W = ACX + B$$

where A, B, and $X \in \mathscr{C}$ are as in Note 7 and $C \in \mathscr{C}$ is an additional causal operator that further restricts the set of W's obtained. Also as in Note 7, this theory can be extended to multivariable systems though no general formulation in a Hilbert resolution space setting is known.

REFERENCES

1. Chang, S. S. L., "Synthesis of Optimal Control Systems." McGraw-Hill, New York, 1961.
2. Cruz, J. B., and Perkins, W. R., A new approach to the sensitivity problem in multivariable feedback system design, *IEEE Trans. Automat. Control* **AC-9** (1964).
3. Damborg, M. J., Ph.D. Dissertation, Univ. of Michigan, Ann Arbor, Michigan (1967).
4. Damborg, M. J., and Naylor, A., Fundamental structure of input-output stability for feedback systems, *IEEE Trans. Systems Sci. Cybernet.* **SSC-6**, 92–96 (1970).
5. DeSantis, R. M., Ph.D. Dissertation, Univ. of Michigan, Ann Arbor, Michigan (1971).
6. DeSantis, R. M., Causality theory in system analysis, *Proc. IEEE* **64**, 36–44 (1976).
7. Desoer, C. A., Liu, R.-W., Murray, J., and Saeks, R., Feedback system design: The fractional representation approach to analysis and synthesis, *IEEE Trans. Automat. Control* **AC-25**, 401–412 (1980).
8. Dolezal, V., "Monotone Operators and Applications in Control and Network Theory." Elsevier, Amsterdam, 1979.

9. Francis, B. A., The multivariate servomechanism problem from the input-output viewpoint, *IEEE Trans. Automat. Control* **AC-22**, 322–328 (1977).
10. Kalman, R. E., Contributions to the theory of optimal control, *Bol. Soc. Mat. Mexicanna* **5**, 102–119 (1960).
11. Kalman, R. E., On the general theory of control systems, *Proc. IFAC Congr. 1st, Moscow* Butterworths, London, 1960.
12. Kou, M. C., and Kazda, L., Minimum energy problems in Hilbert space, *J. Franklin Inst.* **283**, 38–54 (1967).
13. Porter, W. A., "Modern Foundations of System Theory." Macmillan, New York, 1966.
14. Porter, W. A., On sensitivity in multivariate nonstationary systems, *Internat. J. Control* **7** (1968).
15. Porter, W. A., A basic optimization problem in linear systems, *Math. Systems Theory* **5**, 20–44 (1971).
16. Rosenbrock, H. H., "State-Space and Multivariate Theory." Nelson-Wiley, London, 1970.
17. Saeks, R., Synthesis of general linear networks, *SIAM J. Appl. Math.* **16**, 924–930 (1968).
18. Saeks, R., Causality in Hilbert space, *SIAM Rev.* **12**, 357–383 (1970).
19. Saeks, R., "Generalized Networks." Holt, New York, 1972.
20. Saeks, R., and Murray, J., Feedback system design: The tracking and disturbance rejection problems, *IEEE Trans. Automat. Control* **AC-26**, 203–217 (1981).
21. Sandberg, I. W., On the L_2 boundedness of solutions of nonlinear functional equations, *Bell Systems Tech. J.* **43**, 1601–1608 (1965).
22. Schumitzky, A., State feedback control for general linear systems, *Proc. Internat. Symp. Math. Networks and Systems* pp. 194–200. T. H. Delft (1979).
23. Steinberger, M., Ph.D. Dissertation, Univ. of Southern Calif., Los Angeles, California (1977).
24. Willems, J. C., Stability, instability, invertibility, and causality, *SIAM J. Control* **7**, 645–671 (1969).
25. Willems, J. C., "Analysis of Feedback Systems." MIT Press, Cambridge, Massachusetts, 1971.
26. Youla, D. C., Bongiorno, J. J., and Jabr, H. A., Modern Wiener–Hopf design of optimal controllers, parts I and II, *IEEE Trans. Automat. Control* **AC-21**, 3–15, 319–338 (1976).
27. Youla, D. C., Carlin, H. J., and Castriota, L. J., Bounded real scattering matrices and the foundations of linear passive network theory, *IRE Trans. Circuit Theory* **CT-6**, 102–124 (1959).
28. Zames, G., Nonlinear Operators for System Analysis, Tech. Rep. 370. MIT Res. Lab. for Electronics (1960).
29. Zames, G., Functional analysis applied to nonlinear feedback systems, *IEEE Trans. Circuit Theory* **CT-10**, 392–404 (1963).

PART IV

Stochastic Systems

Since a Hilbert resolution space is essentially a Hilbert space into which a time structure has been axiomatically inserted, a resolution space–valued random variable is a natural model for a stochastic process. With this viewpoint in mind this final part of the text is devoted to a study of stochastic system theory. The required results from the theory of Hilbert space–valued random variables are reviewed in Chapter 11 and used to formulate an explicit solution to an appropriate stochastic generalization of the basic optimization problem. The solution of this problem is then used in Chapter 12 as the starting point for the solution of a number of classical estimation and stochastic control problems. These include standard stochastic estimators and predictors, a stochastic servomechanism problem, and a state space theory for stochastic control.

Unlike the feedback system design theory of Part III, much of which could be formulated in a purely algebraic setting, the present section relies heavily on the theory of operators defined on a Hilbert resolution space. Indeed, the operator theoretic concepts formulated in Part I, such as the various forms of hypercausality, operator factorization and decomposition, and causal invertibility theory, etc., play a dominant role in our stochastic system theory. We believe that the material illustrates the essential interrelationship between the theory of operators defined on a Hilbert resolution space and system theory.

11

Stochastic Optimization

A. HILBERT SPACE–VALUED RANDOM VARIABLES

1. Remark: Intuitively, a *stochastic process* is a random function of time. Thus, since a Hilbert resolution space represents an axiomatization of a space of time functions, a resolution space–valued random variable is a natural model for a stochastic process. The purpose of the present section is to review the theory of Hilbert space–valued random variables and to investigate the implication of the resolution structure thereon.

2. Remark: Recall that a *Hilbert space–valued random variable* is a measurable function mapping a probability space (S, Σ, P) to a Hilbert space H. Here S is a set, Σ is a σ field over S, and P is a finitely additive probability measure on Σ. The assumption that P is finitely additive rather than countably additive is a technicality associated with the Hilbert space theory[2] that will be discussed in more detail later. Throughout the development the underlying probability space (S, Σ, P) will be suppressed, appearing only indirectly in the form of the expected value operator

$$E[f(\underline{x}_1, \underline{x}_2, \ldots, \underline{x}_n)] = \int_S f(\underline{x}_1(s), \underline{x}_2(s), \ldots, \underline{x}_n(s))\, dP(s)$$

Here, the $\underline{x}_i$, $i = 1, 2, \ldots, n$, are H-valued random variables defined on (S, Σ, P) and $f: H^n \to C$ is a scalar-valued function of H^n. In general we denote H-valued random variables by a lowercase roman letter with an underbar ($\underline{x}$, $\underline{y}$, etc.) to distinguish them from constant vectors in H.

Although one can formulate a highly sophisticated theory around our Hilbert space–valued random variables, for the present purposes a simple second-order theory suffices. We define a class of random variables with finite second moment whose properties are satisfactorily characterized by an appropriate covariance operator.

3. Definition: An H-valued random variable $\underline{x}$ is said to have *finite second moment* if $E|(x, \underline{x})|^2$ defines a continuous scalar-valued mapping on H.

4. Property: Let $\underline{x}$ be an H-valued random variable with finite second moment. Then there exists a unique vector $m_{\underline{x}} \in H$ such that

$$E(x, \underline{x}) = (x, m_{\underline{x}})$$

for all x in H.

Proof: Since $E|(x, \underline{x})|^2$ defines a continuous map on H, so does $E(x, \underline{x})$. Moreover, $E(x, \underline{x})$ is linear in x and hence defines a continuous linear functional on H. Hence the Riesz representation theorem implies the existence of an appropriate $m_{\underline{x}}$. //

5. Definition: The vector $m_{\underline{x}}$ of Property 4 is termed the *mean* of the random variable $\underline{x}$.

6. Property: Let $\underline{x}$ and $\underline{y}$ be H-valued random variables with finite second moment and let $A \in \mathscr{B}$. Then

(i) $$m_{\underline{x}+\underline{y}} = m_{\underline{x}} + m_{\underline{y}}$$

(ii) $$m_{A\underline{x}} = Am_{\underline{x}}$$

(iii) $$\|m_{\underline{x}}\| \leq E\|\underline{x}\|$$

Proof: (i) follows from the linearity of the inner product and the expectation operator. (ii) follows from the string of equalities

$$(x, m_{A\underline{x}}) = E(x, A\underline{x}) = E(A^*x, \underline{x}) = (A^*x, m_{\underline{x}}) = (x, Am_{\underline{x}})$$

while (iii) follows from the Riesz representation theorem and the fact that $\|m_{\underline{x}}\|$ equals the norm of the functional $E(x, \underline{x})$. As such,

$$\|m_{\underline{x}}\| = \sup_{x \neq 0} \frac{|E(x, \underline{x})|}{\|x\|} \leq \sup_{x \neq 0} \frac{E\|x\| \, \|\underline{x}\|}{\|x\|} = E\|\underline{x}\|$$

where we have invoked the Schwarz inequality. //

7. Remark: Note that if $\underline{x} = \operatorname{col}(\underline{x}_1, \underline{x}_2, \ldots, \underline{x}_n)$ takes its values in R^n, then

$$E(x, \underline{x}) = E\left[\sum_{i=1}^{n} x_i \underline{x}_i\right] = \sum_{i=1}^{n} x_i E[\underline{x}_i] = (x, E[\underline{x}])$$

where $E[\underline{x}] = \operatorname{col}(E[\underline{x}_i])$ is the classical vector of expected values. As such, our abstract definition for $m_{\underline{x}}$ coincides with the classical definition in the finite-dimensional case.

It follows from Property 6(i) that an arbitrary H-valued random variable with finite second moment, $\underline{x}$, can be represented $\underline{x}' + m_x$ where $\underline{x}' = \underline{x} - m_x$ has zero mean. Hence we can greatly simplify our theory by working with zero mean random variables, adding constant vectors where needed to represent nonzero mean random variables. In the sequel we shall assume that all random variables are zero mean unless otherwise stated.

8. Property: Let $\underline{x}$ and $\underline{y}$ be zero mean H-valued random variables with finite second moment. Then there exists a unique operator $Q_{\underline{xy}} \in \mathscr{B}$ such that

$$E(x, \underline{x})(y, \underline{y}) = (x, Q_{\underline{xy}} y)$$

for all x and $y \in H$.

Proof: Since $\underline{x}$ and $\underline{y}$ both have finite second moment, $E(x, \underline{x})(\underline{y}, y)$ is continuous in x and y. Moreover, it is linear in x and sequilinear in y since the inner product has these properties and the expectation operator is linear. $E(x, \underline{x})(\underline{y}, y)$ thus defines a continuous functional on H^2 that is linear in x and sequilinear in y, and hence the Riesz representation theorem implies that there exists a $Q_{\underline{xy}} \in \mathscr{B}$ such that

$$E(x, \underline{x})(y, \underline{y}) = (x, Q_{\underline{xy}} y)$$

as required. //

9. Definition: The operator $Q_{\underline{xy}}$ formulated in Property 8 is termed the *cross-covariance* of $\underline{x}$ and $\underline{y}$. In the special case when $\underline{x} = \underline{y}$ the operator $Q_{\underline{xx}}$ is termed the covariance of $\underline{x}$ and denoted $Q_{\underline{x}}$.

10. Remark: As with the mean our abstract definition for the covariance and cross-covariance operators coincides with the classical definition in the finite-dimensional case. Indeed, if $\underline{x}$, $\underline{y}$, x, $y \in R^n$ and ^{t} denotes transposition,

$$(x, Q_{\underline{xy}}y) = E(x, \underline{x})(\underline{y}, y) = E[x^t\underline{x}\underline{y}^t y] = x^t E[\underline{x}\underline{y}^t]y = (x, E[\underline{x}\underline{y}^t]y)$$

showing that $Q_{\underline{xy}} = E[\underline{x}\underline{y}^t]$ which coincides with the usual definition for the cross-covariance in R^n.

11. Example: Let $\underline{x}$ be an H-valued random variable taking on only finitely many values x_i, $i = 1, 2, \ldots, n$, with probabilities p_i, $i = 1, 2, \ldots, n$. Then for any $x \in H$

$$E(x, \underline{x}) = \int_S (x, \underline{x}(s))\,dP(s) = \sum_{i=1}^{n} (x, x_i)p_i = \left(x, \left[\sum_{i=1}^{n} x_i p_i\right]\right)$$

verifying that

$$m_{\underline{x}} = \sum_{i=1}^{n} x_i p_i$$

Now, taking $m_{\underline{x}} = 0$ we compute

$$E(x, \underline{x})(\underline{x}, y) = \int_S (x, \underline{x}(s))(\underline{x}(s), y)\,dP(s) = \sum_{i=1}^{n} (x, x_i)(x_i, y)p_i$$

$$= \sum_{i=1}^{n} (x, (y, x_i)x_i p_i) = \sum_{i=1}^{n} (x, p_i Q_{x_i} y)$$

$$= \left(x, \left[\sum_{i=1}^{n} p_i Q_i\right]y\right)$$

where Q_i is the *rank-1 operator*, defined by

$$Q_i y = (y, x_i)x_i = [x_i \otimes x_i]y$$

We then have that $Q_{\underline{x}}$ is the finite-rank operator

$$Q = \sum_{i=1}^{n} p_i Q_i = \sum_{i=1}^{n} p_i[x_i \otimes x_i]$$

12. Definition: Let $\underline{x}$ and $\underline{y}$ be zero mean H-valued random variables with finite second moment. Then $\underline{x}$ and $\underline{y}$ are said to be independent if $Q_{\underline{xy}} = 0$.

13. Property: Let $\underline{x}$ and $\underline{y}$ be zero mean H-valued random variables with finite second moment and let A and B be bounded linear operators on H. Then

(i) $W_{\underline{x}+\underline{y}} = Q_{\underline{x}} + Q_{\underline{xy}} + Q_{\underline{yx}} + Q_{\underline{y}}$;
(ii) $Q_{\underline{x}+\underline{y}} = Q_{\underline{x}} + Q_{\underline{y}}$ if and only if $\underline{x}$ and $\underline{y}$ are independent;
(iii) $Q_{(A\underline{x})(B\underline{y})} = AQ_{\underline{xy}}B^*$;
(iv) $Q_{A\underline{x}} = AQ_{\underline{x}}A^*$;
(v) $Q_{\underline{xy}} = Q^*_{\underline{yx}}$;
(vi) $Q_{\underline{x}} = Q^*_{\underline{x}} \geq 0$;
(vii) $E\|\underline{x}\|^2 < \infty$ if and only if $Q_{\underline{x}} \in \mathscr{K}_1$, in which case $E\|\underline{x}\|^2 = \mathrm{tr}[Q_{\underline{x}}]$, i.e., $Q_{\underline{x}}$ is nuclear; and
(viii) if $Q_{\underline{x}} \in \mathscr{K}_1$ then $Q_{\underline{xy}}$ and $Q_{\underline{yx}} \in \mathscr{K}_2$; i.e., they are Hilbert–Schmidt operators.

Proof: (i)

$$\begin{aligned}(x, Q_{\underline{x}+\underline{y}}y) &= E(x, \underline{x} + \underline{y})(\underline{x} + \underline{y}, y)\\ &= E[(x, \underline{x})(\underline{x}, y) + (x, \underline{x})(\underline{y}, y) + (x, \underline{y})(\underline{x}, y) + (x, \underline{y})(\underline{y}, y)]\\ &= E(x, \underline{x})(\underline{x}, y) + E(x, \underline{x})(\underline{y}, y) + E(x, \underline{y})(\underline{x}, y)\\ &\quad + E(x, \underline{y})(\underline{y}, y)\\ &= Q_{\underline{x}} + Q_{\underline{xy}} + Q_{\underline{yx}} + Q_{\underline{y}}\end{aligned}$$

as required.

(ii) This follows from (i) and the definition of independence.

(iii)

$$\begin{aligned}(x, Q_{(A\underline{x})(B\underline{y})}y) = E(x, A\underline{x})(B\underline{y}, y) &= E(A^*x, \underline{x})(\underline{y}, B^*y)\\ &= (A^*x, Q_{\underline{xy}}B^*y) = (x, AQ_{\underline{xy}}B^*y)\end{aligned}$$

verifying (iii).

(iv) This is a special case of (iii) and hence follows from the above.

(v)

$$\begin{aligned}(x, Q_{\underline{xy}}y) &= E(x, \underline{x})(\underline{y}, y) = E\overline{(y, \underline{y})}\,\overline{(\underline{x}, x)} = \overline{E(y, \underline{y})(\underline{x}, x)} = \overline{(y, Q_{\underline{yx}}x)}\\ &= (Q_{\underline{yx}}x, y) = (x, Q^*_{\underline{yx}}y)\end{aligned}$$

verifying that $Q_{\underline{xy}} = Q^*_{\underline{yx}}$.

(vi) The fact that $Q_{\underline{x}} = Q_{\underline{xx}}$ is hermitian follows from the above while its positivity follows from

$$(x. Q_{\underline{x}}x) = E(x, \underline{x})(\underline{x}, x) = E(x, \underline{x})\overline{(x, \underline{x})} = E|(x, \underline{x})|^2 \geq 0$$

(vii) To verify this condition we let x_π denote a complete orthonormal system in H and we expand $E\|x\|^2$ as

$$E\|\underline{x}\|^2 = E\left[\sum_\pi |(x_n, \underline{x})|^2\right] = \sum_\pi E|(x_\pi, \underline{x})|^2 = \sum_\pi E(x_\pi, \underline{x})(\underline{x}, x_\pi)$$
$$= \sum_\pi (x_\pi, Q_{\underline{x}} x_\pi) = \operatorname{tr}[Q_{\underline{x}}]$$

since $\sum_\pi (x_\pi, Q_{\underline{x}} x_\pi)$ defines the trace of an operator independently of the choice of orthonormal system x_π. Hence $E\|\underline{x}\|^2 = \operatorname{tr}[Q_{\underline{x}}]$, and it is therefore finite if and only if $Q_{\underline{x}}$ is *trace class*, i.e., $Q_{\underline{x}} \in \mathscr{K}_1$.

(viii) Recall that for scalar-valued random variables α and β

$$|E[\alpha\beta]| \leq \sqrt{E[|\alpha|^2]}\sqrt{E[|\beta|^2]}$$

Indeed, this is just the Schwarz inequality in a Hilbert space of random variables.[2] Then for any $x \in H$

$$\begin{aligned}(Q_{\underline{yx}} x, Q_{\underline{yx}} x) &= E(Q_{\underline{yx}} x, \underline{y})(\underline{x}, x)\\ &\leq E\sqrt{[|Q_{\underline{yx}} x, \underline{y})|^2]}\sqrt{E[|(\underline{x}, x)|^2]}\\ &= \sqrt{E(Q_{\underline{yx}} x, \underline{y})(\underline{y}, Q_{\underline{yx}} x)}\sqrt{E(x, \underline{x})(\underline{x}, x)}\\ &= \sqrt{(Q_{\underline{yy}} x, Q_{\underline{y}} Q_{\underline{yx}} x)}\sqrt{(x, Q_{\underline{x}} x)}\end{aligned}$$

Now let x_π be an orthonormal system in H and let F denote any finite set of indices. Then

$$\begin{aligned}\sum_F (Q_{\underline{yx}} x_\pi, Q_{\underline{yx}} x_\pi) &\leq \sum_F \sqrt{Q_{\underline{yx}} x_\pi, Q_{\underline{y}} Q_{\underline{yx}} x_\pi)}\sqrt{\textstyle\sum_F (x_\pi, Q_{\underline{x}} x_\pi)}\\ &\leq \sqrt{\textstyle\sum_F (Q_{\underline{yx}} x_\pi, Q_{\underline{x}} Q_{yx} x_\pi)}\sqrt{\textstyle\sum_F (x_\pi, Q_{\underline{x}} x_\pi)}\\ &\leq \sqrt{\|Q_{\underline{y}}\|}\sqrt{\textstyle\sum_F (Q_{\underline{yx}} x_\pi, Q_{\underline{yx}} x_\pi)}\sqrt{\textstyle\sum_F (x_\pi, Q_{\underline{x}} x_\pi)}\end{aligned}$$

As such,

$$\sum_F (Q_{\underline{yx}} x_\pi, Q_{\underline{yx}} x_\pi) \leq \|Q_{\underline{y}}\| \left[\sum_F (x_\pi, Q_{\underline{x}} x_\pi)\right]$$

Since this holds for all F, upon taking limits we have

$$\operatorname{tr}[Q_{\underline{yx}}^* Q_{\underline{yx}}] = \sum_\pi (x_\pi, Q_{\underline{yx}}^* Q_{\underline{yx}} x_\pi) = \sum_\pi (Q_{\underline{yx}} x_\pi, Q_{\underline{yx}} x_\pi)$$
$$\leq \|Q_{\underline{y}}\| \left[\sum_\pi (x_\pi, Q_{\underline{x}} x_\pi)\right] = \|Q_{\underline{y}}\| \operatorname{tr}[Q_{\underline{x}}] < \infty$$

since $Q_y \in \mathscr{B}$ and $Q_{\underline{x}} \in \mathscr{K}_1$. Hence $Q_{\underline{yx}}^* Q_{yx} \in \mathscr{K}_1$, which implies that $Q_{\underline{yx}} \in \mathscr{K}_2$. Finally, $Q_{\underline{xy}} = Q_{\underline{xy}}^* \in \mathscr{K}_2$ since $\mathscr{K}_2$ is closed under adjoints. //

14. Remark: As remarked earlier, a viable theory of Hilbert space–valued random variables should allow for finitely additive probability measures. Indeed, while every positive hermitian operator $Q = Q^* \geq 0$ is the covariance for some finitely additive random variable, such a random variable may be extended to a countably additive probability space if and only if $Q \in \mathscr{K}_1$. Accordingly, we really have three equivalent characterizations for a Hilbert space–valued random variable; its norm squared has finite expected value, its covariance is nuclear, and it has a countably additive extension.[2] Fortunately, this latter property is not required for our theory and will therefore not be investigated further.

15. Example: Let $\underline{x}$ be a zero mean H-valued random variable with covariance $Q_{\underline{x}} \in \mathscr{K}$. Then since $Q_{\underline{x}}$ is compact its eigenvector sequence x_i defines an orthonormal basis and we may expand $\underline{x}$ in this basis, obtaining

$$x = \sum_i \underline{x}_i x_i$$

where $\underline{x}_i = (x_i, \underline{x})$ is a scalar-valued random variable. Now, we may compute

$$E[\underline{x}\,\overline{\underline{x}}_j] = E(x_i, \underline{x})\overline{(x_j, \underline{x})} = E(x_i, \underline{x})(\underline{x}, x_j) = (x_i, Q_{\underline{x}} x_j) = \lambda_j(x_i, x_j)$$
$$= \begin{cases} 0, & i \neq j \\ \lambda_j, & i = j \end{cases}$$

where $\lambda_j \geq 0$ is the eigenvalue associated with x_j. The coefficients in our expansion for x are therefore independent random variables, and we have formulated a version of the Karhunen–Loève expansion.

16. Remark: Thus far the entire development has dealt with Hilbert space–valued random variables independently of any resolution structure. If, however, $\underline{x}$ takes its values in H where $(H, \mathscr{E})$ defines a Hilbert resolution space, then x may be interpreted as a stochastic process and one may study its "time related" behavior. As an example we consider the concept of white noise.

17. Definition: A zero mean H-valued random variable $\underline{x}$ with finite second moment is termed *white noise* if $E^\dagger \underline{x}$ is independent of $E_\dagger \underline{x}$ for all $\dagger \in \$$.

18. Property: Let $\underline{x}$ be a zero mean H-valued random variable with finite second moment. Then $\underline{x}$ is white noise if and only if $Q_{\underline{x}} \in \mathcal{M}$.

Proof: If $Q_x \in \mathcal{M}$, then

$$Q_{(E^\dagger \underline{x})(E_\dagger \underline{x})} = E^\dagger Q_{\underline{x}} E_\dagger^* = E^\dagger Q_{\underline{x}} E_\dagger = E^\dagger E_\dagger Q_{\underline{x}} = 0$$

which implies that $E^\dagger \underline{x}$ and $E_\dagger \underline{x}$ are independent. Conversely, if $E^\dagger \underline{x}$ and $E_\dagger \underline{x}$ are independent for all $\dagger \in \$$,

$$E^\dagger Q_{\underline{x}} E_\dagger = E^\dagger Q_{\underline{x}} E_\dagger^* = Q_{(E^\dagger \underline{x})(E_\dagger \underline{x})} = 0$$

and hence

$$E^\dagger Q_{\underline{x}} = E^\dagger Q_{\underline{x}} E^\dagger + E^\dagger Q_{\underline{x}} E_\dagger = E^\dagger Q_{\underline{x}} E^\dagger$$

showing that $Q_{\underline{x}} \in \mathscr{C}$. Similarly, $E_\dagger Q_{\underline{x}} E^\dagger = (E^\dagger Q_{\underline{x}} E_\dagger)^* = 0^* = 0$ implies that $Q_{\underline{x}} \in \mathscr{C}^*$; hence $Q_{\underline{x}} \in \mathcal{M}$, as was to be shown. //

19. Remark: In the remainder of the chapter a number of much deeper time-related properties of a resolution space–valued random variable will be formulated for our stochastic optimization theory. Interestingly, however, the concept of a stationary process is not encountered. Such a concept is well defined in a uniform resolution space (though not in a general Hilbert resolution space) by requiring that $Q_{\underline{x}}$ be a time-invariant operator. Since the theory of uniform resolution spaces is not covered in detail in the present text (see Note 3 at the end of Part I for an introduction to the concept), we shall not consider the stationary case further.

B. A BASIC STOCHASTIC OPTIMIZATION PROBLEM

1. Remark: The basic stochastic optimization problem parallels the basic optimization problem formulated in Chapter 10 with the vectors z and w replaced by H-valued random variables $\underline{z}$ and $\underline{w}$. Here we are required to minimize

$$J(B) = E\|\underline{z} - AB\underline{w}\|^2$$

over the causal operators $B \in \mathscr{C}$ where A is an arbitrary bounded operator satisfying $A^*A > 0$. As before, once a solution to this abstract optimization problem has been obtained it can be used as a starting point for the derivation of solutions to the various classical stochastic control and estimation problems encountered in system theory.

Basic Stochastic Optimization Theorem: Let $A \in B$, and $\underline{z}$ and $\underline{w}$ be zero mean H-valued random variables with finite second moments and covariance operators $Q_{\underline{z}} \in \mathscr{K}_1$ and $Q_w > 0$, and assume that Q_w and A^*A admit spectral factorizations $A^*A = \bar{C}^*C$ and $Q_{\underline{w}} = DD^*, C, D \in \mathscr{C} \cap \mathscr{C}^{-1}$. Then

$$J(B) = E\|\underline{z} - AB\underline{w}\|^2$$

is minimized over $B \in \mathscr{C}$ by

$$B_0 = C^{-1}[C^{*-1}A^*Q_{\underline{zw}}D^{*-1}]_{\mathscr{C}}D^{-1}$$

Proof: From Property A.13 and the linearity of the trace operator,

$$\begin{aligned} E\|\underline{z} - AB\underline{w}\|^2 &= \mathrm{tr}[Q_{\underline{z}-AB\underline{w}}] \\ &= \mathrm{tr}[ABQ_{\underline{w}}B^*A^*] - \mathrm{tr}[ABQ_{\underline{wz}}] - \mathrm{tr}[Q_{\underline{zw}}B^*A^*] + \mathrm{tr}[Q_{\underline{z}}] \end{aligned}$$

Now since $\mathrm{tr}[XY] = \mathrm{tr}[YX]$ for any operators X and Y for which the appropriate traces exist

$$\begin{aligned} \mathrm{tr}[ABQ_{\underline{w}}B^*A^*] &= \mathrm{tr}[A^*ABQ_{\underline{w}}B^*] = \mathrm{tr}[C^*CBQ_{\underline{w}}B^*] \\ &= \mathrm{tr}[CBQ_{\underline{w}}B^*C^*] = \mathrm{tr}[CBDD^*B^*C^*] \end{aligned}$$

while

$$\begin{aligned} \mathrm{tr}[ABQ_{\underline{wz}}] &= \mathrm{tr}[AC^{-1}CBDD^{-1}Q_{\underline{wz}}] = \mathrm{tr}[C^{-1}CBDD^{-1}Q_{\underline{wz}}A] \\ &= \mathrm{tr}[CBDD^{-1}Q_{\underline{wz}}AC^{-1}] = \mathrm{tr}[CBDX^*] \end{aligned}$$

where $X = C^{*-1}A^*Q_{\underline{zw}}D^{*-1}$. Similarly,

$$\mathrm{tr}[Z_{\underline{zw}}B^*A^*] = \mathrm{tr}[XD^*B^*C^*]$$

Thus

$$\begin{aligned} E\|\underline{z} - AB\underline{w}\|^2 &= \mathrm{tr}[CBDD^*B^*C^*] - \mathrm{tr}[CBDX^*] - \mathrm{tr}[XD^*B^*C^*] \\ &\quad + \mathrm{tr}[Q_{\underline{z}}] \\ &= \mathrm{tr}[CBDD^*B^*C^* - CBDX^* - XD^*B^*C^*] + \mathrm{tr}[Q_{\underline{z}}] \\ &= \mathrm{tr}[(CBD - X)(CBD - X)^*] + \mathrm{tr}[Q_{\underline{z}} - XX^*] \end{aligned}$$

Now since $Q_{\underline{z}} - XX^*$ is independent of B and the trace operator is positive, a minimizing $B \in \mathscr{B}$ can be obtained from the above expression for $E\|\underline{z} - AB\underline{w}\|^2$ by letting $B = C^{-1}XD^{-1}$ since this choice of B zeros the term $\mathrm{tr}[(CBD - X)(CBD - X)^*]$, which would otherwise be positive. Unfortunately, $B \notin \mathscr{C}$, and we must therefore look further for our optimal causal solution. To this end let

$$X = [X]_{\mathscr{C}} + [X]_{\mathscr{R}^*}$$

be decomposed as a causal operator and a strictly anticausal operator. Note that since $Q_{\underline{z}} \in \mathscr{K}_1$, Property A.13 implies that $Q_{\underline{zw}} \in \mathscr{K}_2$ and hence so is X. As such, the decomposition theorem for Hilbert–Schmidt operators guarantees the existence and uniqueness of this decomposition. Now, upon substituting this decomposition into the above expression we have

$$\begin{aligned} E\|\underline{z} - AB\underline{w}\|^2 &= \operatorname{tr}[(CBD + [X]_{\mathscr{C}} - [X]_{\mathscr{R}^*})(CBD + [X]_{\mathscr{C}} + [X]_{\mathscr{R}^*})^*] \\ &\quad + \operatorname{tr}[Q_{\underline{z}} - XX^*] \\ &= \operatorname{tr}[(CBD - [X]_{\mathscr{C}})(CBD + [X]_{\mathscr{C}})^* \\ &\quad - \operatorname{tr}[(CBD - [X]_{\mathscr{C}})[X]^*_{\mathscr{R}^*}] \\ &\quad - \operatorname{tr}[[X]_{\mathscr{R}^*}(CBD - [X]_{\mathscr{C}})^*] \\ &\quad + \operatorname{tr}[Q_{\underline{z}} - xx^* + [X]_{\mathscr{R}^*}[X]^*_{\mathscr{R}^*}] \end{aligned}$$

Interestingly, the two cross terms in this expression cancel out. To see this observe that $(CBD - [X]_{\mathscr{C}}) \in \mathscr{C}$ while $[X]^*_{\mathscr{R}^*} \in \mathscr{R}$. As such, $(BCD - [X]_{\mathscr{C}}) \times [X]^*_{\mathscr{R}^*} \in \mathscr{R}$ since $\mathscr{R}$ is an ideal in $\mathscr{C}$. Finally, the trace of this operator is zero since $\mathscr{R}$ is quasinilpotent. Similarly, $\operatorname{tr}[[X]_{\mathscr{R}^*}(CBD - [X]_{\mathscr{C}})^*] = 0$ since $[X]_{\mathscr{R}^*}(CBD - [X]_{\mathscr{C}})^* \in \mathscr{R}^*$. Thus

$$\begin{aligned} E\|\underline{z} - AB\underline{w}\|^2 &= \operatorname{tr}[(CBD - [X]_{\mathscr{C}})(CBD - [X]_{\mathscr{C}})^*] \\ &\quad + \operatorname{tr}[Q_{\underline{z}} - XX^* + [X]_{\mathscr{R}^*}[X]^*_{\mathscr{R}^*}] \end{aligned}$$

is minimized by

$$B_0 = C^{-1}[X]_{\mathscr{C}}D^{-1} = C^{-1}[C^{*-1}A^*Q_{\underline{zw}}D^{*-1}]_{\mathscr{C}}D^{-1}$$

since

$$\operatorname{tr}[(CBD - [X]_{\mathscr{C}})(CBD - [X]_{\mathscr{C}})^*] \geq 0$$

and

$$\operatorname{tr}[Q_{\underline{z}} - XX^* + [X]_{\mathscr{R}^*}[X]^*_{\mathscr{R}^*}]$$

is independent of B. //

2. Remark: Note that the minimum which results from the optimal solution over $B \in \mathscr{C}$ is

$$\operatorname{tr}[Q_{\underline{z}} - XX^* + [X]_{\mathscr{R}^*}[X]^*_{\mathscr{R}^*}]$$

Thus the price we pay to obtain a causal solution to our optimization problem is $\operatorname{tr}[[X]_{\mathscr{R}^*}[X]^*_{\mathscr{R}^*}]$, which is a natural measure of the deviation of the optimal solution over $\mathscr{B}$ from being causal. Also note that since $Q_{\underline{z}} \in \mathscr{K}_1$ while X and $[X]_{\mathscr{R}^*}$ are in $\mathscr{K}_2$, these minima are finite.

3. Remark: It is interesting to observe that although the proof of the theorem is not overly long we have used in its derivation virtually the entirety of the theory of operators defined on a Hilbert resolution space. Indeed, we have used operator factorizations and decompositions, and the quasinilpotence of the strictly causal operators in addition to the standard results on causality and strict causality.

4. Example: To minimize the complexity of this first example let us consider a case where $H = R^2$. Here $\mathscr{B}$ is represented by 2×2 matrices and $\mathscr{C}$ is represented as 2×2 lower triangular matrices. Now, take $\underline{z}$ to be the R^2-valued random variable with

$$Q_{\underline{z}} = \begin{bmatrix} 1 & 1 \\ 1 & 1 \end{bmatrix}$$

and $\underline{w} = \underline{z} + \underline{v}$ where $\underline{v}$ represents a white noise process $Q_{\underline{v}} = I$, which is independent of $\underline{z}$. Thus

$$Q_{\underline{w}} = I + Q_{\underline{z}} = \begin{bmatrix} 2 & 1 \\ 1 & 2 \end{bmatrix}$$

while $Q_{\underline{zw}} = Q_{\underline{z}}$. Finally, if we take A to be the 2×2 matrix

$$A = \begin{bmatrix} 0 & 1 \\ 1 & 0 \end{bmatrix}$$

then $A^*A = I$ allowing us to take $C = I$. On the other hand,

$$Q_{\underline{w}} = \begin{bmatrix} 2 & 1 \\ 1 & 2 \end{bmatrix} = \begin{bmatrix} \sqrt{2} & 0 \\ 1/\sqrt{2} & \sqrt{3}/\sqrt{2} \end{bmatrix} \begin{bmatrix} \sqrt{2} & 1/\sqrt{2} \\ 0 & \sqrt{3}/\sqrt{2} \end{bmatrix} = DD^*$$

from which it follows that

$$D^{-1} = \begin{bmatrix} 1/\sqrt{2} & 0 \\ -1/\sqrt{6} & \sqrt{2}/\sqrt{3} \end{bmatrix}$$

while

$$D^{*-1} = \begin{bmatrix} 1/\sqrt{2} & -1/\sqrt{6} \\ 0 & \sqrt{2}/\sqrt{3} \end{bmatrix}$$

Therefore

$$X = C^{*-1}A^*Q_{\underline{zw}}D^{*-1} = \begin{bmatrix} 1/\sqrt{2} & 1/\sqrt{6} \\ 1/\sqrt{2} & 1/\sqrt{6} \end{bmatrix}$$

and

$$[X]_{\mathscr{C}} = [C^{*-1}A^*Q_{\underline{zw}}D^{*-1}]_{\mathscr{C}} = \begin{bmatrix} 1/\sqrt{2} & 0 \\ 1/\sqrt{2} & 1/\sqrt{6} \end{bmatrix}$$

yielding

$$B_0 = C^{-1}[C^{*-1}A^*Q_{\underline{zw}}D^{*-1}]_{\mathscr{C}}D^{-1} = \begin{bmatrix} \frac{1}{2} & 0 \\ \frac{1}{3} & \frac{1}{3} \end{bmatrix}$$

as our optimal solution.

Note that without the causality constraint our optimal solution is

$$B = C^{-1}[C^{*-1}A^*Q_{\underline{zw}}D^{*-1}]D^{-1} = \begin{bmatrix} \frac{1}{3} & \frac{1}{3} \\ \frac{1}{3} & \frac{1}{3} \end{bmatrix}$$

Here

$$[X]_{\mathscr{R}^*} = \begin{bmatrix} 0 & 1/\sqrt{6} \\ 0 & 0 \end{bmatrix}$$

and hence $\mathrm{tr}[X]_{\mathscr{R}^*}[X]^*_{\mathscr{R}^*}] = \frac{1}{6}$ defines the cost associated with the causality constraint.

5. Remark: Unlike the deterministic optimization theorem of Chapter 10, in which an interpolation step is required in the stochastic optimization theorem, our performance measure averages over the entire Hilbert space, and as a result that B_0 obtained in the theorem and Example 4 is a globally defined causal operator.

6. Remark: The assumption that $Q_{\underline{w}} > 0$ in the theorem is quite minimal. Indeed, more often than not $\underline{w}$ contains a white noise term with identity covariance that makes $Q_{\underline{w}}$ positive definite. On the other hand, the assumption that $Q_{\underline{z}} \in \mathscr{K}_1$ is nontrivial. In the case that $\underline{z}$ takes on only a finite set of values, this assumption is satisfied as in Example A.11. More generally, if $\underline{z}$ can be satisfactorily approximated by random variables that take on only a finite set of values, a nuclear covariance will result, though the assumption is still somewhat restrictive. A careful inspection of the proof of the theorem will reveal that the primary reason for requiring $Q_{\underline{z}} \in \mathscr{K}_1$ is to guarantee that $E\|\underline{z} - AB\underline{w}\|^2 = \mathrm{tr}[Q_{\underline{z}-AB\underline{w}}]$ has a finite minimum. Thus if one can replace the performance measure $E\|\underline{z} - AB\underline{w}\|^2$ by an alternative performance measure with similar characteristics but less restrictive existence conditions, the restrictions on $\underline{z}$ can be alleviated.

To this end we adopt $[Q_{\underline{z}-AB\underline{w}}]^{\mathscr{M}}$ as an alternative performance measure. Here, $[\ ]^{\mathscr{M}}$ is the expectation operator introduced in Chapter 3, which extends the "memoryless part" operator. An inspection of the properties of this operator, formulated in Property 3.B.7, will reveal that it behaves similarly to the trace operator. Indeed, for our purposes the only major exception is that $[XY]^{\mathscr{M}} \neq [YX]^{\mathscr{M}}$. Thus one would hope to formulate an optimization theory relative to the performance measure

$$K(B) = [Q_{\underline{z}-AB\underline{w}}]^{\mathscr{M}}$$

which is similar in character to that formulated here for the performance measure

$$J(B) = E\|\underline{z} - AB\underline{w}\|^2 = \mathrm{tr}[Q_{\underline{z}-AB\underline{w}}]$$

without the restrictive existence conditions of the latter.

To justify this new performance measure on physical grounds, consider the finite-dimensional case where $\underline{x} = \mathrm{col}(\underline{x}_i)$ is an R^n-valued random variable. Now

$$Q_{\underline{x}} = E\underline{x}\underline{x}^{\mathrm{t}} = \begin{bmatrix} E\underline{x}_1^2 & E\underline{x}_1\underline{x}_2 & E\underline{x}_1\underline{x}_3 & \cdots & E\underline{x}_1\underline{x}_n \\ E\underline{x}_2\underline{x}_1 & E\underline{x}_2^2 & E\underline{x}_2\underline{x}_3 & \cdots & E\underline{x}_2\underline{x}_n \\ E\underline{x}_3\underline{x}_1 & E\underline{x}_3\underline{x}_2 & E\underline{x}_3^2 & \cdots & E\underline{x}_3\underline{x}_n \\ \vdots & \vdots & \vdots & & \vdots \\ E\underline{x}_n\underline{x}_1 & E\underline{x}_n\underline{x}_2 & E\underline{x}_n\underline{x}_3 & \cdots & E\underline{x}_n^2 \end{bmatrix}$$

and hence

$$E\|\underline{x}\|^2 = \mathrm{tr}[Q_{\underline{x}}] = \sum_{i=1}^{n} E\underline{x}_i^2$$

while

$$[Q_{\underline{x}}]^{\mathscr{M}} = \begin{bmatrix} E\underline{x}_1^2 & & & & \\ & E\underline{x}_2^2 & & & \\ & & E\underline{x}_3^2 & & \\ & & & \ddots & \\ & & & & E\underline{x}_n^2 \end{bmatrix}$$

As such, the performance measures $E\|\underline{x}\|^2$ and $[Q_{\underline{x}}]^{\mathscr{M}}$ operate on exactly the same data, though in the former they are summed and thus may lead to existence problems, whereas in the latter case they arrayed and thus must be minimized in the partial ordering of positive hermitian operators as a multiparameter optimization problem. Fortunately, this causes no difficulty and, indeed, leads to an optimal solution formally identical to that formulated in the previous theorem, but with less restrictive existence condition.

7. Property: Let $A \in \mathscr{C} \cap \mathscr{C}^{-1}$, and $\underline{z}$ and $\underline{w}$ be zero mean H-valued random variables with finite second moment and covariance operators $Q_{\underline{z}}$ and $Q_{\underline{w}} > 0$; assume that $Q_{\underline{w}}$ admits a spectral factorization $[Q_{\underline{w}}] = DD^*$, $D \in \mathscr{C} \cap \mathscr{C}^{-1}$. Then if $[Q_{\underline{zw}} D^{*-1}]_{\mathscr{C}}$ exists,

$$K(B) = [Q_{\underline{z} - AB\underline{w}}]^{\mathscr{M}}$$

is minimized (in the partial ordering of positive hermitian operators) over $B \in \mathscr{C}$ by

$$B_0 = A^{-1}[Q_{\underline{zw}} D^{*-1}]_{\mathscr{C}} D^{-1}$$

8. Remark: The proof is essentially the same as that of the previous theorem and will therefore be skipped (see Problem 10). The major difference is that we take $C = A$ and by so doing circumvent the necessity of employing the commutativity property for the trace (which does not hold for $[\ ]^{\mathscr{M}}$) to replace A by C. With this exception, the derivation is identical to that of the theorem, with the properties of the expectation operator formulated in 3.B.7 subsuming those of the trace.

In effect, Property 7 allows us to bypass the requirement that $Q_{\underline{z}}$ be nuclear at the cost of requiring that $A \in \mathscr{C} \cap \mathscr{C}^{-1}$. Since $A = I$ in most estimation problems, this represents a significant improvement. Unfortunately, A does not satisfy this hypothesis in many control problems, wherein we must restrict $\underline{z}$.

9. Example: Let $\underline{w}$ be a white noise process with covariance $Q_{\underline{w}} = I$ and let $\underline{z} = K\underline{w}$, $K \in \mathscr{B}$. Then

$$Q_{\underline{z}} = KQ_{\underline{w}}K^* = KIK^* = KK^*$$

and

$$\begin{aligned}(x, Q_{\underline{zw}} y) &= E(x, \underline{z})(\underline{w}, y) = E(x, K\underline{w})(\underline{w}, y) = E(K^*x, \underline{w})(\underline{w}, y) \\ &= (K^*x, Q_{\underline{w}} y) = (K^*x, y) = (y, Ky)\end{aligned}$$

which implies that $Q_{\underline{zw}} = K$. Moreover, since $Q_{\underline{w}} = I$ we may take $D = I$, obtaining

$$B_0 = [Q_{\underline{zw}} D^{*-1}]_{\mathscr{C}} D^{-1} = [K]_{\mathscr{C}}$$

as the minimizing causal operator for $[Q_{\underline{z} - B\underline{w}}]^{\mathscr{M}}$ so long as $[K]_{\mathscr{C}}$ exists.

Since $\underline{z} = K\underline{w}$, the obvious choice with which to minimize $\underline{z} - B\underline{w}$

would be $B_0 = K$, which is indeed the correct solution when K is causal. Interestingly, however, when K is not causal the optimal solution is just the causal part of K (if it exists). Although one might hope that such a simple solution would carry over to more general optimization problems, this is not the case. Indeed, the present simple solution is a manifestation of the fact that $\underline{w}$ is white noise and holds only in the case where one is working with white noise or in the presence of an appropriate whitening filter.

C. REPRESENTATION, APPROXIMATION, AND IDENTIFICATION

1. Remark: In engineering practice one has little control over the signal and noise processes with which one must deal. It is common, therefore, to filter a given process to make it simulate a desired process. In its simplest form the given process $\underline{v}$ is taken to be white noise (which is readily generated in a laboratory environment); $\underline{v}$ is then passed through a causal filter F to form a new process $\underline{z} = F\underline{v}$ with prescribed covariance Q.

2. Property: Assume that Q admits a spectral factorization $[Q] = FF^*$, $F \in \mathscr{C} \cap \mathscr{C}^{-1}$, and that $Q_{\underline{w}} = I$. Then the H-valued random variable $\underline{z} = F\underline{v}$ has covariance Q.

Proof: From Property A.13

$$Q_{\underline{z}} = FQ_{\underline{v}}F^* = FIF^* = FF^* = Q$$

as required. //

3. Remark: Although the process we have constructed has the prescribed covariance, it may be completely independent of a given process with the same covariance. Hence rather than matching covariances it might be more appropriate to design a filter to minimize $E\|\underline{z} - F\underline{w}\|^2$ given $\underline{w}$ and $\underline{z}$. Indeed, this is just a special case of the basic stochastic optimization theorem with $A = I$.

4. Property: Let $\underline{z}$ and $\underline{w}$ be zero mean H-valued random variables with finite second moment and covariance operators $Q_{\underline{z}} \in \mathscr{K}_1$ and $Q_{\underline{w}} > 0$, and assume that $Q_{\underline{w}}$ admits a spectral factorization $[Q_{\underline{w}}] = DD^*$, $D \in \mathscr{C} \cap \mathscr{C}^{-1}$. Then $E\|\underline{z} - F\underline{w}\|^2$ is minimized by

$$F_0 = [Q_{\underline{zw}}D^{*-1}]_{\mathscr{C}}D^{-1}$$

over $F \in \mathscr{C}$.

5. Remark: Rather than choosing a causal F to make $F\underline{w}$ appropriate a given random variable $\underline{z}$, we may desire to choose $F \in \mathscr{C}$ so that it approximates a given, not necessarily causal, $K \in \mathscr{K}_2$. Although such an approximation problem may be formulated in purely deterministic terms,[1] a stochastic formulation often proves to be more appropriate. Given $K \in \mathscr{K}_2$ and a zero mean H-valued random variable $\underline{w}$ with finite second moment and covariance $Q_{\underline{w}} \in \mathscr{B}$, the problem of finding the causal operator $F \in \mathscr{C}$ that minimizes $E\|(K - F)\underline{w}\|^2$ is termed the *$\underline{w}$-best causal approximation problem.*

6. Property: Assume that $Q_{\underline{w}}$ admits a spectral factorization $[Q_{\underline{w}}] = DD^*$, $D \in \mathscr{C} \cap \mathscr{C}^{-1}$. Then the solution to the $\underline{w}$-best causal approximation problem is given by

$$F_0 = [KQ_{\underline{w}}D^{*-1}]_{\mathscr{C}}D^{-1}$$

Proof: This is just a special case of the basic stochastic optimization problem, and Property B.7 with $A = I$ and $\underline{z} = K\underline{w}$. //

7. Remark: Finally, we consider a system identification problem, illustrated in Fig. 1. Here one observes the input and output processes $\underline{w}$ and $\underline{z}$, respectively, associated with an unknown system. Since the system is "physical," we may assume that it is causal and thus identify it by determining the $F \in \mathscr{C}$ that minimizes $E\|\underline{z} - F\underline{w}\|^2$. Of course, this is just a special case of the basic stochastic optimization theorem, and Property B.7 and admits the following solution.

8. Property: Assume that $Q_{\underline{z}} \in \mathscr{K}_1$, $Q_{\underline{w}} > 0$, and $Q_{\underline{zw}}$ have been measured and that $Q_{\underline{w}}$ admits a spectral factorization $[Q_{\underline{w}}] = DD^{*-1}$, $D \in \mathscr{C} \cap \mathscr{C}^{-1}$. Then the solution to the system identification problem is given by

$$F_0 = [Q_{\underline{zw}}D^{*-1}]_{\mathscr{C}}D^{-1}$$

9. Remark: Note that in all of the above applications $A = I$, and hence we can use the optimization criterion of Property B.7 in lieu of the more

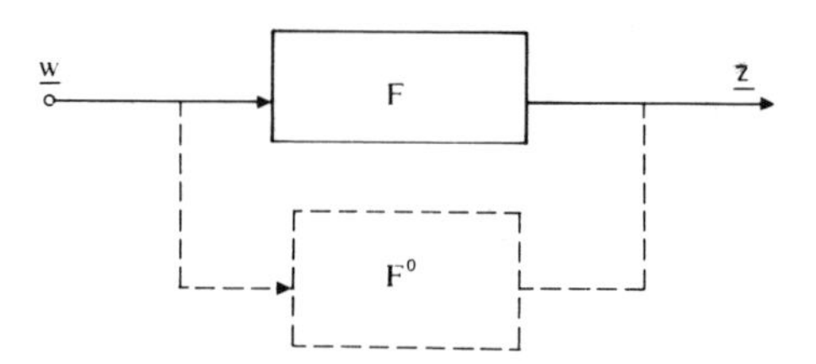

Fig. 1. System identification problem.

restricted criterion of the basic stochastic optimization theorem. Here the expected value is replaced by the expectation operator $[\ ]^{\mathscr{M}}$, and the existence of the causal part term $[\ ]_{\mathscr{C}}$ must be hypothesized, but by so doing the nuclearity assumption is no longer required. In particular, one may take $K \in \mathscr{B}$ for the $\underline{w}$-best causal approximation problem while $Q_{\underline{z}}$ need not be in $\mathscr{K}_1$ for the system identification and stochastic approximation problems.

PROBLEMS

1. Show that $E(x, \underline{x})$ is continuous if $\underline{x}$ is an H-valued random variable with finite second moment.

2. Let $f: H^2 \to C$ be a continuous functional defined on H^2 that is linear in its first coordinate and sequilinear in its second. Then show that there exists a $Q \in \mathscr{B}$ such that $f(x, y) = (x, Qy)$ for all $x, y \in H$.

3. Show that $E(x, \underline{x})(\underline{y}, y)$ is continuous in x and y if $\underline{x}$ and $\underline{y}$ are zero mean H-valued random variables with finite second moment.

4. Assume that a random variable takes on only countably many values x_i, $i = 1, 2, \ldots$; each with probability p_i. Under what conditions does such a random variable have a nuclear covariance?

5. Show that linear combinations of independent random variables are independent.

6. Show that $\mathrm{tr}[XY] = \mathrm{tr}[YX]$ whenever XY *and* YX are nuclear.

7. Show that $A(A^*A)^{-1}A$ is the projection onto the range of A whenever $(A^*A)^{-1}$ exists.

8. Show that $\mathrm{tr}[XX^*] = \mathrm{tr}[P_A Q_{\underline{xy}} Q_{\underline{y}}^{-1} Q_{\underline{yx}} P_A]$ where X is defined as in the proof of the basic stochastic optimization theorem and P_A is the projection onto the range of A.

9. Give an example in R^2 to show that $[XY]^{\mathscr{M}}$ may not equal $[YX]^{\mathscr{M}}$.

10. Give a complete proof for Property B.7.

REFERENCES

1. Arveson, W. A., Interpolation problems in nest algebras, *J. Funct. Anal.* **20**, 208–233 (1975).
2. Balakrishnan, A. V., "Introduction to Optimization Theory in Hilbert Space." Springer-Verlag, Berlin and New York, 1971.

12

Estimation and Control

A. FILTERING AND PREDICTION

1. Remark: The configuration for the filtering and prediction problem is shown in Fig. 1. Here, $\underline{u}$ denotes a signal process of some type that is observed through a sensor S whose output is corrupted by a noise term $\underline{m}$. On the basis of these observations we desire to construct an optimal causal filter F that processes these observations to obtain an estimate of $T\underline{u}$, where $T \in \mathscr{B}$ is a given linear operator. With $T = I$ this reduces to a classical Wiener filter, while $T = \tilde{P}$, the ideal predictor on $L_2(-\infty, \infty)$, corresponds to a Wiener predictor, the case where one desires to estimate the value of $\underline{u}$ at some time in the future based on observation made in the past.[1,2]

2. Property: Let $\underline{u}$ and $\underline{m}$ be zero mean H-valued random variables with finite second moment and covariances $Q_{\underline{u}} \in \mathscr{K}_1$ and $Q_{\underline{m}} > 0$, and assume that $[Q_{\underline{m}} + SQ_{\underline{um}} + Q_{\underline{mu}}S^* + SQ_{\underline{u}}S^*] = DD^*, D \in \mathscr{C} \cap \mathscr{C}^{-1}$. Then $E\|\underline{y}\|^2$ is minimized over $F \in \mathscr{C}$ by

$$F_0 = [T(Q_{\underline{um}} + Q_{\underline{u}}S^*)D^{*-1}]_{\mathscr{C}}D^{-1}$$

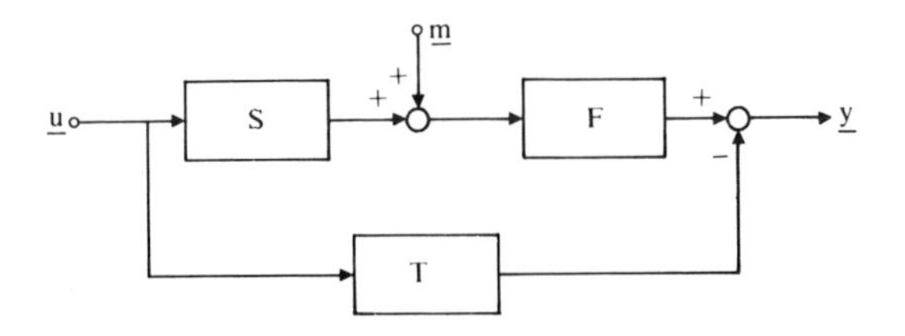

Fig. 1. Configuration for the filtering and prediction problem.

Proof: This filtering and prediction problem is just a special case of the basic stochastic optimization theorem with $\underline{z} = T\underline{u}$ and $\underline{w} = S\underline{u} + \underline{m}$. As such,

$$Q_{\underline{z}} = TQ_{\underline{u}}T^* \in \mathscr{K}_1$$

while

$$Q_{\underline{w}} = Q_{\underline{m}} + SQ_{\underline{um}} + Q_{\underline{mu}}S^* + SQ_{\underline{u}}S^* = DD^* > 0$$

since $Q_{\underline{m}} > 0$. Finally,

$$Q_{\underline{zw}} = Q_{(T\underline{u})(S\underline{u}+\underline{m})} = TQ_{\underline{u}}S^* + TQ_{\underline{um}}$$

Hence it follows from the theorem that

$$F_0 = [T(Q_{\underline{u}}S^* + Q_{\underline{um}})D^{*-1}]_{\mathscr{C}}D^{-1}$$

as was to be shown. //

3. Remark: Note that if $T \in \mathscr{K}_2$ then $Q_{\underline{z}} = TQ_{\underline{u}}T^* \in \mathscr{K}_1$ for arbitrary $Q_{\underline{u}}$, thereby allowing the assumption that $Q_{\underline{u}} \in \mathscr{K}_1$ to be dropped. Indeed, if we minimize $[Q_{\underline{y}}]^{\mathscr{M}}$ rather than $E\|\underline{y}\|^2$ with Property 11.B.7, replacing the basic stochastic optimization theorem, all restrictions on $\underline{u}$ and T may be dropped provided $[T(Q_{\underline{u}}S^* + Q_{\underline{um}})D^{*-1}]_{\mathscr{C}}$ exists.

4. Example: Consider a signal process which is obtained by passing white noise through a filter characterized by the differential operator $G = [D + 2]^{-1}[\sqrt{3}]$ on $L_2(-\infty, \infty)$ with its usual resolution structure. Then $\underline{u} = G\underline{v}$ where $Q_{\underline{v}} = I$, which implies that

$$Q_{\underline{u}} = GQ_{\underline{v}}G^* = GG^* = [D^2 - 4]^{-1}[3]$$

Now, let $\underline{m}$ be white noise independent of $\underline{u}$ and let $S = T = I$. Thus $Q_{\underline{m}} = I$, $Q_{\underline{um}} = 0$, and

$$\begin{aligned}[Q_{\underline{m}} + SQ_{\underline{um}} + Q_{\underline{mu}}S^* + SQ_{\underline{u}}S^*] &= I + Q_{\underline{u}} = [D^2 - 4]^{-1}[3] + I \\ &= [D^2 - 4]^{-1}([3] + [D^2 - 4]) = [D^2 - 4]^{-1}[D^2 - 1] \\ &= ([D + 2]^{-1}[D + 1])([D - 2]^{-1}[D - 1]) = \hat{D}\hat{D}^*\end{aligned}$$

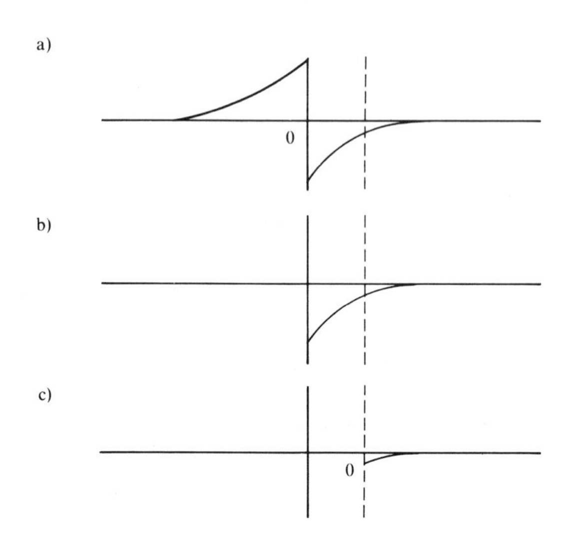

Fig. 2. (a) Green's function for the differential operator D, (b) the Green's function truncated prior to $\dagger$, (c) and the causal part of the operator.

where $\hat{D}$ denotes the desired spectral factor to distinguish it from the differential operator D. The optimal filter is thus given by

$$\begin{aligned} F_0 &= [T(Q_{\underline{u}}S^* + Q_{\underline{um}})D^{*-1}]_{\mathscr{C}}\hat{D}^{-1} \\ &= [([D^2 - 4]^{-1}[3])([D-2]^{-1}[D-1])^{-1}]_{\mathscr{C}}([D+2]^{-1}[D+1])^{-1} \\ &= [([D+2][D-1])^{-1}[3]]_{\mathscr{C}}([D+1]^{-1}[D+2]) \end{aligned}$$

where we have represented the adjoint of the differential operator D by $-D$. Now, the quickest way to compute the causal part of a differential operator is to compute its *Green's function* and truncate it prior to zero. We write

$$([D+2][D-1])^{-1}[3] = [D+2]^{-1}[-1] + [D-1]^{-1}[1]$$

which admits the Green's function

$$g(\dagger) = \begin{cases} -e^{-2\dagger}, & \dagger \geq 0 \\ e^{\dagger}, & \dagger < 0 \end{cases}$$

sketched in Fig. 2a. Truncating the portion of this function prior to $\dagger = 0$ then yields the new Green's function $g_{\mathscr{C}}(\dagger) = -e^{-2\dagger}U(\dagger)$ (sketched in Fig. 2b), where $U(\dagger)$ is the *Heaviside step function.* Now, this is just the Green's function for the differential operator $[D+2]^{-1}[-1]$; hence

$$[([D+2][D-1])^{-1}[3]]_{\mathscr{C}} = [D+2]^{-1}[-1]$$

and

$$F_0 = ([D + 2]^{-1}[-1])([D + 1]^{-1}[D + 2]) = [D + 1]^{-1}[-1]$$

Let us consider the case where $T = \tilde{P}$, the ideal predictor. Now, instead of simply estimating $\underline{u}$ in the presence of noise, our filter is required to predict its value at some time in the future. Since F is causal, however, this prediction can be based only on past measurements. Following the same procedure as before but with $T = \tilde{P}$ we obtain

$$F_0 = [\tilde{P}([D + 2][D - 1])^{-1}[3]]_{\mathscr{C}}([D + 1]^{-1}[D + 2])$$

Now, the Green's function for the differential difference $\tilde{P}([D + 2] \times [D - 1])^{-1}[3]$ is the same as in the previous case except for a shift of one time unit. As such, it is still represented by Fig. 2a but with the origin shifted forward, as indicated by the dotted line. The Green's function for the causal part of this operator is obtained by truncating prior to this new origin, yielding

$$g'_{\mathscr{C}}(\dagger) = (-1/e^2)e^{-2\dagger}U(\dagger)$$

and the differential operator

$$[\tilde{P}([D + 2][D - 1])^{-1}[3]]_{\mathscr{C}} = [D + 2]^{-1}[-1/e^2]$$

This, in turn, results in the optimal predictor

$$F_0 = [D + 1]^{-1}[-1/e^2]$$

Interestingly, the predictor differs from the original estimator only by a reduction in amplitude. In effect, this implies that the system has less confidence in its ability to predict than to estimate.

B. OPEN LOOP SERVOMECHANISMS

1. Remark: The configuration for the open loop stochastic servomechanism sketched in Fig. 3 is identical to that of the deterministic servomechanism problem of Chapter 10 except that the disturbance n is stochastic and our observations of the disturbance are corrupted by a noise term $\underline{m}$. We then observe $\underline{n} + \underline{m}$ and process it through a causal compensator M whose output controls the given plant. Here, the compensator is chosen so as to minimize $E[\|\underline{y}\|^2 + \|\underline{r}\|^2]$ over $M \in \mathscr{C}$.

2. Property: Let $\underline{m}$ and $\underline{n}$ be zero mean H-valued random variables with finite second moment and covariances $Q_{\underline{n}} \in \mathscr{K}_1$ and $Q_{\underline{m}} > 0$.

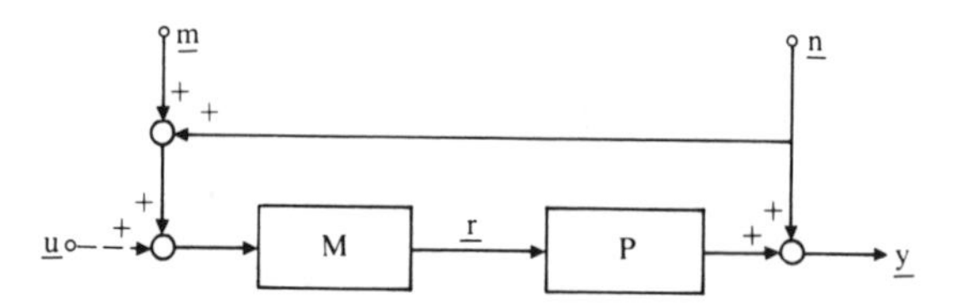

Fig. 3. Configuration for the open loop stochastic servomechanism problem.

Furthermore, assume that $P \in \mathscr{C}$ and $[I + P^*P]$ and $[Q_{\underline{m}} + Q_{\underline{mn}} + Q_{\underline{nm}} + Q_{\underline{n}}]$ admit spectral factorizations

$$[I + P^*P] = C^*C$$

and

$$[Q_{\underline{m}} + Q_{\underline{mn}} + Q_{\underline{nm}} + Q_{\underline{n}}] = DD^*, \ C, D \in \mathscr{C} \cap \mathscr{C}^{-1}$$

Then $E[\|\underline{y}\|^2 + \|\underline{r}\|^2]$ is minimized over $M \in \mathscr{C}$ by

$$M_0 = C^{-1}[C^{*-1}P^*(Q_{\underline{nm}} + Q_{\underline{n}})D^{*-1}]_{\mathscr{C}} D^{-1}$$

Proof: As usual this is just a special case of the basic stochastic optimization theorem with $\underline{w} = \underline{n} + \underline{m}$, $B = M$, $\underline{z} = \text{col}(n, 0) \in H^2$, and $A: H \to H^2$ defined by the 2×1 matrix of operators

$$A = \begin{bmatrix} -P \\ \hdashline -I \end{bmatrix}$$

Then

$$\underline{z} - AB\underline{w} = \begin{bmatrix} \underline{n} \\ \hdashline 0 \end{bmatrix} - \begin{bmatrix} -P \\ \hdashline -I \end{bmatrix} M(\underline{m} + \underline{n}) = \begin{bmatrix} \underline{y} \\ \hdashline \underline{r} \end{bmatrix}$$

and hence $E\|\underline{z} - AB\underline{w}\|^2 = E\|\text{col}(\underline{y}, \underline{r})\|^2 = E[\|\underline{y}\|^2 + \|\underline{r}\|^2]$, verifying that the solution of the basic stochastic optimization theorem with this choice of $\underline{z}$, $\underline{w}$, A, and B yields a solution to the open loop stochastic servomechanism problem. Indeed,

$$\begin{aligned} M_0 &= C^{-1}\left[C^{*-1}[-P^* \ \vdots \ -I]\begin{bmatrix} Q_{\underline{nm}} + Q_{\underline{n}} \\ \hdashline 0 \end{bmatrix} D^{*-1}\right]_{\mathscr{C}} D^{-1} \\ &= -C^{-1}[C^{*-1}P^*(Q_{\underline{nm}} + Q_{\underline{n}})D^{*-1}]_{\mathscr{C}} D^{-1} \end{aligned}$$

where we have used the fact that

$$Q_{\underline{w}} = Q_{\underline{n}+\underline{m}} = [Q_{\underline{m}} + Q_{\underline{mn}} + Q_{\underline{nm}} + Q_{\underline{n}}] = DD^*$$

$$A^*A = [I + P^*P] = C^*C$$

while

$$Q_{\underline{zw}} = \left[\begin{array}{c} Q_{\underline{nm}} + Q_{\underline{n}} \\ \hline 0 \end{array}\right]$$

The proof is therefore complete. //

3. Example: Consider a stochastic servomechanism problem formulated on $l_2[0, \infty)$ with its usual resolution structure and $P = V$, *unilateral shift*. Now, assume that $\underline{n}$ is characterized by the covariance $Q_{\underline{n}} \in \mathscr{K}_1$, represented by the semiinfinite matrix

$$Q_{\underline{n}} = \begin{bmatrix} 1 & \frac{1}{2} & 0 & 0 & 0 & 0 & \cdots \\ \frac{1}{2} & \frac{1}{2} & \frac{1}{4} & 0 & 0 & 0 & \cdots \\ 0 & \frac{1}{4} & \frac{1}{4} & \frac{1}{8} & 0 & 0 & \cdots \\ 0 & 0 & \frac{1}{8} & \frac{1}{8} & \frac{1}{16} & 0 & \cdots \\ 0 & 0 & 0 & \frac{1}{16} & \frac{1}{16} & \frac{1}{32} & \cdots \\ \vdots & \vdots & \vdots & \vdots & \vdots & \vdots & \end{bmatrix}$$

Although infinite in rank, this operator is nuclear since its entries go to zero sufficiently fast toward the "tail" of the semiinfinite matrix. As usual, we assume that the noise term is characterized by $Q_{\underline{m}} = I$ and that $\underline{n}$ and $\underline{m}$ are independent, yielding $Q_{\underline{nm}} = 0$.

To compute our spectral factorizations we have

$$[I + P^*P] = [I + V^*V] = [I + I] = 2I = \sqrt{2}I)^*(\sqrt{2}I) = C^*C$$

since the unilateral shift is an isometry. On the other hand,

$$Q_{\underline{m}} + Q_{\underline{n}} = \begin{bmatrix} 2 & \frac{1}{2} & 0 & 0 & \cdots \\ \frac{1}{2} & \frac{3}{2} & \frac{1}{4} & 0 & \cdots \\ 0 & \frac{1}{4} & \frac{5}{4} & \frac{1}{8} & \cdots \\ 0 & 0 & \frac{1}{8} & \frac{9}{8} & \cdots \\ \vdots & \vdots & \vdots & \vdots & \end{bmatrix}$$

$$= \begin{bmatrix} \sqrt{2} & 0 & 0 & 0 & \cdots \\ 1/2\sqrt{2} & \sqrt{11/8} & 0 & 0 & \cdots \\ 0 & 2/\sqrt{11} & \sqrt{39}/\sqrt{44} & 0 & \cdots \\ \vdots & \vdots & \vdots & \vdots & \end{bmatrix}$$

$$\times \begin{bmatrix} \sqrt{2} & 1/2\sqrt{2} & 0 & 0 & \cdots \\ 0 & \sqrt{11/8} & 2/\sqrt{11} & 0 & \cdots \\ 0 & 0 & \sqrt{39}/\sqrt{44} & 0 & \cdots \\ \vdots & \vdots & \vdots & \vdots & \end{bmatrix} = DD^*$$

Now, the inverse of D may be computed by a back substitution process, yielding

$$D^{-1} = \begin{bmatrix} 1/\sqrt{2} & 0 & 0 & \cdots \\ -2/\sqrt{11} & 8/\sqrt{11} & 0 & \cdots \\ 4\sqrt{44}/11\sqrt{39} & -16\sqrt{44}/11\sqrt{39} & 44/\sqrt{39} & \cdots \\ \vdots & \vdots & \vdots & \end{bmatrix}$$

while D^{*-1} is represented by its transpose. It then follows from Property 2 that

$$\begin{aligned} M_0 &= C^{-1}[C^{*-1}P^*(Q_{\underline{nm}} + Q_{\underline{n}})D^{*-1}]_{\mathscr{C}}D^{-1} \\ &= (1/\sqrt{2})[(1/\sqrt{2})V^*Q_{\underline{n}}D^{*-1}]_{\mathscr{C}}D^{-1} = \tfrac{1}{2}[V^*Q_{\underline{n}}D^{*-1}]_{\mathscr{C}}D^{-1} \end{aligned}$$

where we have used the linearity of $[\ ]_{\mathscr{C}}$ to take the scale factor $1/\sqrt{2}$ out of the causal part of the operator. Now, upon multiplying the matrix representations for V^*, $Q_{\underline{n}}$, and D^{*-1} we obtain a matrix of the form

$$V^*Q_{\underline{n}}D^{*-1} = \begin{bmatrix} \frac{1}{4} & x & x & \cdots \\ 0 & 9/4\sqrt{11} & x & \cdots \\ 0 & 0 & \sqrt{44}/8\sqrt{39} & \cdots \\ \vdots & \vdots & \vdots & \end{bmatrix}$$

where the x's denote nonzero entries. Since $V^*Q_{\underline{n}}D^{*-1}$ is anticausal, $[V^*Q_{\underline{n}}D^{*-1}]_{\mathscr{C}}$ is just its diagonal. Hence

$$M_0 = \tfrac{1}{2}[C^*Q_{\underline{n}}D^{*-1}]_{\mathscr{C}}D^{-1} = \tfrac{1}{2}\begin{bmatrix} \frac{1}{4} & 0 & 0 & \cdots \\ 0 & 9/4\sqrt{11} & 0 & \cdots \\ 0 & 0 & \sqrt{44}/8\sqrt{39} & \cdots \\ \vdots & \vdots & \vdots & \end{bmatrix}$$

$$\times \begin{bmatrix} 1/\sqrt{2} & 0 & 0 & \cdots \\ -2/\sqrt{11} & 8/\sqrt{11} & 0 & \cdots \\ 4\sqrt{44}/11\sqrt{39} & -16\sqrt{44}/11\sqrt{39} & \sqrt{44}/\sqrt{39} & \cdots \\ \vdots & \vdots & \vdots & \end{bmatrix}$$

$$= \begin{bmatrix} 1/8\sqrt{2} & 0 & 0 & \cdots \\ -\frac{9}{44} & \frac{9}{11} & 0 & \cdots \\ \frac{1}{39} & -\frac{4}{39} & \frac{11}{156} & \cdots \\ \vdots & \vdots & \vdots & \end{bmatrix}$$

which is the desired compensator.

C. CLOSED LOOP SERVOMECHANISMS

1. Remark: The configuration for the closed loop stochastic servomechanism problem is sketched in Fig. 4. As with the open loop problem this configuration is identical to that used in the deterministic case except that the disturbance $\underline{n}$ is taken to be a stochastic process and our observations of the system output are corrupted by a noise term. As before the goal is to choose a stabilizing feedback law minimizing $E[\|\underline{y}\|^2 + \|\underline{r}\|^2]$.

To solve the closed loop problem we invoke the feedback system stabilization theorem, which allows us to parametrize the stabilizing compensators by a $W \in \mathscr{C}$ and then apply the basic stochastic optimization theorem to choose an optimal W. Although we shall consider the case of an unstable plant, the solution is greatly simplified in the case of a stable plant and hence we begin with that case.

2. Property: Let $\underline{m}$ and $\underline{n}$ be zero mean H-valued random variables with finite second moment and covariances $Q_{\underline{n}} \in \mathscr{K}_1$ and $Q_{\underline{m}} > 0$. Furthermore, assume that $P \in \mathscr{C}$ and that $[I + P^*P]$ and $[Q_{\underline{m}} + Q_{\underline{mn}} + Q_{\underline{nm}} + Q_{\underline{n}}]$ admit spectral factorizations $[I + P^*P] = C^*C$ and $[Q_{\underline{m}} + Q_{\underline{mn}} + Q_{\underline{nm}} + Q_{\underline{n}}] = DD^*$, $C, D \in \mathscr{C} \cap \mathscr{C}^{-1}$. Then $E[\|\underline{y}\|^2 + \|\underline{r}\|^2]$ is minimized over all stabilizing feedback laws by

$$F_0 = [-C^{-1}[C^{*-1}P^*(Q_{\underline{nm}} + Q_{\underline{n}}D^{*-1}]_{\mathscr{C}}D^{-1} + I]^{-1} \times [C^{-1}[C^{*-1}P^*(Q_{\underline{nm}} + Q_{\underline{n}})D^{*-1}]_{\mathscr{C}}D^{-1}]$$

if $[-C^{-1}[C^{*-1}P^*(Q_{\underline{nm}} + Q_{\underline{n}})D^{*-1}]_{\mathscr{C}}D^{-1} + I]^{-1}$ exists.

Proof: Since $P \in \mathscr{C}$ it admits the doubly coprime fractional representation

$$\begin{bmatrix} I & 0 \\ -P & I \end{bmatrix}^{-1} = \begin{bmatrix} I & 0 \\ P & I \end{bmatrix}$$

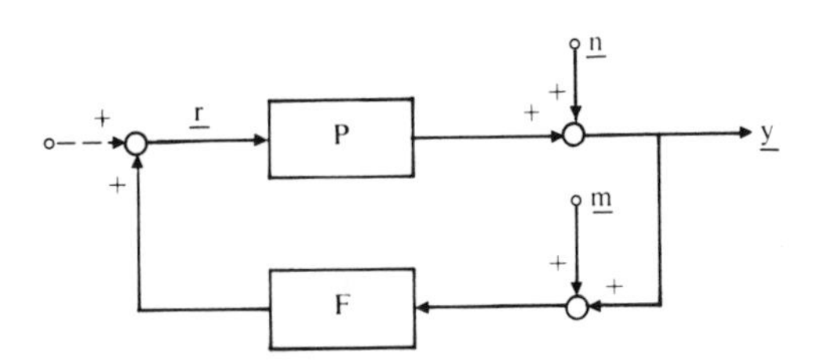

Fig. 4. Configuration for the closed loop stochastic servomechanism problem.

in terms of which the set of stabilizing feedback laws may be parametrized

$$F = (WP + I)^{-1}(-W)$$

where $W \in \mathscr{C}$ and $(WP + I)^{-1}$ is assumed to exist. Moreover, it follows from the feedback system stabilization theorem that

$$\underline{r} = H_{rn}(\underline{n} + \underline{m}) = W(\underline{n} + \underline{m})$$

while

$$\underline{y} = H_{yn}(\underline{n} + \underline{m}) - \underline{m} = [PW + I](\underline{n} + \underline{m}) - \underline{m} = PW(\underline{n} + \underline{m}) + \underline{n}$$

The minimization of $E[\|\underline{y}\|^2 + \|\underline{r}\|^2]$ may now be reduced to an application of the basic stochastic optimization theorem by letting $\underline{w} = \underline{n} + \underline{m}$, $B = W$, $\underline{z} = \mathrm{col}(\underline{n}, 0) \in H^2$, and $A: H \to H^2$ be represented by the 2×1 matrix of operators

$$A = \begin{bmatrix} -P \\ \hdashline -I \end{bmatrix}$$

In this case

$$\underline{z} = AB\underline{w} = \begin{bmatrix} \underline{n} \\ \hdashline 0 \end{bmatrix} - \begin{bmatrix} -P \\ \hdashline -I \end{bmatrix} W(\underline{n} + \underline{m}) = \begin{bmatrix} \underline{y} \\ \hdashline \underline{r} \end{bmatrix}$$

Hence a solution to the basic stochastic optimization theorem is equivalent to the minimization of $E[\|\underline{y}\|^2 + \|\underline{r}\|^2]$ over $W \in \mathscr{C}$. Now,

$$Q_{\underline{w}} = Q_{\underline{n}+\underline{m}} = [Q_{\underline{m}} + Q_{\underline{mn}} + Q_{\underline{nm}} + Q_{\underline{n}}] = DD^*$$

while

$$Q_{\underline{z}} = Q_{\mathrm{col}(\underline{n},\,0)} = \begin{bmatrix} Q_{\underline{n}} & 0 \\ 0 & 0 \end{bmatrix} \in \mathscr{K}_1$$

and

$$Q_{\underline{zw}} = Q_{(\mathrm{col}(\underline{n},\,0))\,(\underline{n}+\underline{m})} = \begin{bmatrix} Q_{\underline{nm}} + Q_{\underline{n}} \\ \hdashline 0 \end{bmatrix}$$

Hence the basic stochastic optimization theorem implies that

$$\begin{aligned} W_0 &= C^{-1}\left[C^{*-1}[-P^* \mid -I]\begin{bmatrix} Q_{\underline{nm}} + Q_{\underline{n}} \\ \hdashline 0 \end{bmatrix} D^{*-1}\right]_{\mathscr{C}} D^{-1} \\ &= -C^{-1}[C^{*-1}P^*(Q_{\underline{nm}} + Q_{\underline{n}})D^{*-1}]_{\mathscr{C}} D^{-1} \end{aligned}$$

Finally, making the assumption that $(W_0 P + I)^{-1}$ exists and substituting W_0 into the formula $F_0 = (W_0 P + I)^{-1}(-W_0)$ we obtain the required optimal stabilizing feedback law. //

3. Remark: Conceptually, the solution to the closed loop servomechanism problem with unstable plant is identical to that described above, except that the simple doubly coprime fractional representation employed in the stable plant case must be replaced by a general doubly coprime fractional representation with its commensurate increase in complexity. As a technicality we must additionally assume that $U_{pr} \in \mathscr{K}_2$ to guarantee that the resultant $Q_{\underline{w}}$ required by the basic stochastic optimization theorem will be nuclear. Since U_{pr} may be taken to be zero when the plant is stable, the assumption that $U_{pr} \in \mathscr{K}_2$ may be interpreted as a slight restriction on the "degree of unstability" permitted the plant.

Stochastic Servomechanism Theorem: Let $\underline{m}$ and $\underline{n}$ be zero mean H-valued random variables with finite second moment and covariances $Q_{\underline{n}} \in \mathscr{K}_1$ and $Q_{\underline{m}} > 0$. Further assume that P admits a doubly coprime fractional representation with $U_{pr} \in \mathscr{K}_2$

$$\begin{bmatrix} V_{pr} & U_{pr} \\ -N_{pl} & D_{pl} \end{bmatrix}^{-1} = \begin{bmatrix} D_{pr} & -U_{pl} \\ N_{pr} & V_{pl} \end{bmatrix}$$

such that $[N_{pr}^* N_{pr} + D_{pr}^* D_{pr}]$ and $D_{pl}[Q_{\underline{m}} + Q_{\underline{mn}} + Q_{\underline{nm}} + Q_{\underline{n}}]D_{pl}^*$ admit spectral factorizations $[N_{pr}^* N_{pr} + D_{pr}^* D_{pr}] = C^*C$ and $D_{pl}[Q_{\underline{m}} + Q_{\underline{mn}}$ $Q_{\underline{nm}} + Q_{\underline{n}}]D_{pl}^* = DD^*$, $C, D \in \mathscr{C} \cap \mathscr{C}^{-1}$. Then $E[\|\underline{y}\|^2 + \|\underline{r}\|^2]$ is minimized over all stabilizing feedback laws by

$$F_0 = (W_0 N_{pl} + V_{pr})^{-1}(-W_0 D_{pl} + U_{pr})$$

if $(W_0 N_{pl} + V_{pr})^{-1}$ exists, where

$$\begin{aligned} W_0 = & -C^{-1}[C^{*-1}(-N_{pr}^* D_{pr}(Q_{\underline{m}} + Q_{\underline{mn}} + Q_{\underline{nm}} + Q_{\underline{n}}) \\ & + D_{pr}^* V_{pl} D_{pl}(Q_{\underline{nm}} + Q_{\underline{n}}) \\ & + D_{pr}^* N_{pr} U_{pr}(Q_{\underline{m}} + Q_{\underline{mn}}))D_{pl}^* D^{*-1}]_{\mathscr{C}} D^{-1} \end{aligned}$$

Proof: Since the proof parallels that of Property 2 we shall give only a sketch of the proof indicating the differences. With our general doubly coprime fractional representation, the stabilizing feedback laws are given by

$$F = (WN_{pl} + V_{pr})^{-1}(-WD_{pl} + U_{pr})$$

where $W \in \mathscr{C}$, and so we optimize over $W \in \mathscr{C}$ using the formulas

$$\underline{r} = H_{rn}(\underline{n} + \underline{m}) = D_{pr} W D_{pl}(\underline{n} + \underline{m}) - D_{pr} U_{pr}(\underline{n} + \underline{m})$$

and

$$\underline{y} = H_{yn}(\underline{n} + \underline{m}) - \underline{m} = N_{pr} W D_{pl}(\underline{n} + \underline{m}) + V_{pl} D_{pl} \underline{n} - N_{pr} U_{pr} \underline{m}$$

The minimization of $E[\|\underline{y}\|^2 + \|\underline{r}\|^2]$ may now be reduced to an application of the basic stochastic optimization theorem by letting $\underline{w} = D_{pl}(\underline{n} + \underline{m})$, $B = W$, $A: H \to H^2$ by

$$A = \begin{bmatrix} -N_{pr} \\ \hdashline -D_{pr} \end{bmatrix}$$

and $\underline{z} = \mathrm{col}(-D_{pr} U_{pr}(n + m), V_{pl} D_{pl} \underline{n} + N_{pr} U_{pr} \underline{m})$. Thus

$$Q_{\underline{w}} = Q_{D_{pl}(\underline{n}+\underline{m})} = D_{pl}[Q_{\underline{m}} + Q_{\underline{mn}} + Q_{\underline{nm}} + Q_{\underline{n}}]D_{pl}^* = DD^*$$

while

$$A^*A = [N_{pr}^* N_{pr} + D_{pr}^* D_{pr}] = C^*C$$

are the required spectral factors. Moreover, a little algebra will reveal that $Q_{\underline{z}} \in \mathscr{K}_1$ since $Q_{\underline{n}} \in \mathscr{K}_1$ (which in turn implies that $Q_{\underline{nm}}$ and $Q_{\underline{mn}} \in \mathscr{K}_2$) and $U_{pr} \in \mathscr{K}_2$. As such, the hypotheses of the basic stochastic optimization theorem are satisfied and yield the claimed W_0 using

$$Q_{\underline{zw}} = \begin{bmatrix} -D_{pr} U_{pr}(Q_{\underline{m}} + Q_{\underline{nm}} + Q_{\underline{mn}} + Q_{\underline{n}})D_{pl}^* \\ \hdashline V_{pl} D_{pl}(Q_{\underline{n}} + Q_{\underline{nm}})D_{pl}^* + N_{pr} U_{pr}(Q_{\underline{m}} + Q_{\underline{mn}})D_{pl}^* \end{bmatrix}$$

in the formula $W_0 = C^{-1}[C^{*-1}A^*Q_{\underline{zw}}D^{*-1}]_{\mathscr{C}}D^{-1}$. //

4. Example: Let us consider the design of a closed loop servomechanism using the same plant and noise processes as in the open loop servomechanism of Example B.3. Recall that the system is defined on $l_2[0, \infty)$ with its usual resolution structure $P = V$, the unilateral shift; $\underline{m}$ is white noise, which is independent of the signal process $\underline{n}$, whose covariance is given in Example B.30. For this special case a careful inspection of the requirements of properties 2 and B.2 will reveal that W_0 coincides with M_0, the optimal open loop compensator designed in Example B.3. As such,

$$W_0 = \begin{bmatrix} 1/8\sqrt{2} & 0 & 0 & \cdots \\ -\frac{9}{44} & \frac{9}{11} & 0 & \cdots \\ \frac{1}{39} & -\frac{4}{39} & \frac{11}{156} & \cdots \\ \vdots & \vdots & \vdots & \end{bmatrix}$$

while

$$W_0 P + I = \begin{bmatrix} 1 & 0 & 0 & \cdots \\ \frac{9}{11} & 1 & 0 & \cdots \\ -\frac{4}{39} & \frac{11}{156} & 1 & \cdots \\ \vdots & \vdots & \vdots & \end{bmatrix}$$

Now, a back-substitution process will reveal that

$$(W_0 P + I)^{-1} = \begin{bmatrix} 1 & 0 & 0 & \cdots \\ -\frac{9}{11} & 1 & 0 & \cdots \\ -\frac{25}{156} & -\frac{11}{156} & 1 & \cdots \\ \vdots & \vdots & \vdots & \end{bmatrix}$$

and thus

$$F_0 = -(W_0 P + I)^{-1} W_0 = \begin{bmatrix} -0.088 & 0 & 0 & \cdots \\ 0.277 & -0.818 & 0 & \cdots \\ 0.162 & 0.160 & -0.071 & \cdots \\ \vdots & \vdots & \vdots & \end{bmatrix}$$

is the optimal feedback law.

5. Remark: In the Example 4 a little algebra will reveal that

$$H_{yn}^{\mathrm{f}} = (I + PF_0)^{-1} = \begin{bmatrix} 1 & 0 & 0 & \cdots \\ 0.088 & 1 & 0 & \cdots \\ -0.201 & 0.818 & 1 & \cdots \\ \vdots & \vdots & \vdots & \end{bmatrix}$$

has norm greater than 1. Contrary to "folklore," an optimal feedback controller may not have sensitivity characteristics superior to an equivalent optimal open loop controller. The sensitivity may, however, be superior over a specified range of inputs.

D. STOCHASTIC REGULATORS

1. Remark: Before proceeding to the stochastic regulator problem we will require the following lemma. In essence this is a variation on the $\underline{w}$-best causal approximation problem of Property 11.C.6 with the requirements that $Q_{\underline{w}} > 0$ and $K \in \mathscr{K}_2$ replaced by the requirement that $Q_{\underline{w}} = DD^*$, $D \in \mathscr{C} \cap \mathscr{K}_2$.

2. Lemma: Let $\underline{w}$ be a zero mean H-valued random variable with finite second moment and covariance $Q_{\underline{w}} = DD^*, D \in \mathscr{C} \cap \mathscr{K}_2$, and assume that $K \in \mathscr{B}$. Then if

$$F_0 D = [KD]_{\mathscr{C}}$$

F_0 minimizes $E\|K\underline{w} - F\underline{w}\|^2$ over $F \in \mathscr{C}$.

Proof:

$$\begin{aligned}
E\|K\underline{w} - F\underline{w}\|^2 &= E\|(F - K)\underline{w}\|^2 = \mathrm{tr}[(F - K)Q_{\underline{w}}(F - K)^*] \\
&= \mathrm{tr}[(F - K)DD^*(F - K)^*] \\
&= \mathrm{tr}[(FD - KD)(FD - KD)^*] \\
&= \mathrm{tr}[(FD - [KD]_{\mathscr{C}} - [KD]_{\mathscr{R}^*}) \\
&\quad \times (FD - [KD]_{\mathscr{C}} - [KD]_{\mathscr{R}^*})^*] \\
&= \mathrm{tr}[(FD - [KD]_{\mathscr{C}^*})(FD - [KD]_{\mathscr{C}^*})^*] - \mathrm{tr}[FD[KD]^*_{\mathscr{R}^*}] \\
&\quad - \mathrm{tr}[[KD]_{\mathscr{R}^*} D^* F^*] + \mathrm{tr}[[KD]_{\mathscr{R}^*}[KD]^*_{\mathscr{R}^*}
\end{aligned}$$

Now since $[KD]_{\mathscr{R}^*} \in \mathscr{R}^*$, $[KD]^*_{\mathscr{R}^*} \in \mathscr{R}$ and hence $FD[KD]^*_{\mathscr{R}^*} \in \mathscr{R}$. As such $\mathrm{tr}[FD[KD]^*_{\mathscr{R}^*}] = 0$ since $\mathscr{R}$ is quasinilpotent, and similarly, $\mathrm{tr}[[KD]_{\mathscr{R}^*} D^* F^*] = 0$ since $[KD]_{\mathscr{R}^*} D^* F^* \in \mathscr{R}^*$. Thus

$$\begin{aligned}
E\|K\underline{w} - F\underline{w}\|^2 = \mathrm{tr}[(FD - [KD]_{\mathscr{C}})(FD - [KD]_{\mathscr{C}})^*] \\
+ \mathrm{tr}[[KD]_{\mathscr{R}^*}[KD]^*_{\mathscr{R}^*}]
\end{aligned}$$

is minimized by any $F_0 \in \mathscr{C}$ such that $F_0 D = [KD]_{\mathscr{C}}$. Indeed, this is the case since the first term in the above expression for $E\|K\underline{w} - F\underline{w}\|^2$ is always greater than or equal to zero, with equality for such an F_0, while the second term is independent of F. //

3. Remark: As with the deterministic regulator problem, we initially formulate our stochastic regulator problem as an open loop problem, later showing that the optimal solution can be implemented using a memoryless state feedback configuration. The configuration for our open loop stochastic regulator is illustrated in Fig. 5. Unlike in the deterministic case, where our disturbance took the form of an initial state at time †, in the present case the disturbance is modeled as a white noise process $\underline{s}$ applied to the plant input. Intuitively, one can think of $\underline{s}$ as representing random inputs (say, gusts of wind that perturb an aircraft control system) driving the plant off of its nominal trajectory (which by linearity may always be taken to be zero). For reasons that will become apparent later, the disturbance $\underline{s}$ is sensed through a model of the plant, the response of which is processed by a compensator M to control the plant. Our goal is to determine an optimal causal compensator M_0 that minimizes $E[\|\underline{y}\|^2 + \|\underline{r}\|^2]$ over all $M \in \mathscr{C}$.

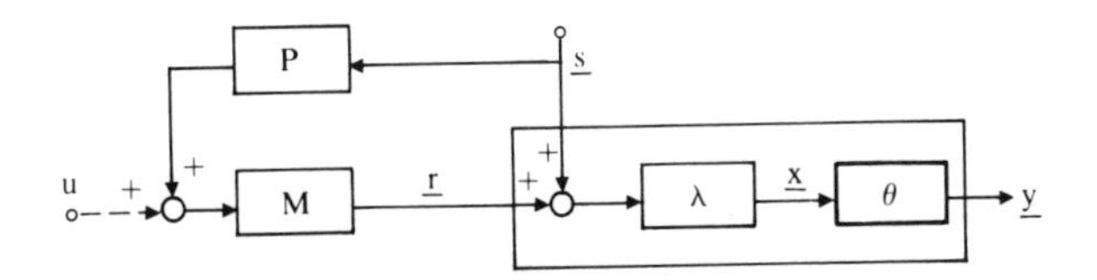

Fig. 5. Configuration for the open loop stochastic regulator problem.

Stochastic Regulator Theorem: Let s be a zero mean H-valued random variable with finite second moment and covariance $Q_{\underline{s}} = I$. Furthermore, let $P \in \mathscr{S} \cap \mathscr{K}_2$ admit a strong minimal decomposition and a factorization $P = \theta\lambda$, and assume that $[I + P^*P]$ admits a radical factorization $[I + P^*P] = (I + V)^*(I + V)$, $V \in \mathscr{R}$. Then $E[\|\underline{y}\|^2 + \|\underline{r}\|^2]$ is minimized by $\underline{r}_0 = PM_0\underline{s} = -V(\underline{r}_0 + \underline{s})$. Furthermore, if $\underline{x}_0 = \lambda(r_0 + \underline{s})$ is the state trajectory resulting from the application of this input, then there exists a memoryless operator F such that $\underline{r}_0 = F\underline{x}_0$.

Proof: From the block diagram of Fig. 5

$$\underline{y} = PMP\underline{s} + P\underline{s} \qquad \text{and} \qquad \underline{r} = MP\underline{s}$$

Thus

$$\begin{aligned} E[\|\underline{y}\|^2 + \|\underline{r}\|^2] &= \mathrm{tr}[(PM + 1)PP^*(PM + 1)^* + MPP^*M^*] \\ &= \mathrm{tr}[PMPP^*M^*P^* + PMPP^* + PP^*M^*P^* \\ &\quad + PP^* + MPP^*M^*] \end{aligned}$$

Now, with use of the linearity of the trace and the fact that $\mathrm{tr}[XY] = \mathrm{tr}[YX]$, this becomes

$$\begin{aligned} E[\|\underline{y}\|^2 + \|\underline{r}\|^2] &= \mathrm{tr}[(I + P^*P)(MPP^*M^*) + PMPP^* + PP^*M^*P^* \\ &\quad + PP^*] \\ &= \mathrm{tr}[(I + V)^*(I + V)(MPP^*M^*) + PMPP^* \\ &\quad + PP^*M^*P^* + PP^*] \\ &= \mathrm{tr}[(I + V)MPP^*M^*(I + V)^* + PMPP^* \\ &\quad + PP^*M^*P^* + PP^*] \\ &= \mathrm{tr}[((I + V)M + (I + V)^{*-1}P^*)PP^*((I + V)M \\ &\quad + (I + V)^{*-1}P^*)^*] \\ &\quad + \mathrm{tr}[PP^* - P^*P(I + P^*P)^{-1}P^*P] \\ &= E\|(I + V)^{*-1}P^*\underline{w} - (-(I + V)M)\underline{w}\|^2 \\ &\quad + \mathrm{tr}[PP^* - P^*P(I + P^*P)^{-1}P^*P] \end{aligned}$$

Since the latter term is independent of M, our performance measure is minimized by the optimal solution to this first term. This, however, falls into the class of optimization problems covered by Lemma 2 with $K = (I + V)^{*-1}P^*$, $F = -(I + V)M$, and $Q_{\underline{w}} = PP^*$. As such, our optimal compensator is characterized by

$$-(I + V)MP_0 = [(I + V)^{*-1}P^*P]_{\mathscr{C}}$$

or equivalently,

$$MP_0 = -(I + V)^{-1}[(I + V)^{*-1}P^*P]_{\mathscr{C}}$$

Now, since $[I + P^*P] = (I + V)^*(I + V) = I + V + V^* + V^*V$,

$$\begin{aligned}[(I + V)^{*-1}P^*P]_{\mathscr{C}} &= [(I + V)^{*-1}(V + V^* + V^*V)]_{\mathscr{C}} \\ &= [(I + V)^{*-1}V^* + (I + V)^{*-1}(I + V)^*V]_{\mathscr{C}} \\ &= [(I + V)^{*-1}V^*]_{\mathscr{C}} + [V]_{\mathscr{C}} = V\end{aligned}$$

Here, the last equality follows from the fact that $(I + V)^{*-1}V^* \in \mathscr{R}^*$ while $V \in \mathscr{C}$. Thus

$$MP_0 = -(I + V)^{-1}V \qquad \text{and} \qquad \underline{r}_0 = M_0P\underline{s} = -(I + V)^{-1}V\underline{s}$$

Accordingly, $(I + V)\underline{r}_0 = -V\underline{s}$, or equivalently,

$$\underline{r}_0 = -V(\underline{r}_0 + \underline{s})$$

Now, if we apply input $\underline{r}_0$ to the plant, once the disturbance is added, the actual input is $\underline{r}_0 + \underline{s}$. Hence the resultant optimal state is

$$\underline{x}_0 = \lambda(\underline{r}_0 + \underline{s})$$

Finally, by the lifting theorem there exists a memoryless operator F such that $V = F\lambda$, and hence

$$\underline{r}_0 = -V(\underline{r}_0 + \underline{s}) = -F\lambda(\underline{r}_0 + \underline{s}) = -F\underline{x}_0$$

as was to be shown. //

4. Remark: The theorem actually suggests three alternative implementations for the optimal stochastic regulator. These are illustrated in Fig. 6. In the obvious open loop implementation, indicated in Fig. 6a, we use the equality

$$M_0P = -(I + V)^{-1}V$$

to implement the cascade of our optimal compensator and the sensor P as a feedback loop with open loop gain V. Since $V \in \mathscr{R}$ the resultant feedback

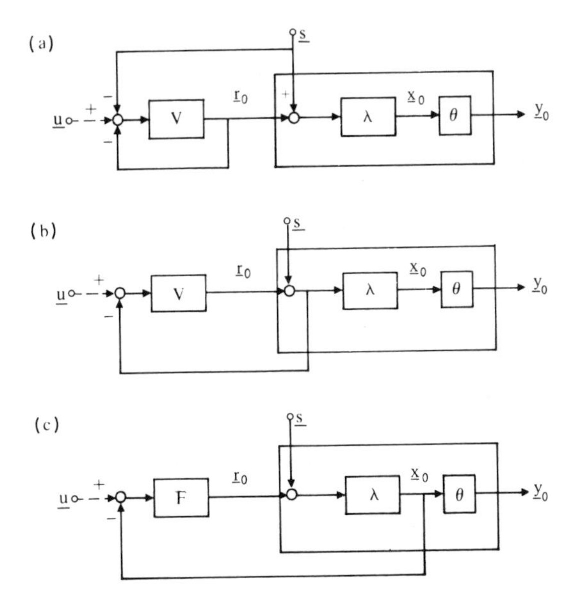

Fig. 6. (a) Open loop, (b) input feedback, and (c) memoryless state feedback implementations of an optimal stochastic regulator.

loop is stable. The second equivalent implementation, illustrated in Fig. 6b, is suggested by the equality

$$\underline{r}_0 = -V(\underline{r}_0 + \underline{s})$$

and is termed an *input feedback* implementation since $\underline{r}_0$ is obtained by processing $\underline{r}_0 + \underline{s}$. Unfortunately, this is not a practical implementation since the disturbance $\underline{s}$ may actually be distributed within the plant rather than added to its input. In such plants the quantity $\underline{r}_0 + \underline{s}$ has no physical realization. Finally, the third implementation, suggested by the equality

$$\underline{r}_0 = -F\underline{x}_0$$

is illustrated in Fig. 6c. Note that the latter two implementations are stable since they are equivalent to the stable open loop implementation. This is the desired realization of our stochastic regulator, since memoryless feedback is much more readily implemented than dynamic feedback and the state is either physically accessible or readily estimated.

Finally, we note that the memoryless state feedback implementation for the stochastic regulator is identical to the corresponding implementation for the deterministic regulator, formulated in Chapter 10, except for the presence of the $E_\dagger$ in the feedback loop of the latter. This term is indicative of the fact that the disturbance in the deterministic regulator has a fixed starting time †, whereas the stochastic regulator is defined at all times.

5. Example: Consider the plant modeled by the differential operator

$$\dot{x}(\dagger) = [-1]x(\dagger) + [1/2]r(\dagger), \qquad x(0) = 0$$
$$y(\dagger) = [2]x(\dagger),$$

defined on $L_2[0, 1]$ with its usual resolution structure. Here, $S = L_2[0, 1]$, $\theta = 2$, and λ is characterized by the differential operator

$$\dot{x}(\dagger) = [-1]x(\dagger) + [\tfrac{1}{2}]r(\dagger), \qquad x(0) = 0$$

Moreover, in operational notation $P = 2\lambda = [D + 1]^{-1}[1]$, so that

$$\begin{aligned}[I + P^*P] &= [D^2 - 1]^{-1}[D^2 - 2] \\ &= (I + [D + 1]^{-1}[\sqrt{2} - 1])(I + [D + 1]^{-1}[\sqrt{2} - 1])^*\end{aligned}$$

and $V = [D + 1]^{-1}[\sqrt{2} - 1]$. On the other hand $\lambda = [D + 1]^{-1}[\tfrac{1}{2}]$, and hence $F = [2\sqrt{2} - 2]$ defines the required feedback law.

6. Remark: Although the feedback configuration for our regulator may be justified from the point of view of convenience of implementation, in fact it also proves to be less sensitive to plant perturbations and modeling errors. Recall from Property 8.D.4 that the return difference for a feedback system defines a measure with which to compare the sensitivity of a feedback system with an equivalent open loop system. In our case, since we feedback the state rather than the system output, the return difference takes the form $(I + F\lambda)^{-1}$, and we desire to show that

$$\|(I + F)^{-1}\| \leq 1$$

Equivalently it suffices to show that

$$I - (I + F\lambda)^{*-1}(I + F\lambda)^{-1} \geq 0$$

with $F\lambda = V$, this becomes

$$\begin{aligned}I - (I + V)^{*-1}(I + V)^{-1} &= (I + V)^{*-1}[(I + V)^*(I + V) - I](I + V)^{-1} \\ &= (I + V)^{*-1}[(I + P^*P) - I](I + V)^{-1} \\ &= (I + V)^{*-1}P^*P(I + V)^{-1} \geq 0\end{aligned}$$

We have verified that the feedback implementation for our optimal stochastic regulator is not only more readily implemented but also less sensitive to plant perturbations than an equivalent open loop regulator.

PROBLEMS

1. Give a solution to the filtering and prediction problem described in Example A.4 with $S = \tilde{D}$, the ideal delay, and $T = \tilde{P}$, the ideal predictor.
2. Verify that the operator $Q_{\underline{n}}$ of Example B.3 is nuclear.
3. Show that $[MA]_{\mathscr{C}} = M[A]_{\mathscr{C}}$ and $[AM]_{\mathscr{C}} = [A]_{\mathscr{C}}M$ if $A \in \mathscr{B}$ and $M \in \mathscr{M}$ whenever the appropriate causal parts exist.
4. In the proof of the closed loop servomechanism theorem verify that $y = H_{yn}(\underline{n} + \underline{m}) - \underline{m} = N_{pr}WD_{pl}(n + \underline{m}) + V_{pl}D_{pl}\underline{n} - N_{pr}U_{pr}\underline{m}$.
5. In the proof of the closed loop servomechanism theorem compute $Q_{\underline{z}}$ and verify that it is nuclear.
6. In the proof of the closed loop servomechanism theorem verify the formula given for $Q_{\underline{zw}}$.
7. Let $\underline{x}$ and $\underline{m}$ be independent zero mean H-valued random variables with finite second moment and covariances $Q_{\underline{x}}$ and $Q_{\underline{m}} = I$, and let $\underline{w} = E_{\dagger}P(E^{\dagger}\underline{x} + \underline{m})$, $P \in \mathscr{K}_2$. Then show that for $K \in \mathscr{B}$ $E\|K\underline{w} - F\underline{w}\|^2$ is minimized over $F \in \mathscr{C}$ by any F_0 satisfying

$$F_0 E_{\dagger} P = [KE_{\dagger}P]_{\mathscr{C}}$$

8. For an operator $A \in \mathscr{B}$ show that $\|A\| \leq 1$ if and only if $I - A^*A$ and $I - AA^*$ are positive.
9. Show that the systems of Fig. 6a and 6b are equivalent.

REFERENCES

1. DeSantis, R. M., Saeks, R., and Tung, L. J., Basic optimal estimation and control problems in Hilbert space," *Math. Systems Theory* **12**, 175–203 (1978).
2. Tung, L. J., Saeks, R., and DeSantis, R. M., Wiener–Hopf filtering in Hilbert resolution space, *IEEE Trans. Circuits and Systems* **CAS-25**, 702–705 (1978).

Notes for Part IV

1. Historical Background: Although the mathematical theory for least squares estimation dates back to the 1930s, the starting point for the present development can be traced to a classic paper by Bode and Shannon[2] wherein a simplified frequency domain derivation of the least squares theory was presented. The explicit formula for the *Wiener filter*

$$F_0 = [Q_{\underline{zw}} D^{*-1}]_{\mathscr{C}} D^{-1}$$

made its first appearance in this paper and, indeed, it was in this classical paper that the fundamental relationship between system theory and the concepts of causality, causal part, and spectral factorization was first exposed. However, the extension of this formula to a frequency domain control problem did not appear until a quarter century after the original work was completed, when Youla *et al.* observed that a second spectral factorization was required and solved a basic stochastic optimization theorem in a frequency domain setting.[15]

Although solid results did not appear until the late 1970s, the extension of these frequency domain formulas for stochastic optimization to a resolution space setting was one of the goals that motivated the initial research in the theory of operators defined on a Hilbert resolution space. Indeed, apparently abstract research on operator decomposition, spectral factorization, and causal invertibility was often motivated by the intuition derived from the Bode–Shannon formula and the goal of formulating a stochastic optimization theory. Although the deterministic optimization theory of Chapter 10 dates back to the late 1960s and early 1970s,[3,9] the first successful generalization of the Bode–Shannon theory to a Hilbert resolution space setting did not appear until 1977 with the thesis of Tung[13] and the follow-up papers by Tung, DeSantis, and Saeks,[4,14] which form the basis for the present development.

A second motivating force behind our theory was the finite-dimensional state space estimation and control theory formulated in the early 1960s by Kalman and Bucy,[7,8] among others. Although this work in a state space setting was actually subsumed by the earlier work of Wiener, Shannon, and Bode in an input/output setting, Kalman and company arrived at the simple implementation presented in Chapter 12, and indeed, it is these implementations rather than the optimization process that has made state space control theory such a powerful tool in modern engineering practice.

2. Causal Approximation: In Chapters 11 and 12 several forms of the $\underline{w}$-best causal approximation were investigated. Although a number of applications of this concept were encountered in our theory, one might desire to formulate a deterministic best causal approximation theory in which an underlying process $\underline{w}$ need not be specified. To this end we define the *distance* between an operator $A \in \mathscr{B}$ and the algebra of causal operators by means of the operator norm

$$d(A, \mathscr{C}) = \inf_{B \in \mathscr{C}} \|A - B\|$$

Although neither an explicit formula for an operator $B_0 \in \mathscr{C}$ such that $\|A - B_0\| = d(A, \mathscr{C})$ nor conditions for its existence are known, the minimizing distance can be computed by the formula

$$d(A, \mathscr{C}) = \sup_{\dagger \in \$} \|E^{\dagger} A E_{\dagger}\|$$

Although rather straightforward, this formula, due to Arveson,[1] is somewhat surprising in that it does not coincide with the intuitive $d(A, \mathscr{C}) = \|[A]_{\mathscr{R}^*}\|$ even when $[A]_{\mathscr{R}^*}$ exists. Indeed, even in the finite-dimensional case, where $\mathscr{C}$ is represented by lower triangular matrices, $d(A, \mathscr{C})$ is computed by taking the supremum over the submatrices lying in the upper right-hand corner of the given array, as indicated in Fig. 1.

$$A = \begin{bmatrix} x & x & x & x & \cdots & x \\ x & x & x & x & \cdots & x \\ x & x & x & x & \cdots & x \\ x & x & x & x & \cdots & x \\ \vdots & \vdots & \vdots & \vdots & \ddots & \vdots \\ x & x & x & x & \cdots & x \end{bmatrix}$$

Fig. 1. Submatrices with which to compute $d(A, \mathscr{C})$ in a finite-dimensional Hilbert resolution space.

3. Reproducing Kernel Resolution Space: Property 11.C.2 considered the problem of generating a random variable $\underline{z}$ with prescribed covariance, Q by filtering white noise $\underline{v}$. Although this goal could be achieved by a simple spectral factorization process, the more general problem of filtering a given process $\underline{w}$ with covariance $Q_{\underline{w}}$ to obtain a new process $\underline{z} = F\underline{w}$ with prescribed covariance is not as readily solved. One approach is to imbed the given process $\underline{w}$ into a renormed Hilbert resolution space wherein $Q_{\underline{w}} = I$ and then use a spectral factorization in this space to compute F. Interestingly, the appropriate resolution space proves to be the reproducing kernel resolution space (introduced in Note 6 at the end of Part I) associated with the operator $Q_{\underline{w}}$. This space is constructed from H but with a new norm and resolution structure, and hence an H-valued $\underline{w}$ may naturally be identified with an $H_{Q_{\underline{w}}}$-valued random variable, while the resultant causal filter on $(H_{Q_{\underline{w}}}, \mathscr{E}_{A_{\underline{w}}})$ may be identified with a causal filter on the original space.[5,6,12]

4. Higher-Order Statistics: Although our stochastic system theory was, in fact, formulated entirely in terms of second-order statistics, it is possible to formulate higher-order statistics for an H-valued random variable.[10] One may define an *nth-order covariance* $Q_{n\underline{x}}$ to be the symmetric n-linear map

$$E(x_1, \underline{x})(x_2, \underline{x})(x_3, \underline{x}) \cdots (x_n, \underline{x}) = Q_{n\underline{x}}(x_1, x_2, \ldots, x_n)$$

$Q_{n\underline{x}} \in \otimes^n H$, with cross-covariances similarly defined. Now rather than minimizing second-order statistics one can formulate a stochastic optimization theory in which higher-order statistics are taken into consideration by minimizing over polynomic rather than linear operators. Indeed, the problem of minimizing

$$E\|\underline{z} - B\underline{w}\|^2$$

over the causal nth-order polynomic operators has an explicit solution in terms of the nth-order statistics of $\underline{z}$ and $\underline{w}$. In fact, this solution can be obtained as a corollary to the basic stochastic optimization theorem simply by lifting the problem into an appropriate tensor product space,[10,11] as described in Note 8 at the end of Part I.

5. Stationary Processes: The concept of a stationary stochastic process is intimately intertwined with time invariance and the various "frequency domain" representations. An H-valued random variable can be identified with a stationary stochastic process only if H is equipped with a uniform resolution space structure. (See Note 3 at the end of Part I for an introduction to this concept.) In that case we say that $\underline{w}$ is *stationary* if $Q_{\underline{w}}$ is time invariant, while the time and frequency domain representations for $Q_{\underline{w}}$ correspond to the autocorrelation and power spectral density functions of $\underline{w}$.

6. Duality: The formulation of an optimal state estimator, unlike that of the stochastic regulator, requires that our state trajectory space admit a Hilbert space structure. This allows one to define a minimum norm problem in the state trajectory space and also paves the way for the construction of the required duality theory. For the purpose of the present discussion on state estimation we shall assume that our plant has been so chosen to permit the state trajectory space S to be imbedded in a Hilbert space in such a manner that the operators λ, θ, and F associated with the state trajectory factorization and the lifting theorem are bounded. Thus, if we are given a plant $P \in \mathscr{S}$ that admits a strong minimal state decomposition, it may be factored through the state trajectory space in the form $P = \theta\lambda$, where λ is an algebraically strictly causal operator mapping H to the state trajectory space S and θ is a memoryless operator mapping S to H. Similarly, by working with P^* in the dual resolution space $(H, \mathscr{E}^*)$, where the direction of time flow has been reversed, we may formulate a state decomposition for P^* that in turn defines a state trajectory factorization $P^* = \beta^*\alpha^*$ where β^* is memoryless and α^* is algebraically strictly causal on $(H, \mathscr{E}^*)$. Finally, on taking adjoints we obtain the *costate trajectory factorization* $P = \alpha\beta$, where β is a memoryless operator mapping H to S' (the state trajectory space for $P^* = \beta^*\alpha^*$) and α is an algebraically strictly causal operator mapping S' to H.

In some sense the state trajectory factorization $P = \theta\lambda$ is ideally suited to the study of the control problem while the costate trajectory factorization, $P = \alpha\beta$ is ideally suited to the state estimation problem. In the present context, however, we would like to investigate the relationship between these two problems in addition to studying them individually, and we thus desire to formulate a single factorization of P that can be employed in both the estimation and control problems. To this end we invoke our minimality assumption to obtain the factorization $P = \theta\pi\beta$, where π is an algebraically strictly causal operator mapping S' to S. Clearly, $\alpha = \theta\pi$ while $\lambda = \pi\beta$. These relationships are illustrated by the commutative diagram

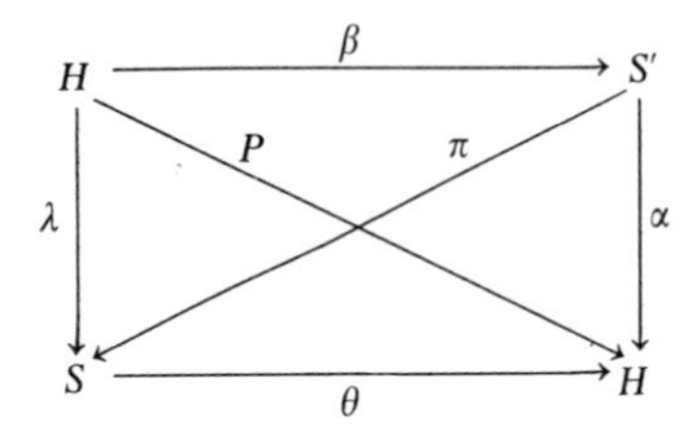

Finally, recall from the lifting theorem of Chapter 7 that if

$$[I + PP^*] = (I + V)^*(I + V)$$

is a radical factorization of $I + P^*P$ with $V \in \mathscr{R}$, then there exists a memoryless operator, $F: S \to H$ such that $V = F\lambda$ $(= F\pi\beta)$. Similarly, it follows from the dual theory that if

$$[I + PP^*] = (I + U)(I + U)^*$$

is a radical factorization of $[I + PP^*]$ with $U \in \mathscr{R}$, then there exists a memoryless operator $G: H \to S'$ such that $U = \alpha G$ $(= \theta\pi G)$. Adding the relationship $V = F\lambda$ and $U = \alpha G$ to the above diagram yields

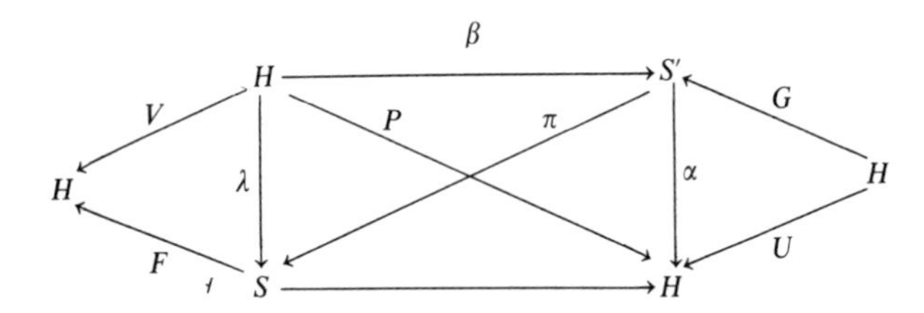

which completely describes the relationship between the various operators required for our estimation theory.

Although the precise condition under which the desired Hilbert space structure can be imposed on our state trajectory space is unknown, the above described dual structure is the natural setting in which to study the state estimation problem. Indeed, the factorization $P = \theta\pi\beta$ is the natural Hilbert resolution space analog of the factorization

$$P = C(D - A)^{-1}B$$

associated with the differential operator

$$\begin{aligned} \dot{x}(\dagger) &= Ax(\dagger) + Bf(\dagger), \\ & \qquad\qquad\qquad\qquad x(0) = 0 \\ g(\dagger) &= Cx(\dagger), \end{aligned}$$

In this respect the present factorization should be compared with the dual structure of Note 4 at the end of Part II wherein the role of the spaces S and S' is interchanged and no lifting theory exists.

7. State Estimation: To implement the stochastic regulator of Section 12.D one must have access to the state of the plant. Although this is often the case, there are systems in which only the output process y is directly measurable, in which case we must estimate the state from observations of the output. Although this can be achieved with a standard Wiener filter if one exploits the special nature of the state model, an especially convenient implementation of the filter can be obtained that employs a model of the plant together with a single memoryless gain. Our goal in this section is the formulation of this particular implementation of the optimal state estimator, which is termed the *Kalman filter.*

The setup for our state estimation problem assumes the dual structure of Note 6 and is illustrated in Fig. 2. Here, one observes the plant output corrupted by a white noise term $\underline{m}$ and desires to estimate the state of the system using a causal estimator that minimizes $E\|\underline{x} - \hat{\underline{x}}\|^2$ over $D \in \mathscr{C}$. As with the regulator problem, the disturbing input $\underline{s}$ is taken to be a white noise.

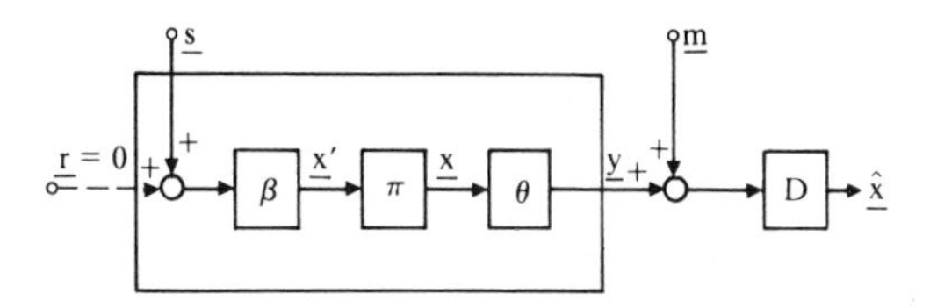

Fig. 2. Configuration for state estimation.

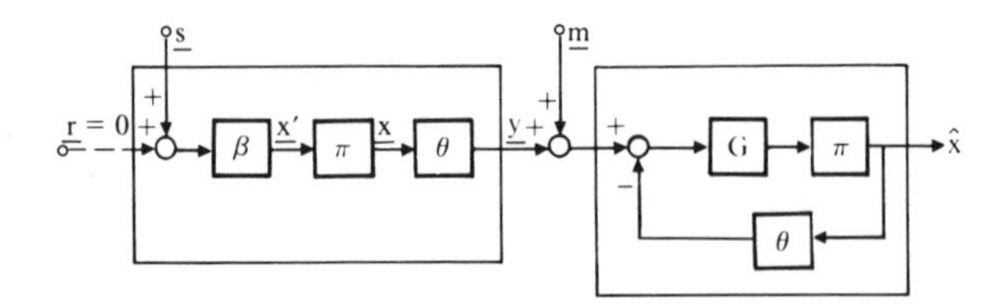

Fig. 3. Feedback implementation for the optimal state estimator.

Now, let $\underline{s}$ and $\underline{m}$ be zero mean H-valued independent random variables with finite second moment and covariances $Q_{\underline{s}} = Q_{\underline{m}} = I$ and assume that $P = \theta\pi\beta \in \mathscr{K}_2$ and that $[I + PP^*]$ admits a radical factorization $[I + PP^*] = (I + U)(I + U)^*$, $U \in \mathscr{R}$. Then an application of the basic stochastic optimization theorem and the lifting theorem reveals that there exists a memoryless operator G such that $E\|\underline{x} - \hat{x}\|^2$ is minimized over $D \in \mathscr{C}$ by

$$D_0 = \pi G(I + \theta\pi G)^{-1}$$

Moreover, $U = \theta\pi G \in \mathscr{R}$ is quasinilpotent, and hence the equality

$$D_0 = \pi G(I + \theta\pi G)^{-1}$$

suggests an implementation of our optimal state estimator by means of the stable feedback loop of Fig. 3.

As formulated here this state estimator assumes that the plant input r is zero. Fortunately, by invoking linearity we may modify our estimator to compensate for the effect of nonzero plant inputs. Indeed, all that need be done is to sense $\underline{r}$ and process it through a model of the plant. The output of this model is then subtracted from the plant output so that the state estimator sees only the zero input response of the plant and produces a zero-input state estimate. Finally, that part of the state that is due to the plant input is added back into the state estimate at the ouput of the estimator. The resulting nonzero-input state estimator is illustrated in Fig. 4.

Although from the point of view of performance the state estimator is just a special case of the Wiener filter, its power lies with the simple feedback implementation of Fig. 3, wherein the entire filter is built from a model of the plant and a single additional memoryless gain G. It is this special implementation of the Wiener filter which is termed the Kalman filter, while the memoryless gain G is often termed the *Kalman gain.*[8]

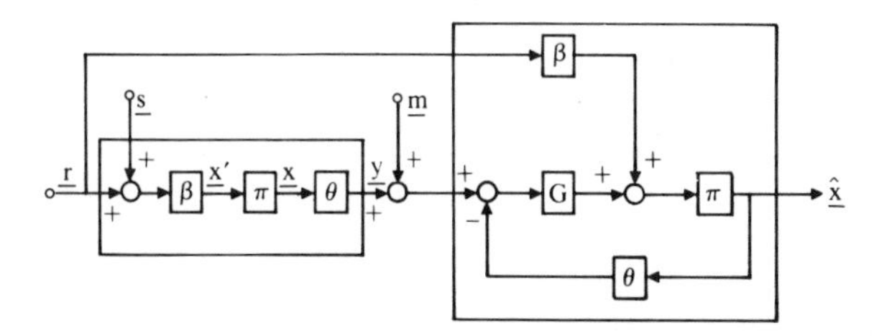

Fig. 4. Implementation of a nonzero-input state estimator.

8. Output Regulation: The implementation of the optimal regulator formulated in Section 12.D requires that the state of the plant be measurable. If not, the obvious choice is to use the Kalman filter to estimate the plant state on the basis of output observations and then to feed this estimate back to the input through the optimal law. The purpose of this note is to show that this is in fact an optimal strategy. As with the state estimator we assume the dual structure of Note 7 and formulate an artificial open loop problem as the starting point for our derivation, eventually showing that the solution to this open loop problem can be implemented using the desired feedback configuration. This problem, which is illustrated in Fig. 5, is identical to that used to derive the memoryless state feedback regulator except for the addition of the noise term $\underline{m}$. As before, our goal is to find a compensator M_0 minimizing $E[\|\underline{y}\|^2 + \|\underline{r}\|^2]$ over $M \in \mathscr{C}$.

To this end we let $\underline{s}$ and $\underline{m}$ be independent zero mean H-valued random variables with finite second moment and covariances $Q_{\underline{s}} = Q_{\underline{m}} = I$. Furthermore, let $P \in \mathscr{S} \cap \mathscr{K}_2$ admit a strong minimal state decomposition and a factorization, $P = \theta\pi\beta$, and assume that $[I + P^*P]$ and $[I + PP^*]$ admit factorizations $[I + P^*P] = (I + V)^*(I + V)$, and $[I + PP^*] = (I + U)(I + U)^*$, $V, U \in \mathscr{R}$. Then an application of the basic stochastic optimization theorem and the lifting theorem will reveal that there exist

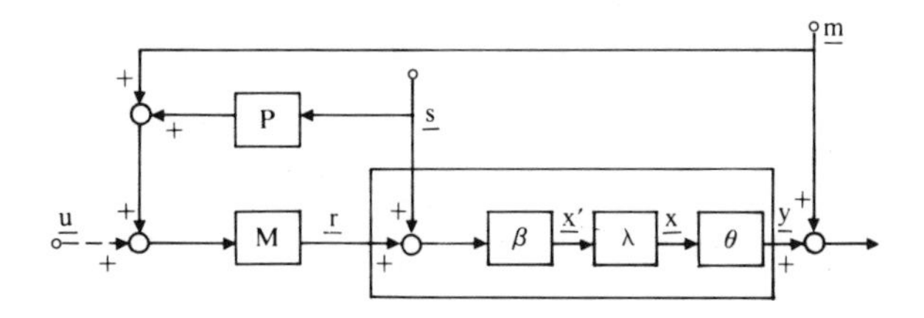

Fig. 5. Open loop regulator with noisy measurements.

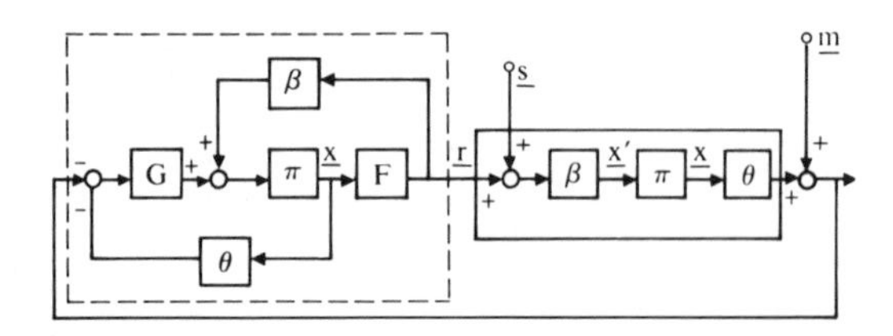

Fig. 6. Feedback implementation of an optimal output regulator.

memoryless operators F and G such that $E[\|\underline{y}\|^2 + \|\underline{r}\|^2]$ is minimized over $M \in \mathscr{C}$ by

$$M_0 = -(I + F\pi\beta)^{-1}F\pi G(I + \theta\pi G)^{-1}$$

Recognizing that $-(I + F\pi\beta)^{-1}F$ and $\pi G(I + \theta\pi G)^{-1}$ have obvious implementations as feedback systems (which are stable, since $V = F\pi\beta$ and $U = \theta\pi G$ are strictly causal), our open loop regulator may be implemented as shown in Fig. 6 with the aid of some diagram chasing. Interestingly, this implementation of the optimal output regulator is just the cascade of the (nonzero input) Kalman filter with the optimal memoryless state feedback law. This classical result of state space optimization theory is termed the *separation principle.*[7,8]

REFERENCES

1. Arveson, W., Interpolation problems in nest algebra, *J. Funct. Anal.* **20**, 208–233 (1975).
2. Bode, H. W., and Shannon, C. E., A simplified derivation of linear least squares smoothing and prediction theory, *Proc. IRE* **38**, 417–425 (1950).
3. DeSantis, R. M., Causality theory in system analysis, *Proc. IEEE* **64**, 36–44 (1976).
4. DeSantis, R. M., Saeks, R., and Tung, L. J., Basic optimal estimation and control problems in Hilbert space, *Math. Systems Theory* **12**, 175–203 (1978).
5. Duttweiler, D., Ph.D. Dissertation, Stanford Univ., Stanford, California (1970).
6. Kailath, T., and Duttweiler, D., An RKHS approach to detection and estimation problems—Part III: Generalized innovations representations and a likelihood ratio formula, *IEEE Trans. Informat. Theory* **IT-18,** 730–745 (1972).
7. Kalman, R. E., Contributions to the theory of optimal control, *Bol. Sci. Math. Mexicanna* **5**, 102–119 (1960).
8. Kalman, R. E., and Bucy, R. S., New results in linear filtering and prediction theory, *Trans. ASME Ser. D J. Basic Eng.* **82**, 95–100 (1961).
9. Porter, W. A., A basic optimization problem in linear systems, *Math. Systems Theory* **5**, 20–44 (1971).
10. Porter, W. A., Multiple signal extraction by polynomic filtering, *Math. Systems Theory* **13**, 237–254 (1980).

11. Porter, W. A., On factoring the polyvariance operator, *Math. Systems Theory* **14**, 67–82 (1981).
12. Saeks, R., Reproducing Kernel resolution space and its application, *J. Franklin Inst.* **302**, 331–355 (1976).
13. Tung, L. J., Ph.D. Dissertation, Texas Tech. Univ., Lubbock, Texas (1977).
14. Tung, L. J., Saeks, R., and DeSantis, R. M., Wiener–Hopf filtering in Hilbert resolution space, *IEEE Trans. Circuits and Systems* **CAS-25**, 702–705 (1978).
15. Youla, D. C., Bongiorno, J. J., and Jabr, H. A., Modern Wiener–Hopf design of optimal controllers, Parts I and II, *IEEE Trans. Automat. Control* **AC-21**, 3–13, 319–337 (1976).

Selected Bibliography

1. RESOLUTION SPACE

Brandon, D., Relativistic Resolution Space, M.S. Thesis, Texas Tech Univ. (1977).

DeCarlo, R. A., Saeks, R., and Strauss, M. J. The Fourier transform of a resolution space and a theorem of Masani, *Proc. Internat. Symp. Operator Theory Networks and Systems, 1st, Montreal* (1975).

DeSantis, R. M., Causality for nonlinear systems in Hilbert space, *Math. Systems Theory* **7** (1973).

DeSantis, R. M., Causality theory in systems analysis, *Proc. IEEE* **64**, 36–44 (1976).

DeSantis, R. M., Causality Structure of Engineering Systems, Ph.D. Thesis, Univ. of Michigan (1971).

DeSantis, R. M., and Feintuch, A., Causality theory and multiplicative transformators, *J. Math. Anal. Appl.* **75**, 411–416 (1980).

DeSantis, R. M., and Porter, W. A., On time related properties of nonlinear systems, *SIAM J. Appl. Math.* **24**, 188–206 (1973).

DeSantis, R. M., and Porter, W. A., Temporal properties of engineering systems, *Proc. Allerton Conf. Circuit and Systems Theory, 9th, Univ. of Illinois* (1971).

Feintuch, A., On diagonal coefficients of C_0 operators, *J. London Math. Soc.* (2) **17**, 507–510 (1978).

Feintuch, A., Causality and C_0 contractions, *Proc. Internat. Symp. Operator Theory and Networks and Systems, 2nd* (1977).

Feintuch, A., Causality and strict causality for symmetric bilinear systems, *J. Math. Anal. Appl.* **65**, 703–710 (1978).

Feintuch, A., Strictly and strongly causal linear operators, *SIAM J. Math. Anal.* **10**, 603–613 (1979).

Porter, W. A., An overview of polynomic system theory, *Proc. IEEE* **64**, 18–23 (1976).

Porter, W. A., Synthesis of polynomic systems, *SIAM J. Math. Anal.* **11**, 308–315 (1980).

Porter, W. A., Nonlinear systems in Hilbert space, *Internat. J. Control.* **13**, 593–602 (1971).

Porter, W. A., Some recent results in nonlinear system theory, *Proc. Midwest Symp. Circuit Theory, 13th, Univ. of Minnesota* (1970).

Porter, W. A., The common causality structure of multilinear maps and their multipower forms, *J. Math. Anal. Appl.* **57**, 667–675 (1977).

Porter, W. A., Causal realization from input–output pairs, *J. Control. Opt.* **15**, 120–128 (1977).

Porter, W. A., Joint interpolation and approximation of causal systems, *J. Math. Anal. Appl.* **72**, 399–412 (1979).

Porter, W. A., Approximation by Bernstein systems, *Math. Systems Theory* **11**, 259–274 (1978).

Porter, W. A., Causal parameter identification, *Internat. J. Control* **9**, 17–29 (1978).

Porter, W. A., Some circuit theory concepts revisited, *Internat. J. Control* **12**, 433–448 (1970).

Porter, W. A., Operator theory of systems, *IEEE Circuits and Systems Newsletter* **7**, 8–12 (1974).

Porter, W. A., Data interpolation, causality structure and system identification, *Inform. and Control* **29**, 217–233 (1975).

Porter, W. A., Clark, T. M., and DeSantis, R. M., Causality structure and the Weierstrass theorem, *J. Math. Anal.* **52**, 351–363 (1975).

Porter, W. A., and DeSantis, R. M., Linear systems with multiplicative control, *Internat. J. Control* **20** (1974).

Porter, W. A., and Zahm, C. L., Basic Concepts in System Theory, Tech. Rep. 33, Systems Eng. Lab., Univ. of Michigan, 1969.

Saeks, R., Causal factorization, shift operators, and the spectral multiplicity function, *in* "Vector and Operator Valued Measures and Applications" (D. Tucker and H. Maynard, eds.). Academic Press, New York, 1973.

Saeks, R., Causality in Hilbert space, *SIAM Rev.* **12**, 357–383 (1970).

Saeks, R., The factorization problem—A survey, *Proc. IEEE* **64**, 90–95 (1976).

Saeks, R., Reproducing kernel resolution space and its applications, *J. Franklin Inst.* **302** (1976).

Saeks, R., Fourier analysis in Hilbert space, *SIAM Rev.* **15**, 604–638 (1970).

Saeks, R., Resolution space—A function analytic setting for control theory, *Proc. Allerton Conf. Circuit and System Theory, 9th, Univ. of Illinois* (1971).

Saeks, R., Finite energy networks, *IEEE Trans.* **CT-17**, 618–619 (1970).

Saeks, R., "Resolution Space, Operators and Systems." Springer-Verlag, Berlin and New York, 1973.

Saeks, R., DeSantis, R. M., and Leake, R. J. On causal decomposition, *IEEE Trans.* **AC-19**, 152–153 (1974).

Saeks, R., and Goldstein, R. A., Cauchy integrals and spectral measures, *Indiana Math. J.* **22**, 367–378 (1972).

Saeks, R., and Leake, R. J., On semi-uniform resolution space, *Proc. Midwest Symp. Circuit Theory, 14th, Univ. of Denver* (1971).
Tung, L. J., and Saeks, R., Reproducing kernel resolution space and its applications II, *J. Franklin Inst.* **306**, 425–447 (1978).
Winslow, L., and R. Saeks, Nonlinear lossless networks, *IEEE Trans.* **CT-19**, 392 (1972).

2. NEST ALGEBRAS

Anderson, N., Ph.D. Dissertation, Univ. of California at Berkeley, Berkeley, California (1979).
Arveson, W., Operator algebras and invariant subspaces, *Ann. of Math.* **120**, 433–532 (1974).
Arveson, W. A., Interpolation in nest algebras, *J. Funct. Anal.* **20**, 208–233 (1975).
Arveson, W. B., Operator algebras and invariant subspaces, *Ann. of Math.* **100**, 433–532 (1974).
Ashton, G., Ph.D. Dissertation, Univ. of London, 1981.
Christensen, E., Derivations of nest algebras, *Math. Ann.* **229**, 155–161 (1977).
Christensen, E., and Peligrad, C., Commutants of nest algebras modulo the compact operators, *Invent. Math.* **56**, 113–116 (1980).
Daniel, V. W., Convolution operators on the lebesque spaces of the half-line, *Trans. Amer. Math. Soc.* **164**, 479–488 (1972).
Deddens, J. A., Every isometry is reflexive, *Proc. Amer. Math. Soc.* **28**, 509–512 (1971).
Erdos, J. A., Non-self-adjoint operator algebras, *Proc. Rep. Ireland Acad. Sci.* **81A**, 127–145 (1981).
Erdos, J. A., The triangular factorization of operators on Hilbert space, *Indiana Univ. Math. J.* **22**, 939–950 (1973).
Erdos, J. A., On some ideals of nest algebras, *Proc. London Math. Soc.* (to appear).
Erdos, J. A., and Powers, S. C., Weakly closed ideals of nest algebras, *J. Operator Theory* (to appear).
Erdos, J., Unitary invariants for nests, *Pacific J. Math.* **23**, 229–256 (1967).
Erdos, J. A., and Longstaff, W. E., The convergence of triangular integrals of operators on Hilbert space, *Indiana Univ. Math. J.* **22**, 929–938 (1973).
Fall, T., Compact Perturbations of Nest Algebras, Ph. D. Dissertation, Univ. of California at Berkeley, Berkeley, California (1977).
Fall, T., Arveson, W., and Muhly, P., Perturbations of Nest algebras, *J. Operator Theory* **1**, 137–150 (1979).
Feintuch, A., Algebras generated by volterra operators, *J. Math. Anal. and Appl.* **56**, 470–476 (1976).
Gilfeather, F., and D. Larson, Nest-subalgebras of von Neumann algebras, *Adv. in Math.* (to appear).
Herrero, D. A., "Operator algebras of finite strict multiplicity, *Indiana Math. J.* **22**, 13–24 (1972).
Hoppenwasser, A., and Larson, D., The carrier space of a reflexive operator algebra, *Pacific J. Math.* **81**, 417–434 (1979).
Halmos, P. R., Quasitriangular operators, *Acta Sci. Math.* (Szegad) **29**, 283–293 (1968).
Hopenwasser, A., Completely isometric maps and triangular operator algebras, *Pacific J. Math.* **9**, 375–392 (1972).

Johnson, B., and Parott, S., Operators commuting with a von Neumann algebra modulo the set of compact operators, *J. Funct. Anal.* **11**, 39–61 (1972).
Kadison, R., and Singer, I., Triangular operator algebras, *Amer. J. Math.* **72**, 227–259 (1960).
Kadison, R. V., and Singer, I. M., Triangular operator algebras, *Amer. J. Math.* **82**, 227–259 (1960).
Lance, E. C., Some properties of nest algebras, *Proc. London Math. Soc.* **19**, 45–68 (1969).
Lance, E. C., Cohomology and perturbations of nest algebras, *Proc. London Math. Soc.* (to appear).
Lance, E. C., Some properties of nest algebras, *Proc. London Math. Soc.* (3), **19**, 47–68 (1969).
Larson, D., Ph.D. Dissertation, Univ. of California at Berkeley, Berkeley, California (1976).
Larson, D., On the structure of certain reflexive operators, *J. Funct. Anal.* **31**, 275–292 (1979).
Loebl, R. I., and Muhly, P. S., Analyticity and flows in von Neumann algebras, *J. Funct. Anal.* **29**, 214–252 (1978).
Plastiras, J., Quasitriangular operator algebras, *Pacific J. Math.* **2**, 543–549 (1976).
Ringrose, J. R., On some algebras of operators, *Proc. London Math. Soc.* **3**, 61–83 (1965).
Ringrose, J., On some algebras of operators II, *Proc. London Math. Soc.* (3), **16**, 385–402 (1966).
Ringrose, J. R., Super-diagonal forms for compact linear operator, *Proc. London Math. Soc.* **12**, 367–384 (1962).
Ringrose, J. R., On the triangular representation of integral operators, *Proc. London Math. Soc.* **12**, 385–399 (1962).
Sarason, D., A remark on the volterra operator, *J. Math. Anal. Appl.* **12**, 244–246 (1965).
Schue, J. R., The structure of hyperreducible triangular algebra, *Proc. Amer. Math. Soc.* **15**, 766–772 (1964).

3. TRIANGULAR OPERATOR MODELS

Barkar, M. A., and Gohberg, I. C., On factorization of operators relative to a discrete chain of projectors in Banach space, *Amer. Math. Soc. Trans.* **90**, 81–133 (1970).
Brodinskii, M. S., "Triangular and Jordan Representation of Linear Operators." American Mathematical Society, Providence, Rhode Island, 1971.
Brodskii, M. S., Gohberg, I. C., and Krein, M. G., General theorems on triangular repre-sensations of linear operators and multiplications of their characteristic functions, *Funct. Anal.* **3**, 1–27 (1969).
Daleckii, J. R., and Krein, S. G., Integration and differentiation of functions of Hermitian operators and their applications to the theory of perturbations, *Amer. Math. Soc. Transl. Ser. 2* **74**, 1–30 (1967).
Gohberg, I. C., The factorization problem for operator functions, *Amer. Math. Soc. Transl.* **49**, 130–161 (1966).
Gohberg, I. C., Triangular representation of linear operators and multiplicative representation of their characteristic functions, *Sov. Math. Dokl.* **8**, 831–834 (1967).
Gohberg, I. C., and Barkar, M. A., On factorization of operators in Banach space, *Amer. Math. Soc. Transl. Ser. 2* **90**, 105–133 (1970).

Gohberg, I. C., and Barkar, M. A., On factorization of operators relative to discrete chain of projections, *Amer. Math. Soc. Transl. Ser. 2* **90**, 81–103 (1970).
Gohberg, I. C., and Krein, M. G., On factorization of operators in Hilbert space, *Sov. Math. Dokl.* **8**, 831–834 (1967).
Gohberg, I. C., and Krein, M. G., "Introduction to the Theory of Linear Non-Self-Adjoint Operators in Hilbert Space." American Mathematical Society, Providence, Rhode Island, 1967.
Gohberg, I. C., and Krein, M. G., "Theory of Volterra Operators in Hilbert Space and Its Applications." American Mathematical Society, Providence, Rhode Island, 1970.
Gohberg, I. C., and Krein, M. G., Systems of integral equations on the half line with kernels depending on the difference of their arguments, *Amer. Math. Soc. Transl. Ser. 2* **14**, 217–287 (1960).
Krein, M. G., Integral equations on the half line with kernels depending on the difference of their arguments, *Amer. Math. Soc. Transl. Ser. 2* **22**, 163–288 (1956).
Livsic, M. S., "Operators, Oscillations, Waves (Open Systems)." American Mathematical Society, Providence, Rhode Island, 1973.

4. STATE SPACE

Balakrishnan, A. V., On the state space theory of nonlinear systems, *in* "Functional Analysis and Optimization" (E. R. Cainanello, ed.), pp. 15–36. Academic Press, New York, 1966.
Balakrishnan, A. V., Foundations of the state space theory of continuous systems, *J. Comp. Systems Sci.* **1**, 91–116 (1967).
Baras, J. S., Brockett, R. W., and Fuhrmann, P. A., State space models for infinite dimensional systems, *IEEE Trans.* **AC-19**, 693–700 (1974).
Curtain, R., and Pritchard, A. J., The infinite-dimensional riccati equations for systems defined by evolution operators, *SIAM J. Control* **14**, 951–983 (1976).
De Santis, R. M., On state realization and causality decomposition for nonlinear systems, *Proc. ORSA Meeting, 41st, New Orleans, Louisiana* (1972).
DeSantis, R. M., On state realization and causality decomposition for nonlinear systems, *SIAM J. Control* **11**, 551–562 (1973).
Feintuch, A., State-space theory for resolution space operators, *J. Math. Anal. Appl.* **74**, 164–191 (1980).
Feintuch, A., On single input controllability for infinite dimensional systems, *J. Math. Anal. Appl.* **62**, 538–546 (1978).
Feintuch, A., Realization theory for symmetric systems, *J. Math. Anal. Appl.* **71**, 131–146 (1979).
Feintuch, A., Strong minimality for infinite dimensional linear systems, *SIAM J. Control Optim.* **14**, 945–950 (1976).
Fuhrmann, P. A., On weak and strong reachability and controllability of infinite dimensional linear systems, *J. Optimization Theory Appl.* **19**, 77–89 (1972).
Fuhrmann, P. A., Exact controllability and observability and realization theory in Hilbert space, *J. Math. Anal. Appl.* (to appear).
Krohn, K., and Rhodes, J., Algebraic theory of machines—I, *Amer. Math. Soc. Transl.* **116**, 450–464 (1966).
Nerode, A., Linear automation transformation, *Proc. Amer. Math. Soc* **9**, 441–444 (1958).
Olivier, P. D., Ph.D. Dissertation, Texas Tech. Univ., Lubbock, Texas (1980).

Olivier, P. D., and Saeks, R., Nonlinear state decomposition, *IEEE Trans. Circuits and Systems* **CAS-25**, 1113–1121 (1980).
Orava, P. J., Causality and state concepts in dynamical systems theory, *Internat. J. Systems Sci.* **4**, 679–691 (1973).
Saeks, R., State in Hilbert space, *SIAM Rev.* **15**, 283–308 (1973).
Salovaara, S., On set theoretic foundations of system theory—A study of the state concept, *Acta Polytech. Scand. Ser. MA*, 1–74 (1967).
Schnure, W. K., System identification: A state space approach, Ph.D. Thesis, Univ. of Michigan (1974).
Schumitzky, A., State space control for general linear systems, *Proc. Internat. Symp. Math. of Networks and Systems* pp. 194–204. T. H. Delft (1979).
Steinberger, M. L., The Realization, Control and Optimal Control of Linear Infinite Dimensional Systems, Ph.D. Dissertation, Dept. of EE, Univ. of Southern California (1977).
Steinberger, M. L., Schumitzky, A., and Silverman, L., Optimal Causal Feedback Control of Linear Infinite Dimensional Systems, Unpublished Notes, Univ. of Southern California (1977).
Zadeh, L., The concepts of system aggregate and state in system theory, *in* "System Theory" (L. Zadeh and E. Polak, eds.), pp. 3–42. McGraw-Hill, New York, 1969.
Zadeh, L., The concept of state in system theory, *in* "Views on General System Theory" (M. Mesarovic, ed.), pp. 39–50. Wiley, New York, 1964.

5. FEEDBACK SYSTEMS

Ahmed, N. U., Optimal control of a class of nonlinear systems on Hilbert space, *IEEE Trans.* **AC-14**, 711–714 (1969).
Balakrishnan, A. V., Optimal control in Banach spaces, *SIAM J. Control* **3**, 152–180 (1965).
Caines, P. E., Causality stability and inverse systems, *Internat J. Systems Sci.* **4**, 825–832 (1973).
Callier, F. M., and Desoer, C. A., L^p-stability ($1 \leq p \leq \infty$) of multivariate non-linear time-varying feedback systems that are open loop unstable, *Internat. J. Control* **20**, 65–72 (1974).
Callier, F. M., and Desoer, C. A., Open loop unstable convolutional feedback systems with dynamical feedback, *Automatica* **12**, 507–518 (1976).
Damborg, M. J., The use of normed linear spaces for feedback system stability, *Proc. Midwest Symp. on Circuit Theory, 14th, Univ. of Denver* (1971).
Damborg, M. J., Stability of the basic nonlinear operator feedback system, Resh, Rep. 37, System Eng. Lab., Univ. of Michigan (1967).
Damborg, M. J., and Naylor, A., Fundamental structure of input–output stability for feedback systems, *IEEE Trans.* **SSC-6**, 92–96 (1970).
Datko, R., A linear control problem in abstract hilbert space, *J. Differential Equations* **9**, 346–359 (1971).
DeCarlo, R. A., and Saeks, R., A new characterization of the Nyquist stability criterion, *Proc. Midwest Symp. Circuits and Systems, 19th, Milwaukee, Wisconsin* (1976).
DeCarlo, R. A., and Saeks, R., Variations on the nonlinear Nyquist criterion, *Proc. Midwest Symp. Circuits and Systems, 18th, Montreal* (1975).

DeSantis, R. M., Causality, strict causality, invertibility for systems in Hilbert resolution spaces, *SIAM J. Control* **12**, 536–553 (1974).

DeSantis, R. M., On the generalization of the Volterra principal of Inversion, *Math. Anal. Appl.* **48** (1974).

DeSantis, R. M., and Porter, W. A., On the Analysis of Feedback Systems with a Multipower Open Loop Chain, Systems Engineering Lab., Tech, Rep. No. 75, Univ. of Michigan (1973).

DeSantis, R. M., On a generalized volterra equation in Hilbert space, *Proc. Amer. Math. Soc.* **38**, 563–570 (1973).

DeSantis, R. M., Causality and stability in resolution space, *Proc. Midwest Symp. on Circuit Theory, 14th, Univ. of Denver* (1971).

DeSantis, R. M., and Porter, W. A., On the analysis of feedback systems with a polynomic plant, *Internat. J. Control* **21**, 159–175 (1975).

Desoer, C. A., Plant perturbations in multivariate systems, *J. Franklin Inst.* **24**, 279–282 (1975).

Desoer, C. A., Liu, R.-W., Murray, J., and Saeks, R., Feedback system design: The fractional representation approach to analysis and synthesis, *IEEE Trans. Automat. Control* **AC-25**, 401–412 (1980).

Dolezal, V., "Monotone Operators and Applications in Control and Network Theory." Elsevier, Amsterdam, 1979.

Falb, P. L., Freedman, M. I., and Zames, G., Input–Output Stability: A General Point of View, Rep. PM-54, NASA Elec. Res. Center (1968).

Feintuch, A., Strictly cyclic linear systems, *SIAM J. Appl. Math.* **34**, 415–422 (1978).

Feintuch, A., On pole assignment for a class of infinite dimensional linear systems, *SIAM J. Control Optim.* **16**, 270–276 (1978).

Feintuch, A., Causal C_0 operators and feedback stability, *Math. Systems Theory* **11**, 283–288 (1978).

Feintuch, A., and Saeks, R., Extension spaces and the resolution topology, *Internat J. Control* **39**, 347–354 (1981).

Freedman, M. I., Falb, P. L., and Zames, G., A Hilbert space stability theory over locally compact Abelian groups, *SIAM J. Control* **7**, 479–493 (1969).

Friedman, A., Optimal control in Banach space, *J. Math. Anal. and Appl.* **19**, 35–55 (1967).

Kuo, M. C. Y., The Application of Functional Analysis to Solve a Class of Linear Control Problems, Ph.D. Thesis, Univ. of Michigan (1964).

Kuo, M. C. Y., and Kazda, L., Minimum energy problems in Hilbert space, *J. Franklin Inst.* **283**, 38–54 (1967).

Porter, W. A., Minimizing system sensitivity through feedback, *J. Franklin Inst.* **286** (1968).

Porter, W. A., A basic optimization problem in linear systems, *Math. Systems Theory* **5**, 20–44 (1971).

Porter, W. A., On sensitivity in multivariate nonstationary systems, *Internat. J. Control* **7** (1968).

Porter, W. A., General sensitivity theory, *IEEE Trans.* **AC-18** (1973).

Porter, W. A., Parameter sensitivity in distributive feedback systems, *Internat. J. Control* **5** (1967).

Porter, W. A., A new approach to the general minimum energy problem, *Joint Automat. Control Conf., Stanford Univ.* (1964).

Porter, W. A., Sensitivity problems in linear systems, *IEEE Trans.* **AC-10** (1965).

Porter, W. A., On the reduction of sensitivity in multivariate systems, *Internat. J. Control* **5**, 1–9 (1967).
Porter, W. A., and DeSantis, R. M., Sensitivity Analysis in Multilinear Systems, SEL TR No. 77, Univ. of Michigan (1973).
Saeks, R., On the encirclement condition and its generalization, *IEEE Trans.* **CAS-22,** 780–785 (1975).
Saeks, R., The index of a nonlinear system, *Proc. IEEE Internat. Symp. Circuits and Systems, New York* (1978).
Saeks, R., Synthesis of general linear networks, *SIAM J. Appl. Math.* **16**, 924–930 (1968).
Saeks, R., and DeCarlo, R. A., Stability and homotopy, *in* "Alternatives for Linear Multivariable Control," pp. 247–252. NEC, Chicago, Illinois, 1978.
Saeks, R., and Murray, J., Feedback system design: The tracking and disturbance rejection problems, *IEEE Trans. Automat. Control* **AC-26,** 203–217 (1981).
Sandberg, J. W., Some stability results related to those of V. M. Popov, *Bell System Tech. J.* **44**, 2133–2148 (1965).
Sandberg, I. W., On the L_2 boundedness of solutions of nonlinear functional equations, *Bell System Tech. J.* **43**, 1601–1608 (1968).
Sandberg, I. W., Some results on the theory of physical systems governed by nonlinear functional equations, *Bell System Tech. J.* **44**, 871 (1965).
Triggiani, R., On the stabilitizability problem in Banach space, *J. Math. Anal. and Appl.* **9**, 383–405 (1975).
Willems, J. C., Stability, instability, invertibility and causality, *SIAM J. Control* **7**, 645–671 (1969).
Willems, J. C., "Analysis of Feedback Systems." MIT Press, Cambridge, Massachusetts, 1971.
Zahm, C., Structure of Sensitivity Reduction, Ph.D. Dissertation, Univ. of Michigan (1969).
Zames, G., Functional analysis applied to nonlinear feedback systems, *IEEE Trans.* **CT-10**, 392–404 (1963).
Zames, G., Nonlinear Operators for System Analysis, Tech. Rep. 370, MIT Research Lab. for Electron. (1960).
Zames, G., and Falb, P. L., Stability conditions for systems with monotone and slope-restricted nonlinearities, *SIAM J. Control* **6**, 89–108 (1968)
Zames, G., On input-output stability of time varying nonlinear feedback systems, *IEEE Trans.* **AC-11**, 228–238, 465–476 (1966).
Zames, G., Nonlinear Operators—Cascading, Inversion and Feedback, Quarterly Progress Rep. No. 53, Research Lab of Electron., MIT, pp. 93–107 (1959).
Zames, G., Realizability conditions for nonlinear feedback systems, *IEEE Trans.* **CT-11**, 186 (1964).
Zames, G., On the stability of nonlinear, time-varying systems, *Proc. Nat. Electron. Conf.* **20**, 725–730 (1964).

6. STOCHASTIC SYSTEMS

Balakrishnan, A. V., Stochastic filtering and control of linear systems: A general theory, "Control Theory of Systems Governed by Partial Differential Equations." Academic Press, New York, 1977.

Balakrishnan, A. V., "Introduction to Optimization Theory in Hilbert Space." Springer-Verlag, Berlin and New York, 1971.
Balakrishnan, A. V., and Lions, J. L., State estimation for infinite dimensional systems, *J. Comparative and Systems Sci.* **11**, 391–403 (1967).
Curtain, R. F., Estimation and stochastic control for linear infinite dimensional systems, "Probabilistic Analysis and Related Topics," (A. Bharucha-Reid, ed.), Vol. 1. Academic Press, New York (1980).
Davis, M. H. A., "Linear Estimation and Stochastic Control." Chapman and Hall, London, 1977.
DeSantis, R. M., Saeks, R., and Tung, L. J., Basic optimal estimation and control problems in Hilbert space, *Math. Systems Theory* **12**, 175–203 (1978).
Duttweiler, D. L., Reproducing Kernel Hilbert Space Techniques for Detection and Estimation Problems, Tech. Rep. 7050-18, Informat. Sci. Lab., Stanford Univ. (1970).
Falb, P., Infinite dimensional Filtering: the Kalman–Bucy Filter in Hilbert space, *Informat. and Control* **11**, 102–137 (1967).
Feintuch, A., Saeks, R., and Neil, C., A new performance measure for stochastic optimization, *Math. Systems Theory* (to appear).
Hagander, P., Linear Control and Estimation Using Operator Factorization, Rep. 7114, Div. of Automatic Control, Lund Inst. of Tech., Lund, Sweden (1971).
Hagander, P., The use of operator factorization for linear control and estimation, *Automatica* **9**, 623–631 (1973).
Kailath, T., Application of a resolvent identity to a linear smoothing problem, *SIAM J. Control* **7**, 68–74 (1969).
Kailath, T., and Duttweiler, D. L., An RKHS approach to detection and estimation problems—Part III: Generalized innovations representations and a likelihood-ratio formula, *IEEE Trans.* **IT-18**, 730–745 (1972).
Kailath, T., and Duttweiler, D. L., Generalized innovation processes and some applications, *Proc. Midwest Symp. on Circuit Theory, 14th, Univ. of Denver* (1971).
Porter, W. A., Multiple signal extraction by polynomial filtering, *Math. Systems Theory* **13**, 237–254 (1980).
Porter, W. A., On factoring the polyvariance operator, *Math. Systems Theory* **14**, 67–82 (1981).
Tung, L. J., Random Variables, Wiener–Hopf Filtering and Control Formulated in Abstract Space, Ph.D. Thesis, Texas Tech. Univ., Lubbock, Texas (1977).
Tung, L. J., and Saeks, R., Wiener–Hopf techniques in resolution space, *Proc. OTNS Internat. Symp., 2nd* (1977).
Tung, L. J., Saeks, R., and DeSantis, R. M., Wiener–Hopf filtering in Hilbert resolution space, *IEEE Trans. Circuits and Systems* **CAS-25**, 702–705 (1978).

7. FREQUENCY DOMAIN METHODS

Baras, J. S., Intrinsic Models for Infinite Dimensional Linear Systems, Ph.D. Dissertation, Harvard Univ., Cambridge, Massachusetts (1973).
Baras, J. S., and Brockett, R. W., H^2-functions and infinite dimensional realization theory, *Proc. IEEE Conf. Decision and Control* (December 1974).
Barrett, J. F., Construction of Linear Quadratic Regulators Using Spectral Factorization and the Return Difference Matrix, Tech. Rep. CN/75/4, Univ. of Cambridge (1975).

Belevitch, V., Factorization of scattering matrices with applications to passive network synthesis, *Phillips Res. Rep.* **18**, 275–317 (1963).
Bode, H. W., "Network Analysis and Feedback Design." Van Nostrand-Reinhold, New York, 1945.
Bode, H. W., and Shannon, C. E., A simplified derivation of linear least square smoothing and prediction theory, *Proc. IRE* **38**, 417–425 (1950).
Bongiorno, J. J., Minimum sensitivity design of linear multivariate feedback control systems by matrix spectral factorization, *IEEE Trans.* **AC-14**, 665–673 (1969).
Chang, S. S. L., "Synthesis of Optimal Control Systems." McGraw-Hill, New York, 1961.
Cruz, J. B., and Perkins, W. R., A new approach to the sensitivity problem in multivariable feedback system design, *IEEE Trans.* **AC-9** (1964).
Douglas, R. G., and Helton, J. W., The precise theoretical limits of causal Darlington synthesis, *IEEE Trans.* **CT-20**, 327 (1973).
Douglas, R. G., and Helton, J. W., Inner dilations of analytic matrix functions and Darlington synthesis, *Acta Sci. Math.* **34**, 61–67 (1973).
Emre, E., and Silverman, L. M., New criteria and system theoretic interpretations for relatively prime polynomial matrices, *IEEE Trans.* **AC-22**, 239–242 (1977).
Falb, P. L., and Freedman, M. I., A generalized transform theory for causal operators, *SIAM J. Control* **7**, 452–471 (1969).
Francis, B. A., The multivariable servomechanism problem from the input-output viewpoint, *IEEE Trans.* **AC-22**, 322–328 (1977).
Fuhrmann, P. A., Realization theory in Hilbert space for a class of transfer functions, *J. Funct. Anal.* **18**, 338–349 (1975).
Helton, J. W., The characteristic functions of operator theory and electrical network realization, *Indiana J. Math.* **22**, 403–414 (1972).
Helton, J. W., Discrete time systems, operator models, and scattering theory, *J. Funct. Anal.* **16**, 15–38 (1974).
Helton, J. W., Passive network realization using abstract operator theory, *IEEE Trans.* **CT-19**, 518–520 (1972).
Hermann, R., and Martin, C., Applications of algebraic geometry to systems theory—Part I, *IEEE Trans.* **AC-22**, 19–25 (1977).
Horowitz, I. M., "Synthesis of Feedback Systems." Academic Press, New York (1963).
Horowitz, I. M., and Shaked, U., Superiority of transfer function over state variable methods in linear time-invariant feedback system design, *IEEE Trans.* **AC-20** (1975).
Jabr, H. A., Modern Analytical Design of Optimal Multivariable Control Systems, Ph.D. Dissertation, Polytechnic Inst. of New York, Farmingdale, New York (1975).
Jury, E. I., "Inners and Stability of Dynamic Systems." Wiley, New York, 1974.
Levan, N., The Nagy–Foias operator models, networks, and systems, *IEEE Trans.* **CAS-23**, 335–343 (1976).
MacFarlane, A. G. J., Return-difference matrix properties for optimal stationary Kalman–Bucy filter, *Proc. IEEE* **118**, 373–376 (1971).
MacFarlane, A. G. J., "Frequency-Response Methods in Control Systems." IEEE Press, New York, 1979.
MacFarlane, A. G. J., and Postlethwaite, I., The generalized Nyquist stability criterion and criterion and multivariable root loci, *Internat. J. Control* **25**, 81–127 (1977).
Martin, C. F., and Hermann, R., Applications of algebraic geometry to systems, Part II, *Proc. IEEE* **65**, 841–848 (1977).

McMillan, B., Introduction to formal realizability theory, *Bell Systems Tech. J.* **31**, 217–279, 541–600 (1952).
Newcomb, R. W., "Linear Multiport Synthesis." McGraw-Hill, New York, 1966.
Newton, G. C., Gould, L. A., and Kaiser, J. F., "Analytical Design of Linear Feedback Controls." Wiley, New York, 1957.
Rosenbrock, H. H., "State-Space and Multivariable Theory." Nelson-Wiley, London, 1970.
Rosenbrock, H. H., Design of multivariable control systems using the inverse Nyquist array, *Proc. IEEE* **116**, 1929–1936 (1969).
Rosenbrock, H. H., Relatively prime polynomial matrices, *Electron. Lett.* **14**, 227–228 (1968).
Rosenbrock, H. H., and Munro, N., The inverse Nyquist array method, *in* "Alternatives for Linear Multivariable Control," pp. 101–137. NEC, Chicago, 1978.
Sain, M. K., Malsa, J. L., and Peczkowski, J. L., "Alternatives for Linear Multivariable Control." NEC, Chicago, Illinois, 1978.
Shaked, U., A general transfer function approach to linear stationary filtering and steady-state optimal control problems, *Internat. J. Control* **23** (1976).
Shaked, U., A general transfer function approach to the steady state linear quadratic Gaussian stochastic control problem, *Internat. J. Control* **23** (1976).
Wiener, N., and Masani, P., The prediction theory of multivariable stochastic processes I and II, *Acta. Math.* **98**, 111–150 (1957); **99**, 93–137 (1958).
Wolovich, W. A., A frequency domain approach to state feedback and estimation, *Proc. IEEE Conf. Decision and Control* (1971).
Youla, D. C., On the factorization of rational matrices, *IRE Trans.* **IT-7**, 172–189 (1961).
Youla, D. C., Bongiorno, J. J., and Lu, C. N., Single-loop feedback stabilization of linear multivariable dynamical plants, *Automatica* **10**, 159–173 (1974).
Youla, D. C., Bongiorno, J. J., and Jabr, H. A., Modern Wiener–Hopf design of optimal controllers—Parts I and II, *IEEE Trans.* **AC-21**, 3–15, 319–338 (1976).
Youla, D. C., Castriota, L. J., and Carlin, H. J., Bounded real scattering matrices and the foundations of linear passive network theory, *IRE Trans.* **CT-6**, 102–124 (1959).

8. TIME DOMAIN METHODS

Anderson, B. D. O., A system theory criterion for positive real matrices, *SIAM J. Control* **5**, 171–182 (1967).
Anderson, B. D. O., An algebraic solution to the spectral factorization problem, *IEEE Trans.* **AC-12**, 410–414 (1967).
Anderson, B. D. O., and Newcomb, R. W., Linear passive networks: Functional theory, *Proc. IEEE* **64**, 72–88 (1976).
Beltrami, E. J., and Wohlers, M. R., Distributional stability criteria, *IEEE Trans.* **CT-12**, 118–119 (1965).
Brockett, R. W., Poles, zeros, and feedback: State space interpretation, *IEEE Trans.* **AC-10**, 129–135 (1965).
Curtain, R. F., and Pritchard, A. J., "Infinite Dimensional Linear Systems," Springer-Verlag, Berlin and New York, 1978.
Desoer, C. A., and Callier, F. M., Convolution feedback systems, *SIAM J. Control* **10**, 737–746 (1972).

Helton, J. W., Systems with infinite-dimensional state space: The Hilbert approach, *Proc. IEEE* **64**, 145–160 (1976).

IEEE Trans. Automat. Control, **AC-16**, 527–869 (1971), Special issue on linear-quadratic Gaussian problem.

Kailath, T., Fredholm resolvants, Wiener–Hopf equations, and Raccati differential equations, *IEEE Trans.* **IT-15**, 665–672 (1969).

Kalman, R. E., Algebraic structure of linear dynamical systems, *Proc. Nat. Acad. Sci. USA* **54**, 1503–1508 (1965).

Kalman, R. E., New methods in Wiener filtering theory, *Proc. Symp. Eng. Appl. Random Function Theory and Probability* (J. L. Bogdanoff, and F. Kozin, F., eds.). Wiley, New York, 1963.

Kalman, R. E., Contributions to the theory of optimal control, *Bol. Soc. Mat. Mexicanna* **5**, 102–119 (1960).

Kalman, R. E., On the general theory of control systems, *Proc. IFAC Congr., 1st, Moscow*. Butterworths, London, 1960.

Kalman, R. E., Arbib, M. A., and Falb, P. L., "Topics in Mathematical System Theory." McGraw-Hill, New York, 1969.

Kalman, R. E., and Bucy, R. S., New results in linear filtering and prediction theory, *Trans. ASME, Ser. J. Basic Eng.* **82**, 95–100 (1961).

Sain, M. K., The growing algebraic presence in systems engineering: An introduction, *Proc. IEEE* **64**, 96–111 (1976).

Schumitzky, A., On the equivalence between matrix Riccati equations and Fredholm resolvents, *J. Comput. System Sci.* **2**, 76–87 (1968).

Solodov, A. V., "Linear Automatic Control Systems with Varying Parameters." American Elsevier, New York, 1966.

Wolovich, W. A., "Linear Multivariable Systems." Springer-Verlag, Berlin and New York, 1974.

Wonham, W. M., On pole assignment in multi-input controllable linear systems, *IEEE Trans.* **AC-12**, 660–665 (1967).

Wonham, W. M., "Linear Multivariable Control: A Geometric Approach," Lecture Notes in Econ. and Math. Systems, Vol. 101. Springer-Verlag, Berlin and New York, 1974.

Zadeh, L., and Desoer, C. A., "Linear System Theory." McGraw-Hill, New York, 1963.

Zamanian, A. H., "Realizability Theory for Continuous Linear Systems." Academic Press, New York, 1972.

Zamanian, A. H., The Hilbert port, *SIAM J. Appl. Math.* **18**, 98–138 (1970).

9. FOUNDATIONS

Balakrishnan, A. V., "Applied Functional Analysis." Springer-Verlag, Berlin and New York, 1976.

Barnett, S., "Matrices in Control Theory." Van Nostrand-Reinhold, Princeton, New Jersey. 1971.

Bart, H., Gohberg, I. C., and Kaashoek, M. A., "Minimal Factorization of Matrix and Operator Functions." Birkhauser-Verlag, Basel, 1979.

Curtain, R. F., and Pritchard, A. J., "Functional Analysis in Modern Applied Mathematics." Academic Press, New York, 1977.

Desoer, C. A., and Vidyasagar, M., "Feedback Systems: Input-Output Properties." Academic Press, New York, 1975.
Douglass, R. G., "Banach Algebra Techniques in Operator Theory." Academic Press, New York, 1972.
Dunford, N., and Schwartz, J. T., "Linear Operators." Wiley (Interscience), New York, 1958.
Duren, P. L., "Theory of H^p Spaces." Academic Press, New York, 1970.
Fures, Y., and Segal, I. E., Causality and analyticity, *Trans. Amer. Math. Soc.* **78**, 385–405 (1955).
Freedman, M. I., Falb, P. L., and Anton, J., A note of causality and analyticity, *SIAM J. Control* **7**, 472–478 (1969).
Holtzman, J., "Nonlinear System Theory: A Functional Analysis Approach." Prentice-Hall, Englewood Cliffs, New Jersey, 1970.
Klir, G., "An Approach to General Systems Theory." Van Nostrand-Reinhold, New York, 1969.
Masani, P., Recent trends in multivariate prediction theory, *in* "Multivariate Analysis." Academic Press, New York, 1966.
Mesarovic, M., Foundations of system theory, *in* "Views of General Systems Theory" (M. Mesarovic, ed.), pp. 1–24. Wiley, New York, 1964.
Paley, R. E. A. C., and Wiener, N., "Fourier Transforms in the Complex Domain," Vol. 19, American Mathematical Society Colloquim Publ., Providence, Rhode Island, 1934.
Porter, W. A., "Modern Foundations of System Theory." Macmillan, New York, 1966.
Ringrose, J. R., "Compact, Non-Selfadjoint Operators." Van Nostrand-Reinhold, New York, 1971.
Sandberg, I., Conditions for the causality of nonlinear operators defined on a linear space, *Quart. Appl. Math.* **23**, 87–91 (1965).
Saeks, R., "Generalized Networks." Holt, New York, 1972.
sz-Nagy, B., and Foias, C., "Analyse Harmonique de Operateurs de L'Espace de Hilbert." Masson, Paris, 1967.
Wiener, N., On the factorization of matrices, *Comment. Math. Helv.* **29**, 97–111 (1955).
Wiener, N., "Extrapolation, Interpolation and Smoothing of Stationary Time Series with Engineering Applications." Wiley, New York, 1949.
Windeknecht, T. C., Mathematical system theory—causality, *Math. Systems Theory* **1**, 279–288 (1976).
Windeknecht, T. C., "General Dynamical Processes." Academic Press, New York, 1971.
Yosida, K., "Functional Analysis." Springer-Verlag, Berlin and New York, 1966.

Index

U

V

W

Pure and Applied Mathematics

A Series of Monographs and Textbooks

Editors **Samuel Eilenberg and Hyman Bass**
Columbia University, New York

RECENT TITLES

CARL L. DEVITO. Functional Analysis

MICHIEL HAZEWINKEL. Formal Groups and Applications

SIGURDUR HELGASON. Differential Geometry, Lie Groups, and Symmetric Spaces

ROBERT B. BURCKEL. An Introduction to Classical Complex Analysis: Volume 1

JOSEPH J. ROTMAN. An Introduction to Homological Algebra

C. TRUESDELL AND R. G. MUNCASTER. Fundamentals of Maxwell's Kinetic Theory of a Simple Monatomic Gas: Treated as a Branch of Rational Mechanics

BARRY SIMON. Functional Integration and Quantum Physics

GRZEGORZ ROZENBERG AND ARTO SALOMAA. The Mathematical Theory of L Systems.

DAVID KINDERLEHRER and GUIDO STAMPACCHIA. An Introduction to Variational Inequalities and Their Applications.

H. SEIFERT AND W. THRELFALL. A Textbook of Topology; H. SEIFERT. Topology of 3-Dimensional Fibered Spaces

LOUIS HALLE ROWEN. Polynominal Identities in Ring Theory

DONALD W. KAHN. Introduction to Global Analysis

DRAGOS M. CVETKOVIC, MICHAEL DOOB, AND HORST SACHS. Spectra of Graphs

ROBERT M. YOUNG. An Introduction to Nonharmonic Fourier Series

MICHAEL C. IRWIN. Smooth Dynamical Systems

JOHN B. GARNETT. Bounded Analytic Functions

EDUARD PRUGOVEČKI. Quantum Mechanics in Hilbert Space, Second Edition

M. SCOTT OSBORNE AND GARTH WARNER. The Theory of Eisenstein Systems

K. A. ZHEVLAKOV, A. M. SLIN'KO, I. P. SHESTAKOV, AND A. I. SHIRSHOV. Translated by HARRY SMITH. Rings That are Nearly Associative

JEAN DIEUDONNÉ. A Panorama of Pure Mathematics; Translated by I. Macdonald

JOSEPH G. ROSENSTEIN. Linear Orderings

AVRAHAM FEINTUCH AND RICHARD SAEKS. System Theory: A Hilbert Space Approach

IN PREPARATION

ROBERT B. BURCKEL. An Introduction to Classical Complex Analysis: Volume 2

HOWARD OSBORN. Vector Bundles: Volume 1, Foundations and Stiefel-Whitney Classes

RICHARD V. KADISON AND JOHN R. RINGROSE. Fundamentals of the Theory of Operator Algebras

BARRETT O'NEILL. Semi-Riemannian Geometry: With Applications to Relatively

ULF GRENANDER. Mathematical Experiments on the Computer

EDWARD B. MANOUKIAN. Renormalization

E. J. MCSHANE. Unified Integration